Lecture Notes in Computer Science 12420

More information about this series at http://www.springer.com/series/7408

Aldeida Aleti · Annibale Panichella (Eds.)

Search-Based Software Engineering

12th International Symposium, SSBSE 2020
Bari, Italy, October 7–8, 2020
Proceedings

Editors
Aldeida Aleti
Monash University
Melbourne, VIC, Australia

Annibale Panichella
Delft University of Technology
Delft, The Netherlands

ISSN 0302-9743 ISSN 1611-3349 (electronic)
Lecture Notes in Computer Science
ISBN 978-3-030-59761-0 ISBN 978-3-030-59762-7 (eBook)
https://doi.org/10.1007/978-3-030-59762-7

LNCS Sublibrary: SL2 – Programming and Software Engineering

This Springer imprint is published by the registered company Springer Nature Switzerland AG
The registered company address is: Gewerbestrasse 11, 6330 Cham, Switzerland

Preface

Message from the General Chair

Welcome to the 12th Symposium on Search-Based Software Engineering (SSBSE 2020). SSBSE is a premium venue dedicated to the discussion of novel ideas and applications of Search-Based Software Engineering, a research area focused on the formulation of software engineering problems as search problems. A wealth of engineering challenges can leverage application of automated approaches and optimization techniques from AI and machine learning research.

This year, SSBSE was organized as a virtual conference due to travel restrictions imposed by the COVID-19 pandemic. We would like to thank the members of the Organizing Committee for their effort in making the virtual event a success. We thank the track chairs for their great work in creating an exciting conference program: Aldeida Aleti and Annibale Panichella (Research Track), Tim Menzies (Journal First Track), Jin Guo and Gregory Gay (New Ideas and Emerging Results Track), Bonita Sharif and Juan Pablo Galeotti (Replications and Negative Results Track), and Ruchika Malhotra and Wesley K. G. Assunção (Challenge Track). We thank Bruno Lima, our publication chair, for his remarkable effort in coordinating the proceedings creation process. Last but not least, we would like to thank our publicity chairs Jinghui Cheng, Rodrigo Morales, Rebecca Moussa, and Nobukazu Yoshioka, for their job in advertising SSBSE through mailing lists and social media, and Luigi Quaranta for setting up our website.

Finally, we would like to thank the University of Bari, Italy, and Polytechnique Montreal, Canada, for the support provided, and our sponsor, Facebook, for the generous support to SSBSE 2020 and the forthcoming 2021 instalment.

We look forward to seeing you all at SSBE 2021.

August 2020

Giuliano Antoniol
Nicole Novielli

Message from the Program Chairs

On behalf of the Program Committee, it is our pleasure to present the proceedings of the 12th International Symposium on Search-Based Software Engineering (SSBSE 2020).

The field of Search-Based Software Engineering has grown tremendously in the last few years, with research covering a wide range of topics in the intersection of software engineering and search algorithms. As in the previous years, SSBSE 2020 continued to bring together the international SBSE community to present innovations, discuss new ideas, and celebrate progress in the field. This year, we faced an unprecedented challenge brought upon us by the global pandemic, which impacted our lives significantly. Hence, it is more important than ever that we stay connected and support each other through these times, even if virtually. Despite these challenges, this year we received a record number of papers; 34 papers in total across four tracks: 24 full research papers, 2 Challenge solutions, 2 New Ideas and Emerging Results (NIER) papers, 5 Replications and Negative Results (RENE) papers, and 1 Challenge case. We would like to thank the authors for their submissions and express our appreciation for their efforts in advancing the SBSE field.

The success of SSBSE depends completely on the effort, talent, and energy of researchers in the field of Search-Based Software Engineering who have written and submitted papers on a variety of topics. Following a strict review process, where each submission received three reviews, we accepted 18 papers: 10 papers in the Research Track, 4 papers in the RENE Track, 1 paper in the NIER Track, 2 papers in the Challenge solutions, and 1 paper in the Challenge cases. Thanks also go to the Program Committee members and external reviewers who invested significant time in assessing multiple papers, and who hold and maintain a high standard of quality for this conference. Their dedicated work and support makes this symposium possible and results in a stronger community.

Finally, the symposium would not have been possible without the efforts of the Organizing Committee, hence we would like to thank:

- NIER Track co-chairs: Jin Guo and Gregory Gay
- Replications and Negative Results Track co-chairs: Bonita Sharif and Juan Pablo Galeotti
- Challenge Track co-chair: Wesley K. G. Assuncao and Ruchika Malhotra
- Publicity co-chairs: Rebecca Moussa, Nobukazu Yoshioka, Jinghui Cheng and Rodrigo Morales
- Virtualization Chair: Marios-Eleftherios Fokaefs
- Journal First Track chair: Tim Menzies
- Publication chair: Bruno Lima
- Web chair: Luigi Quaranta

In addition to a full program of research talks, SSBSE 2020 attendees had the opportunity to listen to an outstanding keynote by Shiva Nejati on "Search-Based Software Testing for Formal Software Verification, and Vice Versa." In this keynote, Shiva presented recent research that combines ideas from SBST research and the formal verification area to improve the analysis of cyber-physical systems (CPS). Next, the keynote focused on an empirical study that compares CPS model testing and verification. While SBST and formal verification communities pursue the common goal of ensuring the correctness of software systems, the interaction between the two communities is sparse. Hence, this keynote provided an exciting opportunity to reflect on the advancements of these two areas and help the software engineering community to reach a common understanding of the problem of testing software systems, to develop common benchmarks and evaluation criteria, and to build more effective software tools.

We hope you enjoy the work contained in this volume and you can apply it to your own work. We are proud of the program assembled this year, and are thankful for the opportunity to present these proceedings to the SBSE research community.

July 2020

Aldeida Aleti
Annibale Panichella

Organization

Organizers

General Chairs

Giuliano Antoniol	Polytechnique Montreal, Canada
Nicole Novielli	University of Bari, Italy

Program Chairs

Aldeida Aleti	Monash University, Australia
Annibale Panichella	Delft University of Technology, The Netherlands

Journal First Chair

Tim Menzies	North Carolina State University, USA

NIER Track Chairs

Gregory Gay	Chalmers and the University of Gothenburg, Sweden
Jin Guo	McGill University, Canada

Replications and Negative Results Track Chairs

Juan Pablo Galeotti	University of Buenos Aires, Argentina
Bonita Sharif	University of Nebraska – Lincoln, USA

Challenge Track Chairs

Ruchika Malhotra	Delhi Technological University, India
Wesley K. G. Assunção	Federal University of Technology, Paraná, Brazil

Publicity Chairs

Jinghui Cheng	Polytechnique Montreal, Canada
Nobukazu Yoshioka	National Institute of Informatics, Japan
Rebecca Moussa	University College London, UK
Rodrigo Morales	Concordia University, Canada

Virtualization Chair

Marios-Eleftherios Fokaefs	Polytechnique Montreal, Canada

Web Chair

Luigi Quaranta	University of Bari, Italy

Publication Chair

Bruno Lima	University of Porto and INESC TEC, Portugal

SSBSE Steering Committee

Gregory Gay	Chalmers and the University of Gothenburg, Sweden
Phil McMinn	The University of Sheffield, UK
Tim Menzies	North Carolina State University, USA
Annibale Panichella	Delft University of Technology, The Netherlands
Federica Sarro	University College London, UK
Shaukat Ali	Simula Research Laboratory, Norway
Shin Yoo	KAIST, South Korea
Mike Papadakis	University of Luxembourg, Luxembourg
Thomas Vogel	Humboldt University Berlin, Germany

Technical Program Committee

Shaukat Ali	Simula Research Laboratory, Norway
Wesley K. G. Assunção	Federal University of Technology, Paraná, Brazil
Carlos Cetina	San Jorge University, Spain
Thelma E. Colanzi	State University of Maringa, Brazil
Xavier Devroey	Delft University of Technology, The Netherlands
Erik M. Fredericks	Oakland University, USA
Fitsum Meshesha Kifetew	Fondazione Bruno Kessler, Italy
Leandro Minku	University of Leicester, UK
Sebastiano Panichella	Zurich University of Applied Science (ZHAW), Switzerland
Mike Papadakis	University of Luxembourg, Luxembourg
Anjana Perera	Monash University, Australia
Pasqualina Potena	RISE SICS Västerås AB, Sweden
José Miguel Rojas	University of Leicester, UK
Paolo Tonella	Università della Svizzera Italiana, Switzerland
Tanja Vos	Universitat Politècnica de València, Spain

NIER Track Committee

Jose Campos	University of Lisbon, Portugal
Junjie Chen	BizSeer, China
James Clause	University of Delaware, USA
Thelma E. Colanzi	State University of Maringa, Brazil
Felix Dobslaw	Chalmers University of Technology, Sweden
Gregory Gay	Chalmers and the University of Gothenburg, Sweden
Jin Guo	McGill University, Canada
Raluca Lefticaru	University of Bradford, UK
Pasqualina Potena	RISE SICS Västerås AB, Sweden

Thomas Vogel	Humboldt University Berlin, Germany
Neil Walkinshaw	The University of Sheffield, UK

RENE Track Committee

Nazareno Aguirre	Universidad Nacional de Río Cuarto, Argentina
Nadia Alshahwan	Facebook, UK
Andrea Arcuri	Kristiania University College, Norway
Earl Barr	University College London, UK
Jose Campos	University of Lisbon, Portugal
Gordon Fraser	University of Passau, Germany
Erik M. Fredericks	Oakland University, USA
Gregory Gay	Chalmers and the University of Gothenburg, Sweden
Fuyuki Ishikawa	National Institute of Informatics, Japan
Fitsum Meshesha Kifetew	Fondazione Bruno Kessler, Italy
Phil McMinn	The University of Sheffield, UK
Shiva Nejati	SnT Centre, University of Luxembourg, Luxembourg
Changhai Nie	Nanjing University, China
Sebastiano Panichella	University of Zurich, Switzerland
Annibale Panichella	Delft University of Technology, The Netherlands
Justyna Petke	University College London, UK
Jose Miguel Rojas	University of Leicester, UK
Shin Yoo	KAIST, South Korea

Challenge Track Program Committee

Earl Barr	University College London, UK
Erik M. Fredericks	Oakland University, USA
Giovani Guizzo	University College London, UK
Gordon Fraser	University of Passau, Germany
Gregory Gay	Chalmers and the University of Gothenburg, Sweden
Jose Miguel Rojas	University of Leicester, UK
Roberto E. Lopez-Herrejon	École de technologie superieuré, Canada
Shaukat Ali	Simula Research Laboratory, Norway
Shin Yoo	KAIST, South Korea
Wasif Afzal	Mälardalen University, Sweden

Sponsoring Institution

Facebook

Contents

Keynote

Search-Based Software Testing for Formal Software Verification – and Vice Versa

Shiva Nejati[1,2](✉)

[1] School of Electrical Engineering and Computer Science, University of Ottawa, Ottawa, Canada
snejati@uottawa.ca

[2] SnT Centre, University of Luxembourg, Esch-sur-Alzette, Luxembourg
https://shnejati.bitbucket.io/

Abstract. "Testing can be quite effective for showing the presence of bugs, but is hopelessly inadequate for showing their absence". This famous remark, which was made by Dijkstra, has often been used to indicate a dichotomy between testing and verification. From a practitioner's point of view, however, there is not much difference in the ways testing and verification techniques may be used in practice. While engineers would try to demonstrate that their systems are correct (verification), they often find themselves in a situation where they have to prioritize bug finding (testing). As a result, the choice to go for formal verification versus testing is largely driven by the practical needs and the context specificities. In this keynote, I will focus on search-based software testing (SBST) and review some recent research that combines ideas from the SBST and the formal verification communities to improve the analysis of models of cyber physical systems (CPS). I will present an empirical study that compares CPS model testing and verification, a search-based testing approach for compute-intensive CPS models that builds on a well-known formal verification framework, and a technique to automatically generate formal environment assumptions for CPS models using search algorithms and genetic programming.

Keywords: Search-based software testing · Formal verification · Model testing · Model checking · Cyber physical systems

1 Introduction

Search-based Software Testing (SBST) and its parent field, i.e., Search-based Software Engineering (SBSE), have significantly expanded since the inception of the Symposium on Search Based Software Engineering (SSBSE) in 2008 [7]. Over the past years, SBST has been successfully applied to many categories of software

This short paper provides an outline of the keynote talk given by Shiva Nejati at SSBSE 2020.

A. Aleti and A. Panichella (Eds.): SSBSE 2020, LNCS 12420, pp. 3–6, 2020.
https://doi.org/10.1007/978-3-030-59762-7_1

testing (e.g., unit testing and regression testing) [6]. SBST solutions provide a number of benefits that are crucial for producing impactful software engineering research [15]: They scale well to large systems, make few assumptions about the structure of their inputs and can be flexibly combined with other methods.

To date, SBST has been largely focused on code testing, and is lesser used to test other artifacts such as software models that are prominently used in the domain of Cyber Physical Systems (CPS) to capture system-level behaviors, facilitate early simulation and enable code generation. CPS models are subject to extensive verification and testing. Historically, formal verification techniques such as model checking [5] have been considered as the main candidate to ensure correctness of CPS requirements and design models. I believe, however, that SBST can be effective in the domain of CPS verification since it can address scalability issues of formal verification, can be combined with formal methods to enhance their effectiveness, and can relax the restrictive assumptions that formal techniques often make about their inputs as they heavily rely on the structure of the artifacts under analysis.

In this keynote, I present recent research that aims to combine ideas from SBST and formal methods to develop testing and verification techniques for industrial CPS models [9,13,14]. I start by presenting an empirical study that compares capabilities of model testing [12] based on SBST and model checking using an industrial benchmark [14]. The models in the benchmark are described in Simulink – a simulation language that is widely used by the CPS industry – and for each model, a set of requirements are specified that must be satisfied by their corresponding model. The experiment aims to compare model testing and model checking to determine how these techniques help assess models against their requirements. Our results show that while model checking is able to prove correctness of Simulink models that capture relatively small and individual components, it fails to analyse Simulink models that are more complex and capture a whole system in a feedback loop (e.g., an autopilot controller and its environment). The study further shows that bounded model checking [3] which is a partial form of model checking and is meant to reveal bugs instead of proving correctness is not as effective as search-based testing in finding bugs in Simulink models.

I then discuss how a formal verification framework, i.e., counterexample guided abstraction and refinement (CEGAR) [4], can be adapted to SBST and help scale model testing to large and complex compute-intensive CPS models from the aerospace domain [13]. CEGAR was originally proposed to help scale model checking and to combat the state explosion problem. The framework was the backbone of early software model checkers such as SLAM [1] developed in early 2000. CEGAR generates a boolean abstraction of a given program using the abstract interpretation framework [8] and iteratively refines the abstract program until a model checker can conclusively verify or refute a given property based on the abstract program. To adapt CEGAR, we replace model checking with search-based model testing and, instead of abstraction interpretation, we use machine learning (more specifically system identification [2]) to approximate a given CPS model.

Finally, I describe how search algorithms and SBST can solve a difficult formal verification problem, i.e., the problem of synthesizing environment assumptions [11], for complex models that cannot typically be handled using formal methods [9]. Specifically, existing assumption synthesis techniques are typically applicable to CPS models expressed as state machines [10], while using search algorithms and machine learning, we can learn assumptions for signal-based formalisms that include both logical and numeric components.

Even though SBST and formal verification communities pursue a common goal, that of ensuring correctness of software systems, there are limited interactions and exchanges between them. Proving correctness which is the ultimate goal of formal verification may allow us to go without testing, but most often, it can only be fully achieved for small-scale programs and components that are assessed as independent units and outside their context and environment. In most other cases, formal verification is applied partially or relies on heuristics (e.g., bounded model checking or symbolic execution techniques), and hence, similar to testing, it can only show the presence of bugs but not their absence. It is, therefore, interesting to study similarities and differences between testing based on statistical optimization (SBST) and testing based on formal verification as they essentially automate the same task (i.e., revealing bugs). This will help the software engineering community to reach a common understanding on the problem of testing software systems, to develop common benchmarks and evaluation criteria and to build more effective software tools.

Acknowledgment. This work is supported by NSERC of Canada under the Discovery program.

References

1. Ball, T., Rajamani, S.K.: The SLAM toolkit. In: Berry, G., Comon, H., Finkel, A. (eds.) CAV 2001. LNCS, vol. 2102, pp. 260–264. Springer, Heidelberg (2001). https://doi.org/10.1007/3-540-44585-4_25
2. Bittanti, S.: Model Identification and Data Analysis. Wiley, New York (2019)
3. Clarke, E.M., Biere, A., Raimi, R., Zhu, Y.: Bounded model checking using satisfiability solving. Formal Methods Syst. Design **19**(1), 7–34 (2001)
4. Clarke, E., Grumberg, O., Jha, S., Lu, Y., Veith, H.: Counterexample-guided abstraction refinement. In: Emerson, E.A., Sistla, A.P. (eds.) CAV 2000. LNCS, vol. 1855, pp. 154–169. Springer, Heidelberg (2000). https://doi.org/10.1007/10722167_15
5. Clarke Jr., E.M., Grumberg, O., Peled, D.A.: Model Checking. MIT Press, Cambridge (1999)
6. Cohen, M.B.: The maturation of search-based software testing: successes and challenges. In: Gorla, A., Rojas, J.M. (eds.) Proceedings of the 12th International Workshop on Search-Based Software Testing, SBST@ICSE 2019, Montreal, QC, Canada, 27 May 2019, pp. 13–14. IEEE/ACM (2019). https://doi.org/10.1109/SBST.2019.00013

7. Colanzi, T.E., Assunção, W.K.G., Farah, P.R., Vergilio, S.R., Guizzo, G.: A review of ten years of the symposium on search-based software engineering. In: Nejati, S., Gay, G. (eds.) SSBSE 2019. LNCS, vol. 11664, pp. 42–57. Springer, Cham (2019). https://doi.org/10.1007/978-3-030-27455-9_4
8. Cousot, P., Cousot, R.: Abstract interpretation: a unified lattice model for static analysis of programs by construction or approximation of fixpoints. In: Graham, R.M., Harrison, M.A., Sethi, R. (eds.) Conference Record of the Fourth ACM Symposium on Principles of Programming Languages, Los Angeles, California, USA, January 1977, pp. 238–252. ACM (1977). https://doi.org/10.1145/512950.512973
9. Gaaloul, K., Menghi, C., Nejati, S., Briand, L., Wolfe, D.: Mining assumptions for software components using machine learning. In: Foundations of Software Engineering ESEC/SIGSOFT FSE 2020. ACM (2020)
10. Giannakopoulou, D., P"s"reanu, C.S., Barringer, H.: Assumption generation for software component verification. In: Proceedings of the 17th IEEE International Conference on Automated Software Engineering, ASE 2002, p. 3. IEEE Computer Society (2002)
11. Li, W., Dworkin, L., Seshia, S.A.: Mining assumptions for synthesis. In: International Conference on Formal Methods and Models, pp. 43–50. IEEE (2011)
12. Matinnejad, R., Nejati, S., Briand, L.C., Bruckmann, T.: Test generation and test prioritization for Simulink models with dynamic behavior. IEEE Trans. Software Eng. **45**(9), 919–944 (2019). https://doi.org/10.1109/TSE.2018.2811489
13. Menghi, C., Nejati, S., Briand, L.C., Parache, Y.I.: Approximation-refinement testing of compute-intensive cyber-physical models: An approach based on system identification. In: International Conference on Software Engineering (ICSE). arXiv (2020)
14. Nejati, S., Gaaloul, K., Menghi, C., Briand, L.C., Foster, S., Wolfe, D.: Evaluating model testing and model checking for finding requirements violations in Simulink models. In: Dumas, M., Pfahl, D., Apel, S., Russo, A. (eds.) Proceedings of the ACM Joint Meeting on European Software Engineering Conference and Symposium on the Foundations of Software Engineering, ESEC/SIGSOFT FSE 2019, Tallinn, Estonia, 26–30 August 2019, pp. 1015–1025. ACM (2019)
15. Zeller, A.: Search-based testing and system testing: a marriage in heaven. In: 10th IEEE/ACM International Workshop on Search-Based Software Testing, SBST@ICSE 2017, Buenos Aires, Argentina, 22–23 May 2017, pp. 49–50. IEEE (2017). https://doi.org/10.1109/SBST.2017.3

Research Papers

Automated Unit Test Generation for Python

Stephan Lukasczyk(✉), Florian Kroiß, and Gordon Fraser

University of Passau, Innstr. 33, 94032 Passau, Germany
{stephan.lukasczyk,gordon.fraser}@uni-passau.de, kroiss@fim.uni-passau.de

Abstract. Automated unit test generation is an established research field, and mature test generation tools exist for statically typed programming languages such as Java. It is, however, substantially more difficult to automatically generate supportive tests for dynamically typed programming languages such as Python, due to the lack of type information and the dynamic nature of the language. In this paper, we describe a foray into the problem of unit test generation for dynamically typed languages. We introduce PYNGUIN, an automated unit test generation framework for Python. Using PYNGUIN, we aim to empirically shed light on two central questions: (1) Do well-established search-based test generation methods, previously evaluated only on statically typed languages, generalise to dynamically typed languages? (2) What is the influence of incomplete type information and dynamic typing on the problem of automated test generation? Our experiments confirm that evolutionary algorithms can outperform random test generation also in the context of Python, and can even alleviate the problem of absent type information to some degree. However, our results demonstrate that dynamic typing nevertheless poses a fundamental issue for test generation, suggesting future work on integrating type inference.

Keywords: Dynamic Typing · Python · Random Test Generation · Whole Suite Test Generation

1 Introduction

Unit tests can be automatically generated to support developers and the dynamic analysis of programs. Established techniques such as feedback-directed random test generation [15] or evolutionary algorithms [7] are implemented in mature research prototypes, but these are based on strong assumptions on the availability of static type information, as is the case in statically typed languages like Java. Dynamically typed languages such as Python or JavaScript, however, have seen increased popularity within recent years. Python is the most popular programming language in the category of dynamically typed languages, according to, for example, the IEEE Spectrum Ranking[1]. It is heavily used in the fields of machine

[1] https://spectrum.ieee.org/computing/software/the-top-programming-languages-2019, accessed 2020–07–25.

A. Aleti and A. Panichella (Eds.): SSBSE 2020, LNCS 12420, pp. 9–24, 2020.
https://doi.org/10.1007/978-3-030-59762-7_2

learning and data analysis, and it is also popular in other domains. This can be seen, for example, from the Python Package Index (PyPI), which contains more than 200 000 packages at the time of writing. In languages like Python, the type information that automated unit test generators require is not available.

An automated unit test generator primarily requires type information in order to select parameters for function calls and to generate complex objects. If type information is absent, the test generator can only guess which calls to use to create new objects, or which existing objects to select as parameters for new function calls. Existing test generators for dynamically typed languages therefore resort to other means to avoid having to make such choices in the first place, for example by using the document object model of a web browser to generate tests for JavaScript [14], or by targeting the browser's event handling system rather than APIs [3,12]. However, there is no general purpose unit test generator at API level yet for languages like Python.

In order to allow test generation research to expand its focus from statically to dynamically typed languages, in this paper we introduce Pynguin, a new automated test generation framework for Python. Pynguin takes as input a Python module and its dependencies, and aims to automatically generate unit tests that maximise code coverage. In order to achieve this, Pynguin implements the established test generation techniques of whole-suite generation [9] and feedback-directed random generation [15]. Pynguin is available as open source to support future research on automated test generation for dynamically typed programming languages. Pynguin is designed to be extensible; in this paper we focus on established baseline algorithms for foundational experiments, we will add further algorithms such as DynaMOSA [16] in future work.

Using Pynguin, we empirically study the problem of automated unit test generation for Python using ten popular open source Python projects taken from GitHub, all of which contain type information added by developers in terms of type annotations. This selection allows us to study two central questions: (1) Do previous findings, showing that evolutionary search achieves higher code coverage than random testing [5], also generalise to dynamically typed languages? (2) What is the influence of the lack of type information in a dynamically typed language like Python on automated unit test generation?

In detail, the contributions of this paper are the following:

1. We introduce Pynguin, a new framework for automated unit test generation for the Python programming language.
2. We replicate experiments previously conducted only in the context of statically typed languages to compare test generation approaches.
3. We empirically study the influence of type information on the effectiveness of automated test generation.

Our experiments confirm that the whole-suite approach generally achieves higher code coverage than random testing, and that the availability of type information also leads to higher resulting coverage. However, our experiments reveal several new technical challenges such as generating collections or iterable input types. Our findings also suggest that the integration of current research on type inference is a promising route forward for future research.

2 Background

The main approaches to automatically generate unit tests are either by creating random sequences, or by applying metaheuristic search algorithms. Random testing assembles sequences of calls to constructors and methods randomly, often with the objective to find undeclared exceptions [6] or violations of general object contracts [15], but the generated tests can also be used as automated regression tests. The effectiveness of random test generators can be increased by integrating heuristics [13,17]. Search-based approaches use a similar representation, but apply evolutionary search algorithms to maximize code coverage [1,4,9,19].

As an example to illustrate how type information is used by existing test generators, consider the following snippets of Java (left) and Python (right) code:

```
class Foo {
  Foo(Bar b) { ... }
  void doFoo(Bar b) { ... } }
class Bar {
  Bar() { ... }
  Bar doBar(Bar b) { ... } }
```

```
class Foo:
  def __init__(self, b): ...
  def do_foo(self, b): ...
class Bar:
  def __init__(self): ...
  def do_bar(self, b): ...
```

Assume Foo of the Java example is the class under test. It has a dependency on class Bar: in order to generate an object of type Foo we need an instance of Bar, and the method doFoo also requires a parameter of type Bar.

Random test generation would typically generate tests in a forward way. Starting with an empty sequence $t_0 = \langle\rangle$, all available calls for which all parameters can be satisfied with objects already existing in the sequence can be selected. In our example, initially only the constructor of Bar can be called, since all other methods and constructors require a parameter, resulting in $t_1 = \langle o_1 = \texttt{new Bar()}\rangle$. Since t_1 contains an object of type Bar, in the second step the test generator now has a choice of either invoking doBar on that object (and use the same object also as parameter), or invoking the constructor of Foo. Assuming the chosen call is the constructor of Foo, we now have $t_2 = \langle o_1 = \texttt{new Bar()}; o_2 = \texttt{new Foo}(o_1);\rangle$. Since there now is also an instance of Foo in the sequence, in the next step also the method doFoo is an option. The random test generator will continue extending the sequence in this manner, possibly integrating heuristics to select more relevant calls, or to decide when to start with a new sequence, etc.

An alternative approach, for example applied during the mutation step of an evolutionary test generator, is to select necessary calls in a backwards fashion. That is, a search-based test generator like EvoSuite [9] would first decide that it needs to, for example, call method doFoo of class Foo. In order to achieve this, it requires an instance of Foo and an instance of Bar to satisfy the dependencies.

To generate a parameter object of type `Bar`, the test generator would consider all calls that are declared to return an instance of `Bar`—which is the case for the constructor of `Bar` in our example, so it would prepend a call to `Bar()` before the invocation of `doFoo`. Furthermore, it would try to instantiate `Foo` by calling the constructor. This, in turn, requires an instance of `Bar`, for which the test generator might use the existing instance, or could invoke the constructor of `Bar`.

In both scenarios, type information is crucial: In the forward construction type information is used to inform the choice of call to append to the sequence, while in the backward construction type information is used to select generators of dependency objects. Without type information, which is the case with the Python example, a forward construction (1) has to allow all possible functions at all steps, thus may not only select the constructor of `Bar`, but also that of `Foo` with an arbitrary parameter type, and (2) has to consider all existing objects for all parameters of a selected call, and thus, for example, also `str` or `int`. Backwards construction without type information would also have to try to select generators from all possible calls, and all possible objects, which both result in a potentially large search space to select from.

Type information can be provided in two ways in recent Python versions: either in a stub file that contains type hints or directly annotated in the source code. A stub file can be compared to C header files: they contain, for example, method declarations with their according types. Since Python 3.5, the types can also be annotated directly in the implementing source code, in a similar fashion known from statically typed languages (see PEP 484[2]).

3 Search-Based Unit Test Generation

3.1 Python Test Generation as a Search Problem

As the *unit* for unit test generation, we consider Python *modules*. A module is usually identical with a file and contains definitions of, for example, functions, classes, or statements; these can be nested almost arbitrarily. When the module is loaded the definitions and statements at the top level are executed. While generating tests we do not only want all definitions to be executed, but also all structures defined by those definitions, for example, functions, closures, or list comprehensions. Thus, in order to apply a search algorithm, we first need to define a proper representation of the valid solutions for this problem.

We use a representation based on prior work from the domain of testing Java code [9]. For each statement s_j in a test case t_i we assign one value $v(s_j)$ with type $\tau(v(s_j)) \in \mathcal{T}$, with the finite set of types $\mathcal{T}$ used in the subject-under-test (SUT) and the modules imported by the SUT. We define four kinds of statements: *Primitive statements* represent `int`, `float`, `bool`, and `str` variables, for example, `var0 = 42`. Value and type of a statement are defined by the primitive variable. Note that although in Python everything is an object, we treat

[2] https://python.org/dev/peps/pep-0484/, accessed 2020-07-25.

these values as primitives because they do not require further construction in Python's syntax. Other simple types, such as lists, require the construction of the list and its elements, which we do not yet handle. *Constructor statements* create new instances of a class, for example, `var0 = SomeType()`. Value and type are defined by the constructed object; any parameters are satisfied from the set $V = \{v(s_k) \mid 0 \leq k < j\}$. *Method statements* invoke methods on objects, for example, `var1 = var0.foo()`. Value and type are defined by the return value of the method; source object and any parameters are satisfied from the set V. *Function statements* invoke functions, for example, `var2 = bar()`. They do not require a source object but are otherwise identical to method statements. This representation is of variable size; we constrain the size of test cases $l \in [1, L]$ and test suites $n \in [1, N]$. In contrast to prior work on testing Java [9], we do not define field or assignment statements; fields of objects are not explicitly declared in Python but assigned dynamically, hence it is non-trivial to identify the existing fields of an object and we leave it as future work.

The search operators for this representation are based on those used in EvoSuite [9]: Crossover takes as input two test suites P_1 and P_2, and generates two offspring O_1 and O_2. Individual test cases have no dependencies between each other, thus the application of crossover always generates valid test suites as offspring. Furthermore, the operator decreases the difference in the number of test cases between the test suites, thus $\text{abs}(|O_1| - |O_2|) \leq \text{abs}(|P_1| - |P_2|)$. Therefore, no offspring will have more test cases than the larger of its parents.

When mutating a test suite T, each of its test cases is mutated with probability $\frac{1}{|T|}$. After mutation, we add new randomly generated test cases to T. The first new test case is added with probability σ. If it is added, a second new test case is added with probability σ^2; this happens until the i-th test case is not added (probability: $1 - \sigma^i$). Test cases are only added if the limit N has not been reached, thus $|T| \leq N$. The mutation of a test case can be one of three operations: *remove*, which removes a statement from the test case, *change*, which randomly changes values in a statement—for example, by adding random values to numbers, adding/replacing/deleting characters, or changing method calls—and *insert*, which adds new statements at random positions in the test case. Each of these operations can happen with the same probability of $\frac{1}{3}$. A test case that has no statements left after the application of the mutation operator is removed from the test suite T. For constructing the initial population, a random test case t is sampled by uniformly choosing a value r with $1 \leq r \leq L$, and then applying the insertion operator repeatedly starting with an empty test case t', until $|t'| \geq r$.

3.2 Covering Python Code

A Python module contains various control structures, for example, `if` or `while` statements, which are guarded by logical predicates. The control structures are represented by conditional jumps at the bytecode level, based on either a unary or binary predicate. We focus on *branch coverage* in this work, which requires that each of those predicates evaluates to both true and false.

Let B denote the set of branches in the SUT—two for each conditional jump in the byte code. Everything executable in Python is represented as a *code object*. For example, an entire module is represented as a code object, a function within that module is represented as another code object. We want to execute all code objects C of the SUT. Therefore, we keep track of the executed code objects C_T as well as the minimum *branch distance* $d_{\min}(b, T)$ for each branch $b \in B$, when executing a test suite T. $B_T \subseteq B$ denotes the set of taken branches. We then define the branch coverage $\text{cov}(T)$ of a test suite T as $\text{cov}(T) = \frac{|C_T| + |B_T|}{|C| + |B|}$.

The fitness function required by the genetic algorithm of our whole-suite approach is constructed similar to the one used in EvoSuite [9] by incorporating the branch distance. *Branch distance* is a heuristic to determine how far a predicate is away from evaluating to true or false, respectively. In contrast to previous work on Java, where most predicates at the bytecode level operate only on Boolean or numeric values, in our case the operands of a predicate can be any Python object. Thus, as noted by Arcuri [2], we have to define our branch distance in such a way that it can handle arbitrary Python objects.

Let $\mathbb{O}$ be the set of possible Python objects and let $\Theta := \{\equiv, \not\equiv, <, \leq, >, \geq, \in, \notin, =, \neq\}$ be the set of binary comparison operators (remark: we use '$\equiv$', '$=$', and '$\in$' for Python's `==`, `is`, and `in` keywords, respectively). For each $\theta \in \Theta$, we define a function $\delta_\theta : \mathbb{O} \times \mathbb{O} \to \mathbb{R}_0^+ \cup \{\infty\}$ that computes the branch distance of the true branch of a predicate of the form $a \,\theta\, b$, with $a, b \in \mathbb{O}$ and $\theta \in \Theta$. By $\delta_{\bar{\theta}}(a, b)$ we denote the distance of the false branch, where $\bar{\theta}$ is the complementary operator of θ. Let further k be a positive number, and let $\text{lev}(x, y)$ denote the Levenshtein distance [11] between two strings x and y. The predicates is_numeric(z) and is_string(z) determine whether the type of their argument z is numeric or a string, respectively.

$$\delta_{\equiv}(a,b) = \begin{cases} 0 & a \equiv b \\ |a - b| & a \not\equiv b \wedge \text{is\_numeric}(a) \wedge \text{is\_numeric}(b) \\ \text{lev}(a,b) & a \not\equiv b \wedge \text{is\_string}(a) \wedge \text{is\_string}(b) \\ \infty & \text{otherwise} \end{cases}$$

$$\delta_{<}(a,b) = \begin{cases} 0 & a < b \\ a - b + k & a \geq b \wedge \text{is\_numeric}(a) \wedge \text{is\_numeric}(b) \\ \infty & \text{otherwise} \end{cases}$$

$$\delta_{\leq}(a,b) = \begin{cases} 0 & a \leq b \\ a - b + k & a > b \wedge \text{is\_numeric}(a) \wedge \text{is\_numeric}(b) \\ \infty & \text{otherwise} \end{cases}$$

$$\delta_{>}(a,b) = \delta_{<}(b,a)$$

$$\delta_{\geq}(a,b) = \delta_{\leq}(b,a)$$

$$\delta_{\theta}(a,b) = \begin{cases} 0 & a \,\theta\, b \\ k & \text{otherwise} \end{cases} \qquad \theta \in \{\not\equiv, \in, \notin, =, \neq\}$$

Note that every object in Python represents a Boolean value and can therefore be used as a predicate. We assign a distance of 0 to the true branch of such a unary predicate, if the object represents a true value, otherwise k. Future work shall refine the branch distance for different operators and operand types.

The fitness function estimates how close a test suite is to covering *all* branches of the SUT. Thus, every predicate has to be executed at least twice, which we enforce in the same way as existing work [9]: the actual branch distance $d(b,T)$ is given by

$$d(b,T) = \begin{cases} 0 & \text{if the branch has been covered} \\ \nu(d_{min}(b,T)) & \text{if the predicate has been executed at least twice} \\ 1 & \text{otherwise} \end{cases}$$

with $\nu(x) = \frac{x}{x+1}$ being a normalisation function [9].

Finally, we can define the resulting fitness function f of a test suite T as

$$f(T) = |C| - |C_T| + \sum_{b \in B} d(b,T)$$

3.3 The PYNGUIN Framework

PYNGUIN is a framework for automated unit test generation written in and for the Python programming language. The framework is available as open-source software licensed under the GNU Lesser General Public License from its GitHub repository[3]. It can also be installed from the Python Package Index (PyPI)[4] using the `pip` utility.

PYNGUIN takes as input a Python module and allows the generation of unit tests using different techniques. For this, it parses the module and extracts information about available methods in the module and types from the module and its imports. So far, PYNGUIN focuses on test-input generation and excludes the generation of oracles. A tool run emits the generated test cases in the style of the widely-used PYTEST[5] framework or for the `unittest` module from the Python standard library.

PYNGUIN is built to be extensible with other test generation approaches and algorithms. For experiments in this paper, we implemented a feedback-directed random approach based on RANDOOP [15] in addition to the whole-suite test-generation approach. Feedback-directed test generation starts with two empty test suites, a passing and a failing test suite, and adds statements randomly to an empty test case. After each addition, the test case is executed and the execution result is retrieved. Successful test cases, that is, test cases that do not raise exceptions are added to the passing test suite; a test case that raises an exception is added to the failing test suite. In the following, the algorithm

[3] https://github.com/se2p/pynguin, accessed 2020–07–27.
[4] https://pypi.org/project/pynguin/, accessed 2020–07–25.
[5] https://www.pytest.org, accessed 2020–07–25.

randomly chooses a test case from the passing test suite or an empty test case and adds statements to it. We refer the reader to the description of RANDOOP [15] for details on the algorithm; the main differences of our approach are that it does not yet check for contract violations, and does not require the user to provide a list of relevant classes and methods, which RANDOOP does.

4 Experimental Evaluation

Using our PYNGUIN test generator, we aim to empirically study automated unit test generation on Python. First, we are interested in determining whether previous findings on the performance of test generation techniques established in the context of statically typed languages generalise also to Python:

Research Question 1 (RQ1) *How do whole-suite test generation and random test generation compare on Python code?*

A central difference between prior work and the context of Python is the type information: Previous work evaluated test-generation techniques mainly for statically typed languages, such as Java, where information on parameter types is available at compile time, that is, without running the program. This is not the case for many programs written in dynamically typed languages, such as Python. Therefore, we want to explicitly evaluate the influence of the type information for the test-generation process:

Research Question 2 (RQ2) *How does the availability of type information influence test generation?*

4.1 Experimental Setup

In order to answer the two research questions, we created a dataset of Python projects for experimentation. We used the 'typed' category of the PyPI package index of Python projects, and selected ten projects by searching for projects that contain type hints in their method signatures, and that do not have dependencies to native-code libraries, such as `numpy`. Details of the chosen projects are shown in Table 1: the column *Project Name* gives the name of the project on PyPI; the lines of code were measured with the CLOC[6] utility tool. The table furthermore shows the absolute average number of code objects, predicates, and detected types per module of each project. The former two measures give an insight on the project's complexity; higher numbers indicate larger complexity. The latter provides an overview how many types PYNGUIN was able to parse (note that PYNGUIN may not be able to resolve all types).

The central metric we use to evaluate the performance of a test generation technique is code coverage. In particular, we measure branch coverage at the level of bytecode; like in Java bytecode, complex conditions are compiled to

[6] https://github.com/AlDanial/cloc, accessed 2020-07-25.

Table 1. Projects used for evaluation

Project name	Version	LOCs	Modules	CodeObjs.	Preds.	Types
apimd	1.0.2	316	1	35.0	83.0	11.0
async_btree	1.0.1	284	6	9.0	8.7	6.3
codetiming	1.2.0	85	2	18.0	8.0	6.0
docstring_parser	0.7.1	608	6	12.0	15.7	9.5
flutes	0.2.0.post0.	1085	9	19.0	26.0	5.0
flutils	0.6	1715	13	10.2	22.3	8.4
mimesis	4.0.0	1663	34	12.3	5.7	9.2
pypara	0.0.22	1305	6	47.2	23.5	12.0
python-string-utils	1.0.0	476	4	21.0	29.5	6.5
pytutils	0.4.1	1108	23	8.2	6.6	6.1
Total		8645	104	191.9	229.0	79.9

nested branches with atomic conditions also in Python code. In addition to the final overall coverage, we also keep track of coverage over time to shed light on the speed of convergence. In order to statistically compare results we use the Mann-Whitney U-test and the Vargha and Delaney effect size $\hat{A}_{12}$.

We executed PYNGUIN in four different configurations: First, we executed PYNGUIN using random test generation and whole test suite generation; second, we ran PYNGUIN with the developer-written type annotations contained in the projects, and without them. To answer RQ1, we compare the performance of random test generation and whole test suite generation; to answer RQ2 we compare the performance of each of these techniques for the case with and without type information.

For each project, PYNGUIN was run on each of the constituent modules in sequence. We executed PYNGUIN in GIT revision 5f538833 in a Docker container that is based on Debian 10 and utilises Python 3.8.3. In line with previous work, we set the maximum time limit for the test-generation algorithms, that is, the time without analysing the module-under-test and without export of the results, to 600 s per module. We ran PYNGUIN 30 times on each module and configuration to minimise the influence of randomness. All experiments were executed on dedicated compute servers equipped with Intel Xeon E5-2690v2 CPUs and 64 GB RAM, running Debian 10. All scripts and the raw data are available as supplementary material[7].

4.2 Threats to Validity

Internal Validity. The standard coverage tool for Python is COVERAGE.PY, which offers the capability to measure branch coverage. However, it measures branch coverage by comparing which transitions between source lines have

[7] https://github.com/se2p/artifact-pynguin-ssbse2020, accessed 2020-07-27.

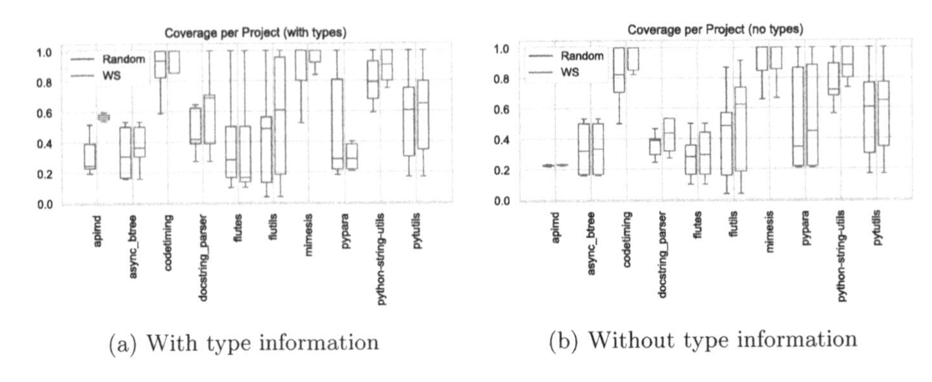

(a) With type information

(b) Without type information

Fig. 1. Coverage per project and configuration

occurred and which are possible. This method of measuring branch coverage is imprecise, because not every branching statement necessarily leads to a source line transition, for example, `x = 0 if y > 42 else 1337`. We thus implemented our own coverage measurement. We tried to mitigate possible errors in our implementation, by providing sufficient unit tests for it.

External Validity. We used 104 modules from ten different Python projects for our experiments. It is conceivable that the exclusion of projects without type annotations or native-code libraries leads to a selection of smaller projects, and the results may thus not generalise to other Python projects. However, besides the two constraints listed, no others were applied during the selection.

Construct Validity. Methods called with wrong input types may still cover parts of the code before possibly raising exceptions due to the invalid inputs. We conservatively included all coverage in our analysis, which may improve coverage for configurations that ignore type information, and thus reduce the effect we observed. However, it does not affect our general conclusions. Further, we cannot measure fault finding capability as our tool does not generate assertions, which is explicitly out of scope of this work.

4.3 RQ1: Whole-Suite Test Generation Vs. Random Testing

Figure 1 provides an overview over the achieved coverage per project in box plots. Each data point in the plot is one achieved coverage value for one of the modules of the project. Figure 1a reports the coverage values for whole-suite and random test generation with available type hints, whereas Fig. 1b reports the same without the usage of type hints to guide the generation.

Coverage values range from 0% to 100% depending on the project. The coverage achieved varies between projects, with some projects achieving generally high coverage (for example, `python-string-utils`, `mimesis`, `codetiming`), and others posing challenges for PYNGUIN (for example, `apmid`, `async_btree`, `pypara`,

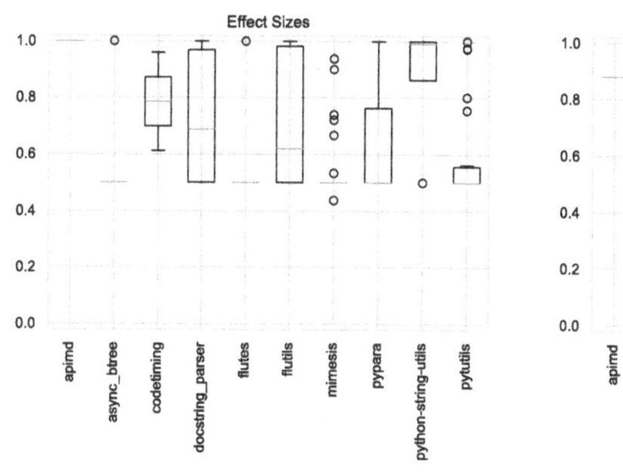

(a) $\hat{A}_{12}$ effect sizes with type information

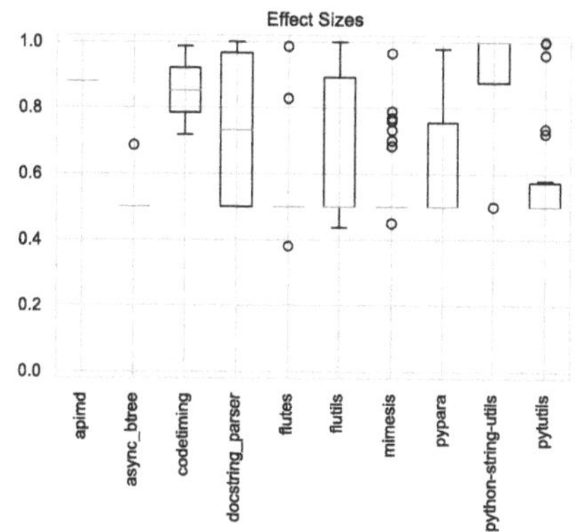

(b) $\hat{A}_{12}$ effect sizes without type information

Fig. 2. Effect sizes of whole suite versus random generation. Values greater than 0.500 indicate whole suite is better than random.

`flutes`). For example, for the `apimd` project without type information the coverage is slightly above 20%, which is the coverage achieved just by importing the module. In Python, when a module is imported, the import statements of the module, as well as class and method definitions are executed, and thus covered. Note that this does not execute the method bodies. For other projects with low coverage, PYNGUIN is not able to generate reasonable inputs, for example, higher-order functions or collections, due to technical limitations.

To better understand whether whole-suite test generation performs better than random test generation, Fig. 2a reports the $\hat{A}_{12}$ effect sizes for the per-module comparison of the two with available type information, whereas Fig. 2b reports the same without available type information. In both box plots, a value greater than 0.500 means that whole-suite performs better than random test generation, that is, yields higher coverage results. Both plots show that on average whole-suite does not perform worse than random and, depending on the project, is able to achieve better results in terms of coverage (average $\hat{A}_{12}$ with type information: 0.618, without type information: 0.603). The effect of these improvements is significant ($p < 0.05$) for six out of ten projects, most notably for `apimd` ($\hat{A}_{12} = 1.00, p{-}\text{value} < 0.001$), `python-string-utils` ($\hat{A}_{12} = 0.705, p{-}\text{value} < 0.001$), and `codetiming` ($\hat{A}_{12} = 0.636, p{-}\text{value} = 0.00531$).

For the other projects the effect is negligible. In case of `mimesis` ($\hat{A}_{12} = 0.530$) this is due to high coverage values in all configurations—most method parameters expect primitive types, which are also used for input generation if no type information is given. Other projects require specific technical abilities, for example, most methods in `async_btree` ($\hat{A}_{12} = 0.535$) are coroutines, which require special calls that cannot currently be generated by PYNGUIN. The consequence of this technical limitations is that PYNGUIN cannot reach higher coverage independent of the used algorithm in these cases. We observed that

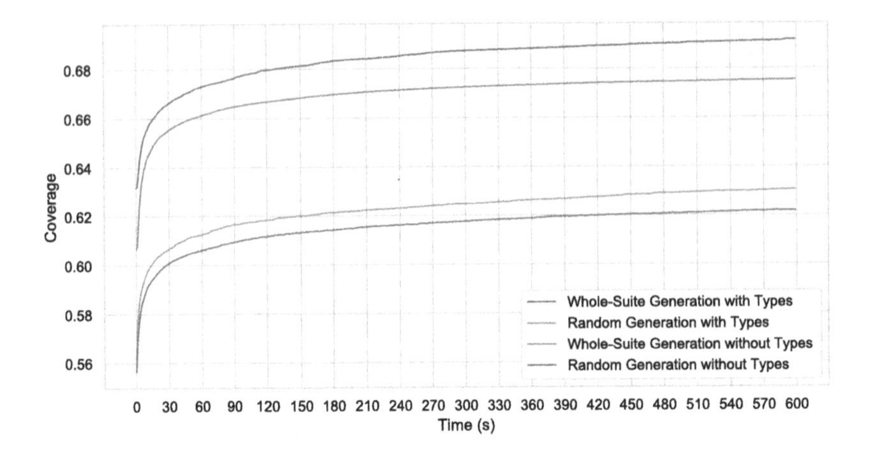

Fig. 3. Average coverage development over time

methods under test often require collection types as inputs, in Python prevalently lists and dictionaries. Also generating these input types would allow us to execute more parts of the code which would lead to higher coverage and thus better results. We leave this, however, as future work.

A further current limitation of our framework lies in how the available type information is processed. PYNGUIN can currently only generate inputs for concrete types, unions of types, and the `Any` type—for which it attempts to use a random type from the pool of available types in the SUT. Future work shall handle sub-typing relations as well as generic types [10]. Another prevalent parameter type that limits our current tool are callables, that is, higher order functions that can be used, for example, as call backs. Previous work has shown that generating higher-order functions as input types is feasible for dynamically typed languages and beneficial for test generation [18]. Furthermore, PYNGUIN currently only has a naive static seeding strategy for constants that incorporates all constant values from the project under test into test generation, whereas seeding has been show to have a positive influence on the quality of test generation [8] since it allows better-suited input values.

Figure 3 shows the development of the average coverage over all modules over the available generation time of 600 s. The line plot clearly indicates that whole-suite generation achieves higher coverage than random generation, which again supports our claim. Overall, we can answer our first research question as follows:

Summary (*RQ1*): Whole-suite test generation achieves at least as high coverage as random test generation. Depending on the project it achieves moderate to strongly higher coverage.

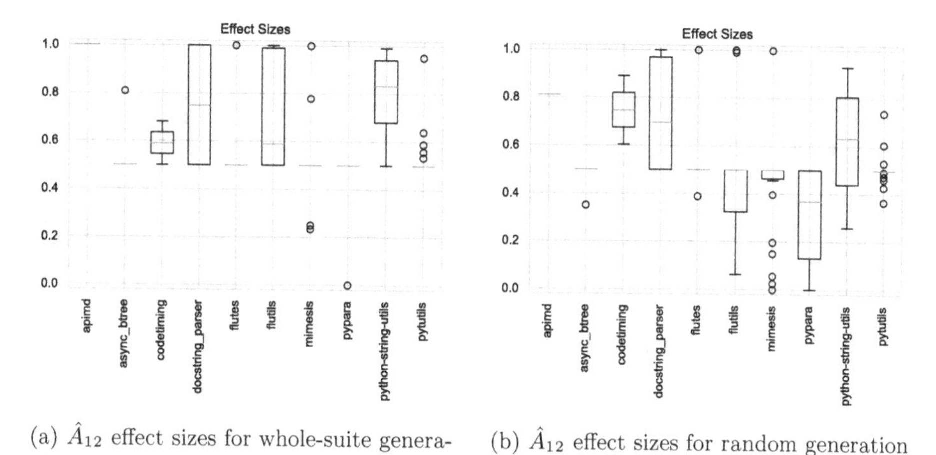

(a) $\hat{A}_{12}$ effect sizes for whole-suite generation

(b) $\hat{A}_{12}$ effect sizes for random generation

Fig. 4. Effect sizes for type influences

4.4 RQ2: Influence of Type Information

To answer RQ2 we compare the coverage values between the configurations with and without type annotations, again using the per-module $\hat{A}_{12}$ effect sizes on the coverage values. This time, we show the effect of type information for whole-suite generation in Fig. 4a and for random generation in Fig. 4b. For whole-suite generation, we observe a large positive effect on some modules, and barely any effect for other modules when type information is incorporated; we report an average $\hat{A}_{12}$ value of 0.578 in favour of type information. For random generation, we note similar effects, except for the `pypara` project ($\hat{A}_{12} = 0.411, p\text{−value} = 0.00937$); inspecting the `pypara` source code reveals that it uses abstract-class types as type annotations. Pynguin tries to instantiate the abstract class, which fails, and is thus not able to generate method inputs for large parts of the code because it cannot find an instantiable subtype. Overall, however, we report an average $\hat{A}_{12}$ value of 0.554 in favour of random generation with type information.

The box plots in Fig. 4 indicate similar conclusions for both whole-suite and random testing: the availability of type information is beneficial for some projects while its effect is negligible for other projects. The `docstring_parser` project, for example, requires their own custom types as parameter values for many methods. Without type information, Pynguin has to randomly choose types from all available types, with a low probability of choosing the correct one, whereas with available type information it can directly generate an object of correct type. Another effect comes in place for the `python-string-utils` projects: most of its methods only require primitive input types but very specific input values. Pynguin utilises a simple static constant seeding heuristic for input-value generation. Due to many values in the constant pool the chance of picking the correct value is smaller when not knowing the requested type, thus leading to lower coverage without type information.

On the other hand, projects such as `flutes` require iterables and callables as parameters in many cases or need special treatment of their methods to execute them properly (see coroutines in `async_btree`, for example). PYNGUIN currently lacks support to generate these required types, which prevents larger effects but does not limit the general approach. Thus, the type information cannot be used effectively, which results in negligible effects between the compared configurations.

The line plot in Fig. 3 shows the average coverage per evaluated configuration over the available time for test generation. It shows that both for whole-suite and random generation the configuration that incorporates type information yields higher coverage values over the full runtime of the generation algorithms, compared to ignoring type information. This again supports our claim that type information is beneficial when generating unit tests for Python programs. Overall, we therefore conclude for our second research question:

Summary (*RQ2*): Incorporating type information supports the test generation algorithms and allows them to cover larger parts of the code. The strength of this effect, however, largely depends on the SUT. Projects that require specific types from a large pool of potential types benefit more, and thus achieve larger effect sizes, than projects only utilising simple types.

5 Related Work

Closest to our work is whole-suite test generation in EVOSUITE [9] and feedback-directed random test generation in RANDOOP [15]. Both of these approaches target test generation for Java, a statically typed language, whereas our work adapts these approaches to Python.

To the best of our knowledge, little has been done in the area of automated test generation for dynamically typed languages. Approaches such as SYMJS [12] or JSEFT [14] target specific properties of JavaScript web applications, such as the browser's DOM or the event system. Feedback-directed random testing has also been adapted to web applications with ARTEMIS [3]. Recent work proposes LAMBDATESTER [18], a test generator that specifically addresses the generation of higher-order functions in dynamic languages. Our approach, in contrast, is not limited to specific application domains.

For automated generation of unit tests for Python we are only aware of AUGER[8]; it generates test cases from recorded SUT executions, while our approach does the generation automatically.

6 Conclusions

In this paper we presented PYNGUIN, an automated unit test generation framework for Python that is available as an open source tool, and showed that PYNGUIN is able to emit unit tests for Python that cover large parts of existing code

[8] https://github.com/laffra/auger, accessed 2020-07-25.

bases. PYNGUIN provides a whole-suite and a random test generation approach, which we empirically evaluated on ten open source Python projects. Our results confirm previous findings from the Java world that a whole-suite approach can outperform a random approach in terms of coverage. We further showed that the availability of type information has an impact on the test generation quality. Our investigations revealed a range of technical challenges for automated test generation, which provide ample opportunities for further research, for example, the integration of further test-generation algorithms, such as (Dyna)MOSA [16], the generation of assertions, or the integration of type inference approaches.

References

1. Andrews, J.H., Menzies, T., Li, F.C.: Genetic algorithms for randomized unit testing. IEEE Trans. Software Eng. **37**(1), 80–94 (2011)
2. Arcuri, A.: It really does matter how you normalize the branch distance in search-based software testing. Softw. Test. Verif. Reliab. **23**(2), 119–147 (2013)
3. Artzi, S., Dolby, J., Jensen, S.H., Møller, A., Tip, F.: A framework for automated testing of JavaScript web applications. In: Proceedings of the ICSE, pp. 571–580. ACM (2011)
4. Baresi, L., Miraz, M.: Testful: automatic unit-test generation for java classes. In: Proceedings of the ICSE, vol. 2, pp. 281–284. ACM (2010)
5. Campos, J., Ge, Y., Albunian, N., Fraser, G., Eler, M., Arcuri, A.: An empirical evaluation of evolutionary algorithms for unit test suite generation. Inf. Softw. Technol. **104**, 207–235 (2018)
6. Csallner, C., Smaragdakis, Y.: JCrasher: an automatic robustness tester for java. Softw. Pract. Exp. **34**(11), 1025–1050 (2004)
7. Fraser, G., Arcuri, A.: Evosuite: automatic test suite generation for object-oriented software. In: Proceedings of the ESEC/FSE, pp. 416–419. ACM (2011)
8. Fraser, G., Arcuri, A.: The seed is strong: seeding strategies in search-based software testing. In: Proceedings of the ICST, pp. 121–130. IEEE Computer Society (2012)
9. Fraser, G., Arcuri, A.: Whole test suite generation. IEEE Trans. Software Eng. **39**(2), 276–291 (2013)
10. Fraser, G., Arcuri, A.: Automated test generation for java generics. In: Winkler, D., Biffl, S., Bergsmann, J. (eds.) SWQD 2014. LNBIP, vol. 166, pp. 185–198. Springer, Cham (2014). https://doi.org/10.1007/978-3-319-03602-1_12
11. Levenshtein, V.I.: Binary codes capable of correcting deletions, insertions, and reversals. Soviet Physics Doklady **10**, 707–710 (1966)
12. Li, G., Andreasen, E., Ghosh, I.: SymJS: automatic symbolic testing of JavaScript web applications. In: Proceedings of the FSE, pp. 449–459. ACM (2014)
13. Ma, L., Artho, C., Zhang, C., Sato, H., Gmeiner, J., Ramler, R.: GRT: program-analysis-guided random testing (T). In: Proceedings of the ASE, pp. 212–223. IEEE Computer Society (2015)
14. Mirshokraie, S., Mesbah, A., Pattabiraman, K.: JSEFT: automated JavaScript unit test generation. In: Proceedings of the ICST, pp. 1–10. IEEE Computer Society (2015)
15. Pacheco, C., Lahiri, S.K., Ernst, M.D., Ball, T.: Feedback-directed random test generation. In: Proceedings of the ICSE, pp. 75–84. IEEE Computer Society (2007)

16. Panichella, A., Kifetew, F.M., Tonella, P.: Automated test case generation as a many-objective optimisation problem with dynamic selection of the targets. IEEE Trans. Software Eng. **44**(2), 122–158 (2018)
17. Sakti, A., Pesant, G., Guéhéneuc, Y.G.: Instance generator and problem representation to improve object oriented code coverage. IEEE Trans. Softw. Eng. **41**(3), 294–313 (2014)
18. Selakovic, M., Pradel, M., Karim, R., Tip, F.: Test generation for higher-order functions in dynamic languages. Proc. ACM Prog. Lang. **2**(OOPSLA), 16:11–16:127 (2018)
19. Tonella, P.: Evolutionary testing of classes. In: Proceedings of the ISSTA, pp. 119–128. ACM (2004)

Do Quality Indicators Prefer Particular Multi-objective Search Algorithms in Search-Based Software Engineering?

Shaukat Ali[1], Paolo Arcaini[2], and Tao Yue[1,3](✉)

[1] Simula Research Laboratory, Oslo, Norway
taoyue@nuaa.edu.cn
[2] National Institute of Informatics, Tokyo, Japan
[3] Nanjing University of Aeronautics and Astronautics, Nanjing, China

Abstract. In Search-Based Software Engineering (SBSE), users typically select a set of Multi-Objective Search Algorithms (MOSAs) for their experiments without any justification, or they simply choose an MOSA because of its popularity (e.g., NSGA-II). On the other hand, users know certain characteristics of solutions they are interested in. Such characteristics are typically measured with Quality Indicators (QIs) that are commonly used to evaluate the quality of solutions produced by an MOSA. Consequently, these QIs are often employed to empirically evaluate a set of MOSAs for a particular search problem to find the best MOSA. Thus, to guide SBSE users in choosing an MOSA that represents the solutions measured by a specific QI they are interested in, we present an empirical evaluation with a set of SBSE problems to study the relationships among commonly used QIs and MOSAs in SBSE. Our aim, by studying such relationships, is to identify whether there are certain characteristics of a QI because of which it prefers a certain MOSA. Such preferences are then used to provide insights and suggestions to SBSE users in selecting an MOSA, given that they know which quality aspects of solutions they are looking for.

Keywords: Search-based software engineering · Quality indicator · Multi-objective search algorithm

1 Introduction

Researchers and practitioners (i.e., users) in Search-Based Software Engineering (SBSE) dealing with multi-objective problems, need to select multi-objective search algorithms (MOSAs) to solve them. However, the use of a specific MOSA

This work is supported by the Co-evolver project (No. 286898/F20) funded by the Research Council of Norway, National Natural Science Foundation of China under Grant No. 61872182, and ERATO HASUO Metamathematics for Systems Design Project (No. JPMJER1603), JST, Funding Reference number: 10.13039/501100009024 ERATO.

A. Aleti and A. Panichella (Eds.): SSBSE 2020, LNCS 12420, pp. 25–41, 2020.
https://doi.org/10.1007/978-3-030-59762-7_3

in a particular SBSE context is rarely justified [14]. At the same time, users are aware of certain characteristics of solutions that they are interested to find with MOSAs. Such characteristics are measured with Quality Indicators (QIs) that are commonly employed to assess the quality of solutions produced by MOSAs from different perspectives. Consequently, these QIs are often used by users to select the best MOSA for their particular applications. To this end, the motivation of this paper is to help SBSE users in selecting an MOSA, i.e., if they are interested in the qualities of solutions preferred by a given QI, we identify the MOSA that will likely produce solutions having these qualities.

This application context is common in practical applications, where one knows which qualities in solutions they prefer and only wants to select one suitable MOSA without conducting extensive experiments to compare multiple MOSAs, which are often expensive. Indeed, in practical contexts, users have limited time budget to run experiments involving multiple MOSAs and would like to select one that produces the "best" solutions in terms of preferred QIs. Thus, in this paper, we aim to provide evidence that is useful for users to choose an MOSA that is highly likely to give solutions satisfying qualities measured by a preferred QI.

There exist surveys and studies about investigating various quality aspects of QIs and their relationships with MOSAs in the context of SBSE. For example, the survey of Sayyad et al. [14] reports that most of the publications in SBSE that were included in the survey do not provide justifications on why one or more particular MOSAs have been chosen in their experiments. The survey also reports that, in some of the investigated publications, researchers compared MOSAs against each other with certain QIs. This indicates that researchers were aware of the quality aspects provided by the selected QIs, and were looking for MOSAs that can produce solutions entailing such quality aspects. The survey also concluded that researchers sometimes chose MOSAs only based on their "popularity", i.e., selecting commonly used MOSA(s).

In the literature, there are also studies on investigating relationships of QIs and their characteristics. For instance, in our previous paper [1], we analyzed agreements among QIs commonly used in SBSE, with the aim to provide users with a set of guidelines on selecting QIs for their SBSE applications. Similarly, Li and Yao [10] surveyed 100 QIs that have been used in the evolutionary computation domain with the aim of studying their strengths and weaknesses. In the current SBSE literature, however, relationships between QIs and MOSAs are not well studied. In particular, the question of whether there are certain MOSAs that are preferred by a given QI is not answered. Answering such research question can help in guiding users to select an MOSA, on the basis of their preferences of QIs.

In this paper, we present an empirical evaluation in SBSE to study relationships between the QIs and the MOSAs to provide evidence indicating which MOSA(s) are highly likely to produce solutions that entail the given quality aspects represented by a specific QI. The empirical evaluation was performed with various industrial, real-world, and open source SBSE problems available online with commonly used MOSAs and QIs in SBSE. Our results reveal that certain QIs prefer specific MOSAs (e.g., Hypervolume prefers NSGA-II), whereas some QIs (e.g., the ones involving the quality aspect of *Cardinality* [10]) do not have any strong preference. Based on our results, we present a set of suggestions and insights to select an MOSA based on a particular QI or a category of quality aspects. The rest of the paper is organized as follows: Sect. 2 relates our work with the existing works in the literature. Section 3 shows the design of our empirical evaluation, and Sects. 4 and 5 describe analyses and results. Section 6 presents the discussion and our recommendations, whereas threats to validity are presented in Sect. 7. Finally, Sect. 8 concludes the paper.

2 Related Work

Sayyad and Ammar [14] presented a survey on SBSE papers that use MOSAs for solving software engineering optimization problems, from the perspectives of the chosen algorithms, QIs, and used tools. The paper concludes that more than half of the 51 surveyed papers do not provide justifications on the selection of a specific MOSA for a specific problem or simply state that an MOSA is selected because it is often applied by others. This observation, to a certain extent, implies that in the SBSE research community, there is no evidence showing which MOSA(s) to apply, in particular in the context in which researchers do know which QI(s) they prefer. Our current study provides evidence for guiding researchers selecting MOSAs based on the preferences of QIs.

The most relevant work, though not in the SBSE context, was presented by Ravber et al. in [13]. The work studied the impact of 11 QIs on the rating of 5 MOSAs: IBEA, MOEA/D, NSGA-II, PESA-II, and SPEA2, and concluded that QIs even with the same optimization goals (*convergence*, *uniformity*, and/or *spread*) might generate different and contradictory results in terms of ranking MOSAs. The authors analyzed the 11 QIs using a Chess Rating System for Evolutionary Algorithms [16], with 10 synthetic benchmark problems from the literature and 3 systems for a real-world problem. Based on the results of the analysis, the studied QIs were categorized into groups that had insignificant differences in ranking MOSAs. A set of guidelines were briefly discussed, considering preferred optimization aspects (e.g., *convergence*) when selecting QIs for a given search problem and selecting a robust (achieving the same rankings of MOSAs for different problems) and big enough set of QIs. To compare with our work reported in this paper, our study differentiates itself from [13] in the following two aspects. First, our study focuses exclusively on SBSE problems, whereas their study was conducted in a general context, and therefore the sets of MOSAs and QIs used in the two studies are different. The MOSAs and QIs we selected in our study are commonly applied ones in the context of SBSE. Second, our study aims to provide evidence on selecting an MOSA for solving an SBSE problem, in the context in which the user is aware of the desired quality aspects in the final solutions, and has limited time budget (in terms of running experiments). Instead, their study aims to suggest which QI(s) to select for assessing

MOSAs. When looking at the results of both studies on ranking MOSAs for each QI, there are similarities and dissimilarities, details of which will be discussed in Sect. 5.

Li and Yao reported a survey [10] on 100 QIs from the literature, discussed their strengths and weaknesses, and presented application scenarios for a set of QIs. In this survey, only two studies [9,18] related to SBSE were included, which are about understanding QIs from various aspects. Wang et al. [18] proposed a guide for selecting QIs in SBSE based on the results of an experiment with 8 QIs, 6 MOSAs, and 3 industrial and real-world problems. Their guide helps to determine a category of the QIs (*Convergence*, *Diversity*, *Combination of convergence and diversity*, or *Coverage*). In our previous work [1], we conducted an extensive empirical evaluation with 11 SBSE search problems from industry, real-world ones, and open source ones, and automatically produced 22 observations based on the results of the statistical tests for studying QI agreements, by considering different ways in which SBSE researchers typically compare MOSAs. We also provided a set of guidelines in the form of a process that can be used by SBSE researchers. To compare with our previous work [1], in this paper, we aim to suggest which MOSA to select given a QI that is preferred, while previously, we aimed at suggesting which QI(s) to use for evaluating a given MOSA.

3 Design of Empirical Evaluation

The design of our empirical evaluation is shown in Fig. 1. It consists of five steps. Steps 1–3 were performed by the authors of the papers [1,17], who afterwards made the data publicly available. We used this public data for our empirical evaluation and performed Steps 4–5. These steps will be described in following subsections.

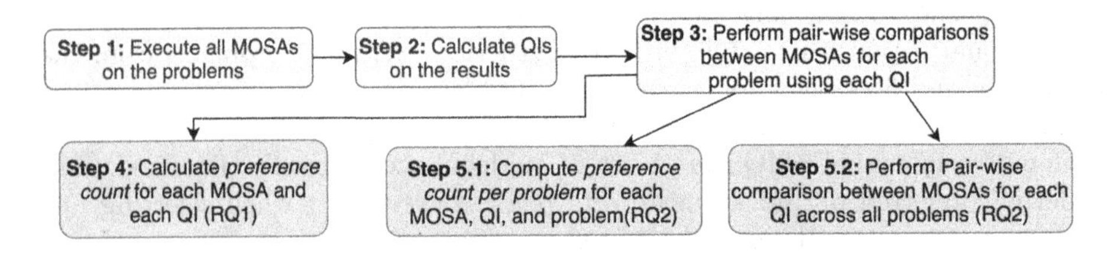

Fig. 1. Design of the experiment

3.1 Selection of Search Problems, MOSAs, and QIs

For our empirical evaluation, we chose data from SBSE search problems that are described in details in [1]. The data consists of a mix of industrial, real-world, and open source SBSE problems. In total, there were data available for 11 SBSE search problems. The data also had the results of 100 runs of commonly used MOSAs in SBSE (Step 1 in Fig. 1), i.e., NSGA-II [5], MoCell [12], SPEA2 [20], PAES [8], SMPSO [11], CellDE [7], and Random Search (RS). CellDE was not applicable to one of the SBSE problems that requires Integer solutions, because CellDE works for Real type solutions only [6]. The chosen MOSAs were run using appropriate parameter settings of the MOSAs based on the previous experiments provided in [1]. These parameter settings can be found in [1]. In addition, the available data also had computed QIs for the commonly used QIs in SBSE (Step 2 in Fig. 1), i.e., Generational Distance (GD), Euclidean Distance (ED), Epsilon (EP), Generalized Spread (GS), Pareto Front Size (PFS), Inverted Generational Distance (IGD), Hypervolume (HV), and Coverage (C). The definition of these QIs can be consulted in [10]. Moreover, the authors performed relevant statistical tests to compare each pair of MOSAs using each QI (Step 3 in Fig. 1). The results of the tests reveal which MOSA performed significantly better than the other one with respect to a particular QI. Note that all the MOSAs performed significantly better than RS; therefore, we did not include the results of RS. If we include the results of RS in our experiments, it will always be the least preferred by all the QIs.

The results of these statistical tests were used in our empirical evaluation reported in this paper: namely, we used them to perform Step 4 and Step 5, in order to answer the RQs defined in the next subsection.

3.2 Research Questions

Our overall objective is to study the relationships between the QIs and the MOSAs with the aim of finding whether there are specific characteristics of MOSAs that are preferred by a specific QI. To this end, we define the following Research Questions (RQs):

- RQ1: How frequently a QI prefers a particular MOSA? This RQ studies the percentage of times that a QI prefers a particular MOSA by ignoring the differences of the SBSE problems when studying pairs of MOSAs. This RQ helps in understanding the overall preferences of a QI.
- RQ2: How frequently a QI prefers a particular MOSA across the different SBSE problems? This RQ studies the preferences of QIs across the problems when studying pairs of MOSAs, whereas in RQ1 we aim to study preferences while ignoring the differences of the problems.

3.3 Evaluation Metrics, Statistical Analyses, and Parameter Settings

We define a set of evaluation metrics to answer the two RQs. First of all, we use the predicate *pref* to indicate the preference relationship between MOSAs for a given quality indicator and a given problem. Let Q be a quality indicator, A and B two MOSAs, and P a search problem. $pref(A, B, Q, P) = true$ iff Q prefers MOSA A to B when these are applied to the search problem P. The preference relation has been computed in our previous work [1], where we compared 100 runs of MOSA A and MOSA B over problem P using the *Mann-Whitney U* test and *Vargha and Delaney* $\hat{A}_{12}$ statistic. Note that $pref(A, B, Q, P) = true$ implies $pref(B, A, Q, P) = false$. If $pref(A, B, Q, P) = false$ and $pref(B, A, Q, P) = false$, it means that Q does not have any significant preference among the two MOSAs.

RQ1. In order to answer RQ1 (*Step 4* in Fig. 1), we introduce the following measure. Let $MOSAs$ be the set of MOSAs, $Problems$ the set of search problems, and Q a quality indicator. We define the *preference count* as the percentage of times Q prefers MOSA A when compared to another MOSA in any problem, formally:

$$PC(A, Q) = \frac{|\bigcup_{P \in Problems} \{B \in (MOSAs \setminus \{A\}) \mid pref(A, B, Q, P)\}|}{(|MOSAs| - 1) \times |Problems|} \tag{1}$$

The rationale is that if an MOSA A is consistently preferred by a QI Q (when compared with other MOSAs and for different problems), it means that A tends to produce solutions that have the quality aspects assessed by Q. The higher $PC(A, Q)$ is, the higher the probability is that, also on new problems, A will produce solutions preferred by Q.

RQ2. In order to answer RQ2, first, we compute the *preference count per problem* defined as follows (*Step 5.1* in Fig. 1)[1]:

$$PC(A, Q, P) = \frac{|\{B \in (MOSAs \setminus \{A\}) \mid pref(A, B, Q, P)\}|}{|MOSAs| - 1} \tag{2}$$

Second, we also perform, for each Q, pair-wise comparisons of the selected MOSAs across search problems (*Step 5.2* in Fig. 1). We choose the *Mann-Whitney U* test to determine the statistical significance of results, whereas we choose the *Vargha and Delaney* $\hat{A}_{12}$ statistics as the effect size measure. These statistical tests were chosen based on the published guidelines reported in [3]. When comparing two algorithms A and B with respect to a QI Q, if the p-value computed by the Mann-Whitney U test is less than 0.05 and $\hat{A}_{12}$ is greater than 0.5, then it means that A is significantly better than B with respect to Q. Similarly, when a p-value is less than 0.05 and $\hat{A}_{12}$ is less than 0.5, it means that B is significantly better than A. Finally, a p-value greater than or equal to 0.05 implies no significant differences between A and B with respect to Q.

The results of these tests give a more trustworthy definition of preference between MOSAs. In order to distinguish it from the one used in RQ1, we will call it *significant preference*, i.e., we will say that MOSA A is *significantly preferred* over MOSA B. Note that in Step 5.1 of Fig. 1, we study preferences per problem, whereas in Step 5.2 we study preferences across the problems. Moreover, in Step 4 we count preferences, while in RQ2 we determine the significance of preferences with the statistical tests.

[1] Note that CellDE is not applicable to one of the search problems, and so the formulations of Eq. 1 and Eq. 2 should be slightly more complicated. We report the simplified versions here, but we use the correct versions in the experiments.

4 Results and Analyses

In this section, we present the results and analyses for our RQs. Section 4.1 presents the results of RQ1, whereas Sect. 4.2 presents the results of RQ2.

4.1 RQ1

Recall that RQ1 aims to study the percentage of times that a QI Q prefers a particular MOSA A across all the problems when comparing pairs of MOSAs using the relevant statistical tests (see Eq. 1 in Sect. 3.3). Answering this RQ helps us in understanding the overall preferences of QIs.

The results are reported in Table 1. The *Preferred (%)* columns show the percentages calculated with the formula $PC(A, Q)$ presented in Sect. 3.3 (Eq. 1). A percentage determines the *preference count* of a QI for each MOSA across all the SBSE problems.

Table 1. RQ1 – *Preference count*

QI	MOSA	Preferred (%)	QI	MOSA	Preferred (%)	QI	MOSA	Preferred (%)	QI	MOSA	Preferred (%)
HV	NSGA-II	75.93%	EP	NSGA-II	74.07%	GS	SMPSO	74.07%	PFS	NSGA-II	62.96%
HV	SPEA2	68.52%	EP	SPEA2	70.37%	GS	CELLDE	64%	PFS	SPEA2	57.41%
HV	SMPSO	55.56%	EP	SMPSO	62.96%	GS	SPEA2	61.11%	PFS	SMPSO	35.19%
HV	CELLDE	42%	EP	CELLDE	46%	GS	NSGA-II	57.41%	PFS	PAES	31.48%
HV	MOCELL	29.63%	EP	MOCELL	27.78%	GS	MOCELL	22.22%	PFS	CELLDE	28%
HV	PAES	7.41%	EP	PAES	5.56%	GS	PAES	5.56%	PFS	MOCELL	18.52%
IGD	NSGA-II	77.78%	GD	SPEA2	77.78%	ED	SPEA2	66.67%	C	SPEA2	46.3%
IGD	SPEA2	77.78%	GD	NSGA-II	75.93%	ED	NSGA-II	62.96%	C	NSGA-II	38.89%
IGD	SMPSO	46.3%	GD	MOCELL	42.59%	ED	SMPSO	44.44%	C	CELLDE	22%
IGD	MOCELL	35.19%	GD	CELLDE	36%	ED	CELLDE	42%	C	MOCELL	18.52%
IGD	CELLDE	34%	GD	SMPSO	33.33%	ED	MOCELL	33.33%	C	SMPSO	16.67%
IGD	PAES	16.67%	GD	PAES	24.07%	ED	PAES	22.22%	C	PAES	7.41%

Based on the results, we can see that some QIs seem to prefer particular MOSAs. For example, HV prefers NSGA-II (75.93%) and SPEA2 is preferred by GD the most (77.78%). This observation suggests that NSGA-II may have some characteristics that are preferred by HV, and SPEA2 has some characteristics that are preferred by GD.

From the table, we can also observe that some QIs have low preference for some MOSAs, e.g., EP with PAES (5.56%). This means that such MOSAs do not usually produce solutions that have the qualities preferred by these QIs. Some QIs don't have strong preferences for any MOSA. For example, C has low percentages for all the MOSAs, thus suggesting that C assesses qualities that are not peculiar of any MOSA.

Moreover, in Fig. 2, we present preferences of each QI for all the selected MOSAs, sorted based on the percentages. For instance, by looking at HV and EP, we can see that NSGA-II was the most preferred by HV and EP, followed by SPEA2, whereas PAES was the least preferred by HV and EP. Similarly, GD, ED, and C prefer SPEA2 the most, followed by NSGA-II, and they least prefer PAES.

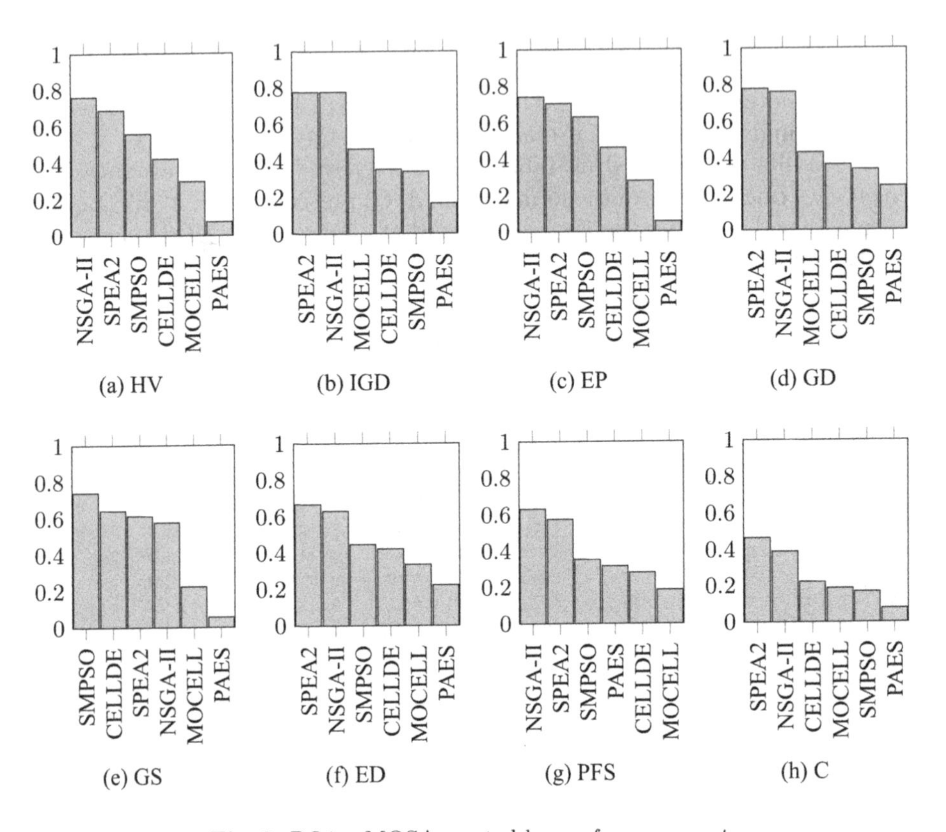

Fig. 2. RQ1 – MOSAs sorted by *preference count*

We provide a summary of the results in terms of which MOSAs are most preferred across all the QIs in Table 2. We can see from the table that NSGA-II and SPEA2 are on the top rankings (i.e., 1 and 2), meaning that these two MOSAs are the most preferred ones by most of the QIs, except for the cases of SMPSO for GS for rank 1 and CellDE for rank 2. We can also observe that PAES is least preferred by seven out of eight studied QIs. For PFS, MOCELL was the least preferred.

Table 2. RQ1 – Overall ranking of MOSAs

Rank	Instances for each MOSA
1	NSGA-II (3), SPEA2 (3), NSGA-II/SPEA2 (1), SMPSO (1)
2	NSGA-II (3), SPEA2 (3), CELLDE (1)
3	SMPSO (4), MOCELL (2), SPEA2 (1), CELLDE (1)
4	CELLDE (5), NSGA-II (1), PAES (1), MOCELL (1)
5	MOCELL (4), SMPSO (3), CELLDE (1)
6	PAES (7), MOCELL (1)

4.2 RQ2

Figure 3 reports, for each quality indicator Q and each MOSA A, the distribution of the metric *preference count per problem* $PC(A, Q, P)$ across search problems P (see Eq. 2). Of course, for MOSAs having a general low *preference count* (see Table 3), the variance of metric *preference count per problem* across problems is low. On the other hand, for MOSAs that in general are preferred, the variance is higher. This means that problem characteristics can influence the effectiveness of a particular MOSA A and so some quality indicator Q may prefer MOSA A on some problems, but not on some others. Note that the influence of the problem characteristics on the results of QIs has been discovered in a different setting. In our previous work [1], we discovered that the agreement between pairs of QIs, i.e., whether they prefer the same MOSA, sometimes depends on problems solved by MOSAs.

Note that we cannot perform an analysis on the base of the different problem characteristics (e.g., number of objectives), as this would require many more problems to have enough problems for each given characteristic. For the number of objectives, for example, we have four problems with two objectives, three problems with three objectives, and four problems with four objectives. This is not enough to draw any conclusion about the influence of the number of objectives.

Table 3 reports the overall results from the statistical test we performed. Recall from Sect. 3.3 that we performed the Mann-Whitney U test and the $\hat{A}_{12}$ statistics. A number in a cell of the table means the number of times that an MOSA (e.g., NSGA-II) was *significantly preferred* over other MOSAs, i.e., a p-value with the U Test was less than 0.05 and the $\hat{A}_{12}$ value greater than 0.5. For example, with respect to GD, NSGA-II was significantly preferred over other four MOSAs.

Based on the results, we can see that, for all the QIs, NSGA-II is significantly preferred by all the selected QIs, since the columns for NSGA-II has either the higher numbers (e.g., PFS) or equal numbers (e.g., EP) as compared to other MOSAs. Followed by NSGA-II, SPEA2 is the most preferred, having all the results the same as NSGA-II except for PFS, where NSGA-II has value 4, whereas SPEA2 has value 3.

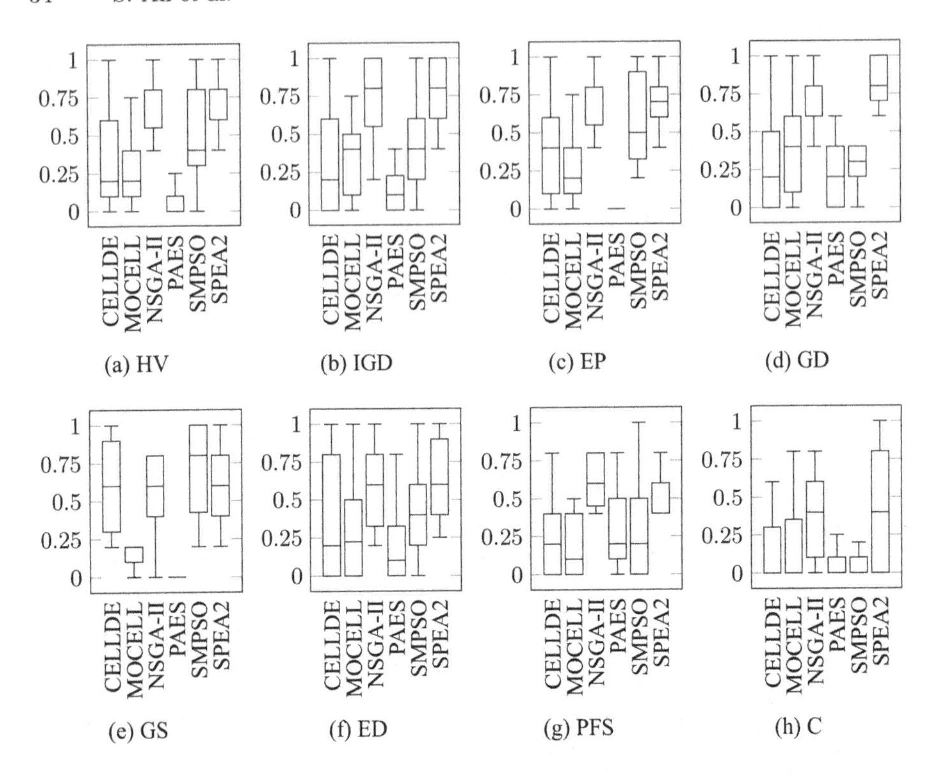

Fig. 3. RQ2 – Distribution of *preference count per problem* over search problems

Table 3. RQ2 – Overall preferences of QIs

	MOSA					
	CELLDE	MOCELL	NSGA-II	SMPSO	SPEA2	PAES
C	0	0	2	0	2	0
ED	0	0	2	0	2	0
EP	1	1	2	2	2	0
GD	0	0	4	0	4	0
GS	2	1	2	2	2	0
HV	1	1	3	1	2	0
IGD	0	0	4	1	4	0
PFS	0	0	4	0	3	0

Figure 4 shows a more detailed representation of the significant preference relation. For each QI, it shows which MOSAs are significantly preferred over others. An arrow from MOSA A to MOSA B means that A is significantly preferred over B. We observe that some MOSAs are consistently significantly preferred over some others, as NSGA-II and SPEA2 that are always preferred over PAES. The most preferred MOSAs (i.e., those with the highest numbers in Table 3) are usually preferred over the same other MOSAs. Moreover, there are some MOSAs that, although are worst than some MOSAs, are better than some others. For example, MOCELL in HV, EP, and GS. These are MOSAs that, although cannot produce optimal solutions, can still produce good ones.

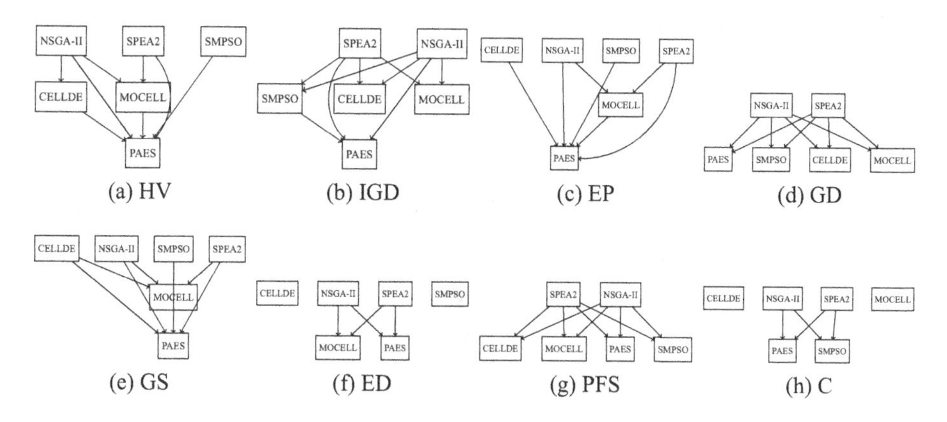

Fig. 4. RQ2 – Significant preference relation between MOSAs pairs

5 Analyses Based on Quality Aspects of QIs

At a higher level, each QI covers certain *quality aspects* of solutions produced by an MOSA. Thus, we analyzed our results from a different perspective, i.e., we checked whether an MOSA is preferred by QIs that cover certain quality aspects.

To this end, we present results of overall preferences for various categories of quality aspects measured by QIs. Li and Yao [10] categorized such quality aspects into four categories. The first category is *Convergence* that focuses on measuring the quality of a set of solutions on a Pareto front based on how close these solutions are to a reference Pareto front. The second category is *Spread*, which measures the quality by measuring the spread of solutions. The third category is *Uniformity* that measures the quality by looking at the distribution of solutions in terms of how evenly these solutions are distributed. Finally, the *Cardinality* category focuses on counting the number of solutions on a Pareto front. We chose this classification since it is published in the recent survey by Li and Yao [10] that studied 100 QIs from the literature. Table 4 shows which of our selected QIs fully or partially cover which categories of quality aspects.

Table 4. Relation between QIs and categories of quality aspects [10]

	QI							
	C	ED	EP	GD	GS	HV	IGD	PFS
Convergence	−	−	+	+		+	+	
Spread			+		−	+	+	
Uniformity			+		+	+	+	+
Cardinality			−			−	−	+

A cell with a "+" signifies that a particular quality aspect is fully represented by a QI, where "−" signifies a partial representation.

Table 5. Overall preferences of QI categories

	MOSA					
	CELLDE	MOCELL	NSGA-II	SMPSO	SPEA2	PAES
Convergence	2	2	17	4	16	0
Spread	4	3	11	6	10	0
Uniformity	4	3	11	6	10	0
Cardinality	2	2	13	4	11	0

Table 5 summarizes our results. For each MOSA A and each category, it reports how many times A is preferred (as reported in Table 3) by a QI belonging to the considered category. For example, for the *Convergence* category, NSGA-II has the highest value, i.e., 17. In general, we can see that NSGA-II is the most preferred one in all the four categories. However, note that these results are based on a particular set of SBSE problems, and once additional SBSE problems are added, these results may change.

We also checked whether QIs that cover the same quality aspects have also the same preferences for MOSAs. Looking at Table 4, we can see that C and ED are partially represented by the *Convergence* quality aspect. Now, looking at the results of C and ED in Fig. 2, we can see partial similarity, i.e., they both preferred SPEA2 and NSGA-II at the first and second place, whereas they both least preferred PAES. However, the other three MOSAs were in different positions. In addition, we can see that for C the preference count is in general lower for all the MOSAs as compared to ED.

As a further observation, we note that the quality indicators EP, HV, and IGD fully represent *Convergence*, *Spread* and *Uniformity*, and partially represent *Cardinality* as shown in Table 4. For HV and EP, we observe the same results in Fig. 2. However, for HV/EP and IGD, we see some differences. For example, SPEA2 and NSGA-II are equally at the first place for IGD, whereas NSGA-II is at the first place followed by SPEA2 for HV/EP. Also, MOCELL, CELLDE, and SMPSO are at different places.

Based on these observations, we can conclude that the QIs covering the same quality aspects don't necessarily have the same preferences for MOSAs. This observation is consistent with what has been reported in [13], in which the authors concluded that QIs with the same quality aspect(s) do not necessarily yield the same rankings of MOSAs.

6 Overall Discussion

Here, we present an overall discussion both for the results based on individual QIs and for the QI categories. Moreover, we provide suggestions to users for selecting an MOSA that will likely produce solutions preferred by a given QI or a given quality aspect.

When we look at the results based on QIs (see Table 3), we observe that both NSGA-II and SPEA2 are highly preferred by each QI. However, one must understand that some results provide more confidence than others when we want to suggest to pick a specific MOSA on the base of these results. For example, for HV, NSGA-II has a value of 3 out of 5 (i.e., 60% of the times NSGA-II was preferred over the other MOSAs), whereas, for PFS, NSGA-II has a value of 4 out of 5 (i.e., 80%). In both cases in which a user wants solutions that are represented by HV or PFS, our results suggest to use NSGA-II. However, in the latter case, the user will be more confident to follow the suggestion, because this is based on a stronger result.

Sometimes there is a tie among the MOSAs. For example, for EP, NSGA-II, SMPSO, and SPEA2 all have a value of 2. In these situations, we suggest the following options:

(1) selecting any MOSA, or
(2) selecting the MOSA from the category table (i.e., Table 5) by checking which quality aspect(s) is(are) represented by the selected QI (i.e., Table 4). For example, suppose that a user selected C. From Table 3, we see that there is a tie between NSGA-II and SPEA2 for C (value 2). In this case, the user may consult Table 4 and find that C represents solutions with the *Convergence* quality aspect. Then, the user can see from Table 5 that, for *Convergence*, NSGA-II is the preferred MOSA. Note that a QI could represent more than one category of quality aspects (e.g., HV). In our context, the preferred MOSA is always NSGA-II. However, when more data are available, the preferred MOSAs may be different for the different categories of quality aspects. In this case, more complicated guidelines could be provided (e.g., taking the MOSA scoring the highest in the majority of categories), or the user could decide to select the MOSA associated with the quality aspect they prefer.

Now, looking at our results for QI categories (Table 5), we note that NSGA-II is the clear option; however, once the results are updated based on additional experimental results of other problems, the preferences may change and we may have ties between two or more MOSAs. If such a case arises, we suggest selecting any of the preferred MOSAs of the tie or taking into account other aspects of MOSAs (e.g., their time performance).

Note that, as also observed in Sect. 4.2, in our experiments we did not study relationships between QI preferences and characteristics of the SBSE problems (e.g., search objective types, data distributions). Such characteristics could help us to provide a better guidance for SBSE users based on different characteristics of SBSE problems. Please note that conducting such an experiment requires a complete and well-planned experiment of its own, involving controlling various characteristics of SBSE problems in a systematic way. Finding publicly available SBSE problems that systematically cover various characteristics is challenging and one may resort to creating synthetic problems. We plan to conduct such an experiment in the future, where we could also study characteristics of QIs and SBSE problems together to suggest appropriate MOSAs.

7 Threats to Validity

We here discuss threats that may affect the validity of our experiments, namely *internal validity*, *conclusion validity*, *construct validity*, and *external validity* [19].

Internal Validity: Many MOSAs have been proposed in the literature, and a threat to the internal validity is that we may have not considered an MOSA that is consistently better than those considered in this paper. In order to address such a threat, we selected the MOSAs that are commonly used in SBSE [1,17].

Another threat is related to the settings of the parameters of the selected MOSAs. an MOSA A may perform better (i.e., preferred by a given QI) than another MOSA B, because it has been configured better. In order to address such threat, we have configured the selected MOSAs by following the commonly applied guides [3,15]. Note that these same settings were used in the papers from which we obtained the case studies, and in those papers these settings have been proven to give good results.

In terms of selection of QIs, one may argue that we did not cover enough QIs, given that there exist 100's of QIs [10]. However, note that we selected the most commonly used QIs in the SBSE literature [14], since our empirical evaluation was focused on SBSE problems. When presenting our results based on QI categories, one may wonder why we did not choose other QI categories, such as the ones proposed by Wang et al. [17]. We chose the QI categories instead from a recent survey [10], which is based on the study of 100 QIs, since it performs extensive evaluation of the existing QIs. Moreover, the QI categories in Wang et al. [17] are not precise as argued in [9].

Finally, we would like to mention that QIs may have different preferences thresholds to determine the significance of preferences. Such thresholds weren't studied in this paper and will require additional experiments.

Conclusion Validity: One such a threat is that the input data that we used in our experiment may not be sufficient to draw conclusions between the application of an MOSA A on a given problem P, and its evaluation with a given QI.

To mitigate such a threat, we have selected benchmarks in which each MOSA has been run 100 times, in order to reduce the effect of random variations. The conclusion whether a QI prefers an MOSA A to an MOSA B for a given problem P, is decided using the Mann-Whitney U test over the distribution of 100 QI values for A and B. Note that, in order to mitigate another threat related to wrong assumption for the tests, we selected such tests by following guidelines for conducting experiments in SBSE [3].

Construct Validity: One construct validity threat is that the measures we used for drawing our conclusions may not be adequate. As first measure, we computed the percentage of times that an MOSA A is preferred over another MOSA B by a given QI Q, since our aim is to suggest an MOSA that will likely produce solutions preferred by Q. Hence, we believe that this metric is adequate. Moreover, to draw more stable conclusions, we also assessed the statistical significance of the results with the Mann-Whitney U test and the $\hat{A}_{12}$ statistics. More specifically, we compared the *preference counts per problem* (see Eq. 2) of the two MOSAs across the problems with the statistical tests for each QI. Note that, in rare cases, the p-value with the Mann-Whitney U test may be less than 0.05, but the $\hat{A}_{12}$ gives a value still close to 0.5. This means that differences are not representing an actual preference. We need to look into such cases more carefully in the future.

External Validity: A major threat is that the results may not be generalizable to other case studies. In order to address such a threat, we selected as many SBSE problems as possible and ended up with 11 problems in total, trying to cover different types of SBSE problems: rule mining in product line engineering, test optimization, and requirements engineering. However, we are aware that such a selection is inherently partial, and we need more case studies from more SBSE problems to generalize the results. The lack of real-world case studies to be used in empirical studies is recognized to be a common threat to external validity [2, 4]. Note that the work presented in this paper does not aim at giving ultimate results, but at providing a methodology that should be followed to build a body of knowledge about the relationship between MOSAs and QIs. To this aim, we make our implementation publicly available[2] and invite SBSE researchers to share with us their empirical studies, so to derive more reliable conclusions.

8 Conclusion and Future Work

We were motivated by the observation that, in the community of search-based software engineering (SBSE), users (researchers and practitioners) often need to select a multi-objective search algorithm (MOSA) for their application, especially in the situation that the users do not have sufficient time budget to conduct experiments to compare multiple MOSAs. Though in the literature there exist

[2] Data and scripts are available at https://github.com/ERATOMMSD/QIsPreferences.

works studying QIs (their characteristics and relationships), the relationships between QIs and MOSAs are however not sufficiently studied. Motivated by this, in this paper, we presented an empirical evaluation and provided evidence to help users to choose an MOSA that is highly likely to produce solutions satisfying qualities measured by a QI preferred by the users. Specifically, we observed that NSGA-II and SPEA2 are preferred by most of the QIs we investigated; PAES is not preferred by most of the QIs. However, we would like to point out that, when selecting an MOSA, in addition to the quality aspects covered in each QI, other aspects such as time performance of MOSA should be considered as well. In the future, we would like to include such aspects in our study.

In addition, we would also like to design experiments for studying each specific quality aspect (e.g., Convergence). Furthermore, when more data will be available, we will conduct more analyses and update our findings. Finally, we would also like to study the preferences of QIs together with various characteristics of search problems. Such study can help users to select MOSAs based on characteristics of SBSE problems.

References

1. Ali, S., Arcaini, P., Pradhan, D., Safdar, S.A., Yue, T.: Quality indicators in search-based software engineering: an empirical evaluation. ACM Trans. Softw. Eng. Methodol. **29**(2) (2020). https://doi.org/10.1145/3375636
2. Ali, S., Briand, L.C., Hemmati, H., Panesar-Walawege, R.K.: A systematic review of the application and empirical investigation of search-based test case generation. IEEE Trans. Softw. Eng. **36**(6), 742–762 (2010)
3. Arcuri, A., Briand, L.: A practical guide for using statistical tests to assess randomized algorithms in software engineering. In: Proceedings of the 33rd International Conference on Software Engineering, ICSE 2011, pp. 1–10. ACM, New York (2011)
4. Barros, M., Neto, A.: Threats to validity in search-based software engineering empirical studies. RelaTe-DIA 5, January 2011
5. Deb, K., Pratap, A., Agarwal, S., Meyarivan, T.: A fast and elitist multiobjective genetic algorithm: NSGA-II. Trans. Evol. Comp **6**(2), 182–197 (2002)
6. Durillo, J.J., Nebro, A.J.: jMetal: a Java framework for multi-objective optimization. Adv. Eng. Softw. **42**(10), 760–771 (2011)
7. Durillo, J.J., Nebro, A.J., Luna, F., Alba, E.: Solving three-objective optimization problems using a new hybrid cellular genetic algorithm. In: Rudolph, G., Jansen, T., Beume, N., Lucas, S., Poloni, C. (eds.) PPSN 2008. LNCS, vol. 5199, pp. 661–670. Springer, Heidelberg (2008). https://doi.org/10.1007/978-3-540-87700-4_66
8. Knowles, J.D., Corne, D.W.: Approximating the nondominated front using the pareto archived evolution strategy. Evol. Comput. **8**(2), 149–172 (2000)
9. Li, M., Chen, T., Yao, X.: A critical review of: "a practical guide to select quality indicators for assessing Pareto-based search algorithms in search-based software engineering": essay on quality indicator selection for SBSE. In: Proceedings of the 40th International Conference on Software Engineering: New Ideas and Emerging Results, ICSE-NIER 2018, pp. 17–20. ACM, New York (2018)
10. Li, M., Yao, X.: Quality evaluation of solution sets in multiobjective optimisation: a survey. ACM Comput. Surv. **52**, 1–38 (2019)

11. Nebro, A.J., Durillo, J.J., Garcia-Nieto, J., Coello Coello, C.A., Luna, F., Alba, E.: SMPSO: a new PSO-based metaheuristic for multi-objective optimization. In: IEEE Symposium on Computational Intelligence in Multi-Criteria Decision-Making (MCDM), pp. 66–73, March 2009
12. Nebro, A.J., Durillo, J.J., Luna, F., Dorronsoro, B., Alba, E.: MOCell: a cellular genetic algorithm for multiobjective optimization. Int. J. Intell. Syst. **24**(7), 726–746 (2009)
13. Ravber, M., Mernik, M., Črepinšek, M.: The impact of quality indicators on the rating of multi-objective evolutionary algorithms. Appl. Soft Comput. **55**(C), 265–275 (2017)
14. Sayyad, A.S., Ammar, H.: Pareto-optimal search-based software engineering (POSBSE): a literature survey. In: 2nd International Workshop on Realizing Artificial Intelligence Synergies in Software Engineering (RAISE), pp. 21–27, May 2013
15. Sheskin, D.J.: Handbook of Parametric and Nonparametric Statistical Procedures, 5th edn. Chapman & Hall/CRC, Boca Raton (2011)
16. Veček, N., Mernik, M., Črepinšek, M.: A chess rating system for evolutionary algorithms: a new method for the comparison and ranking of evolutionary algorithms. Inf. Sci. **277**, 656–679 (2014)
17. Wang, S., Ali, S., Gotlieb, A.: Cost-effective test suite minimization in product lines using search techniques. J. Syst. Softw. **103**(C), 370–391 (2015)
18. Wang, S., Ali, S., Yue, T., Li, Y., Liaaen, M.: A practical guide to select quality indicators for assessing Pareto-based search algorithms in search-based software engineering. In: Proceedings of the 38th International Conference on Software Engineering, ICSE 2016, pp. 631–642. ACM, New York (2016)
19. Wohlin, C., Runeson, P., Hst, M., Ohlsson, M.C., Regnell, B., Wessln, A.: Experimentation in Software Engineering. Springer, Boston (2012). https://doi.org/10.1007/978-1-4615-4625-2
20. Zitzler, E., Laumanns, M., Thiele, L.: SPEA2: improving the strength Pareto evolutionary algorithm for multiobjective optimization. In: Evolutionary Methods for Design Optimization and Control with Applications to Industrial Problems, Athens, Greece, pp. 95–100 (2001)

It Is Not Only About Control Dependent Nodes: Basic Block Coverage for Search-Based Crash Reproduction

Pouria Derakhshanfar(✉), Xavier Devroey, and Andy Zaidman

Delft University of Technology, Delft, The Netherlands
{p.derakhshanfar,x.d.m.devroey,a.e.zaidman}@tudelft.nl

Abstract. Search-based techniques have been widely used for white-box test generation. Many of these approaches rely on the *approach level* and *branch distance* heuristics to guide the search process and generate test cases with high line and branch coverage. Despite the positive results achieved by these two heuristics, they only use the information related to the coverage of explicit branches (*e.g.*, indicated by conditional and loop statements), but ignore potential implicit branchings within basic blocks of code. If such implicit branching happens at runtime (*e.g.*, if an exception is thrown in a branchless-method), the existing fitness functions cannot guide the search process. To address this issue, we introduce a new secondary objective, called Basic Block Coverage (*BBC*), which takes into account the coverage level of relevant basic blocks in the control flow graph. We evaluated the impact of *BBC* on *search-based crash reproduction* because the implicit branches commonly occur when trying to reproduce a crash, and the search process needs to cover only a few basic blocks (*i.e.*, blocks that are executed before crash happening). We combined *BBC* with existing fitness functions (namely *STDistance* and *WeightedSum*) and ran our evaluation on 124 hard-to-reproduce crashes. Our results show that *BBC*, in combination with *STDistance* and *WeightedSum*, reproduces 6 and 1 new crashes, respectively. *BBC* significantly decreases the time required to reproduce 26.6% and 13.7% of the crashes using *STDistance* and *WeightedSum*, respectively. For these crashes, *BBC* reduces the consumed time by 44.3% (for *STDistance*) and 40.6% (for *WeightedSum*) on average.

Keywords: Automated crash reproduction · Search-based software testing · Evolutionary algorithm · Secondary objective

1 Introduction

Various search-based techniques have been introduced to automate different white-box test generation activities (*e.g.*, unit testing [9,10], integration testing [7], system-level testing [2], *etc.*). Depending on the testing level, each of

Electronic supplementary material The online version of this chapter (https://doi.org/10.1007/978-3-030-59762-7_4) contains supplementary material, which is available to authorized users.

A. Aleti and A. Panichella (Eds.): SSBSE 2020, LNCS 12420, pp. 42–57, 2020.
https://doi.org/10.1007/978-3-030-59762-7_4

these approaches utilizes dedicated fitness functions to guide the search process and produce a test suite satisfying given criteria (*e.g.*, line coverage, branch coverage, *etc.*).

Fitness functions typically rely on *control flow graphs (CFGs)* to represent the source code of the software under test [12]. Each node in a CFG is a *basic block* of code (*i.e.*, maximal linear sequence of statements with a single entry and exit point without any internal branch), and each edge represents a possible *execution flow* between two blocks. Two well-known heuristics are usually combined to achieve high line and branch coverages: the *approach level* and the *branch distance* [12]. The former measures the distance between the execution path of the generated test and a target basic block (*i.e.*, a basic block containing a statement to cover) in the CFG. The latter measures, using a set of rules, the distance between an execution and the coverage of a *true* or *false* branch of a particular predicate in a branching basic block of the CFG.

Both *approach level* and *branch distance* assume that only a limited number of basic blocks (*i.e.*, *control dependent* basic blocks [1]) can change the execution path away from a target statement (*e.g.*, if a target basic block is the true branch of an conditional statement). However, basic blocks are not atomic due to the presence of **implicit branches** [4] (*i.e.*, branches occurring due to the exceptional behavior of instructions). As a consequence, any basic block between the entry point of the CFG and the target baic block can impact the execution of the target basic block. For instance, a generated test case may stop its execution in the middle of a basic block with a runtime exception thrown by one of the statements of that basic block. In these cases, the search process does not benefit from any further guidance from the approach level and branch distance.

Fraser and Arcuri [11] introduced testability transformation, which instruments the code to guide the unit test generation search to cover implicit exceptions happening in the class under test. However, this approach does not guide the search process in scenarios where an implicit branch happens in the other classes called by the class under test. This is because of the extra cost added to the process stemming from the calculation and monitoring of the implicit branches in all of the classes, coupled with the class under test. For instance, the class under test may be heavily coupled with other classes in the project, thereby finding implicit branches in all of these classes can be expensive.

However, for some test case generation scenarios, like **crash reproduction**, we aim to cover a limited number of paths, and thereby we only need to analyse a limited number of basic blocks [5,13,16,19,21]. Current crash reproduction approaches rely on information about a reported crash (*e.g.*, stack trace, core dump *etc.*) to generate a **crash reproducing test case (CRT)**.

Among these approaches, search-based crash reproduction [16,19] takes as input a **stack trace** to guide the generation process. More specifically, the statements pointed by the stack trace act as target statements for the approach level and branch distance. Hence, current search-based crash reproduction techniques suffer from the lack of guidance in cases where the involved basic blocks contain implicit branches (which is common when trying to reproduce a crash).

Listing 1.1. Method **fromMap** from XWIKI version 8.1 [17]

```
402   public BaseCollection fromMap(Map<[...]> map, BaseCollection object){
403     for (PropertyClass property : (Collection<[...]>) getFieldList()) {
404       String name = property.getName();
405       Object formvalues = map.get(name);
406       if (formvalues != null) {
407         BaseProperty objprop;
408         if (formvalues instanceof String[]) {
409           [...]
410         } else if (formvalues instanceof String) {
411           objprop = property.fromString(formvalues.toString());
412         } else {
413           objprop = property.fromValue(formvalues);
414         }
415         [...]
416     }}
417     return object;}
```

This paper introduces a novel secondary objective called **Basic Block Coverage (*BBC*)** to address this guidance problem in crash reproduction. *BBC* helps the search process to compare two generated test cases with the same distance (according to approach level and branch distance) to determine which one is closer to the target statement. In this comparison, *BBC* analyzes the coverage level, achieved by each of these test cases, of the basic blocks in between the closest covered control dependent basic block and the target statement.

To assess the impact of *BBC* on search-based crash reproduction, we re-implemented the existing *STDistance* [16] and *WeightedSum* [19] fitness functions and empirically compared their performance with and without using *BBC* (4 configurations in total). We applied these four crash reproduction configurations to 124 hard-to-reproduce crashes introduced as JCrashPack [17], a crash benchmark used by previous crash reproduction studies [8]. We compare the performances in terms of *effectiveness in crash reproduction ratio* (*i.e.*, percentage of times that an approach can reproduce a crash) and *efficiency* (*i.e.*, time required by for reproducing a crash).

Our results show that *BBC* significantly improves the crash reproduction ratio over the 30 runs in our experiment for respectively 5 and 1 crashes when compared to using *STDistance* and *WeightedSum* without any secondary objective. Also, *BBC* helps these two fitness functions to reproduce 6 (for *STDistance*) and 1 (for *WeightedSum*) crashes that they could not be reproduced without secondary objective. Besides, on average, *BBC* increases the crash reproduction ratio of *STDistance* by 4%. Applying *BBC* also significantly reduces the time consumed for crash reproduction guided by *STDistance* and *WeightedSum* in 33 (26.6% of cases) and 14 (13.7% of cases) crashes, respectively, while it was significantly counter productive in only one case. In cases where *BBC* has a significant impact on efficiency, this secondary objective improves the average efficiency of *STDistance* and *WeightedSum* by 40.6% and 44.3%, respectively.

2 Background

2.1 Coverage Distance Heuristics

Many structural-based search-based test generation approaches mix the *branch distance* and *approach level* heuristics to achieve a high line and branch cover-

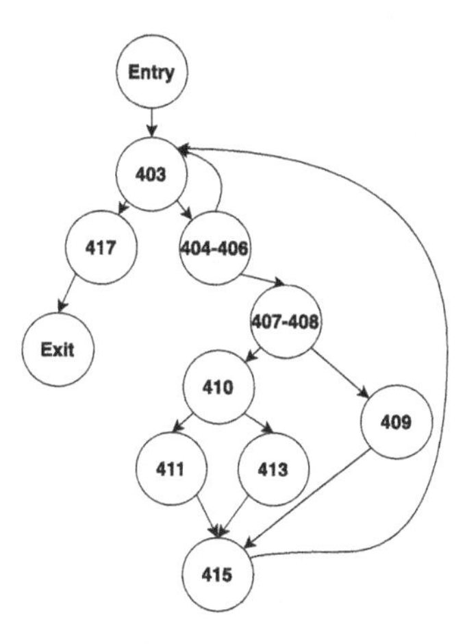

Fig. 1. CFG for method `fromMap`

age [12]. These heuristics measure the distance between a test execution path and a specific statement or a specific branch in the software under test. For that, they rely on the coverage information of *control dependent basic blocks*, *i.e.*, basic blocks that have at least one outgoing edge leading the execution path toward the *target basic block* (containing the targeted statement) and at least another outgoing edge leading the execution path away from the target basic block. As an example, Listing 1.1 shows the source code of method `fromMap` in XWIKI[1], and Fig. 1 contains the corresponding CFG. In this graph, the basic block `409` is control dependent on the basic block `407--408` because the execution of line 409 is dependent on the satisfaction of the predicate at line 408 (*i.e.*, line 409 will be executed only if elements of array `formvalues` are `String`).

The *approach level* is the number of uncovered control dependent basic blocks for the target basic block between the closest covered control dependent basic block and the target basic block. The *branch distance* is calculated from the predicate of the closest covered control dependent basic block, based on a set of predefined rules. Assuming that the test t covers only line `403` and `417`, and our target line is `409`, the approach level is 2 because two control dependent basic blocks (`404--406` and `407--408`) are not covered by t. The branch distance the predicate in line `403` (the closest covered control dependency of node `409`) is measured based on the rules from the established technique [12].

To the best of our knowledge, there is no related work studying the extra heuristics helping the combination of approach level and branch distance to improve the coverage. Most related to our work, Panichella *et al.* [14] and Rojas *et al.* [15] introduced two heuristics called *infection distance* and *propagation distance*, to improve the weak mutation score of two generated test cases. However, these heuristics do not help the search process to improve the general statement coverage (*i.e.*, they are effective only after covering a mutated statement).

[1] https://github.com/xwiki.

In this paper, we introduce a new secondary objective to improve the statement coverage achieved by fitness functions based on the approach level and branch distance, and analyze the impact of this secondary objective on **search-based crash reproduction**.

Listing 1.2. XWIKI-13377 crash stack trace [17]

```
0  java.lang.ClassCastException: [...]
1      at [...].BaseStringProperty.setValue(BaseStringProperty.java:45)
2      at [...].PropertyClass.fromValue(PropertyClass.java:615)
3      at [...].BaseClass.fromMap(BaseClass.java:413)
4      [...]
```

2.2 Search-Based Crash Reproduction

After a crash is reported, one of the essential steps of software debugging is to write a **Crash Reproducing Test case (CRT)** to make the crash observable to the developer and help them in identifying the root cause of the failure [22]. Later, this CRT can be integrated into the existing test suite to prevent future regressions. Despite the usefulness of a CRT, the process of writing this test can be labor-intensive and time-taking [19]. Various techniques have been introduced to automate the reproduction of a crash [5,13,16,19,21], and search-based approaches (EvoCrash [19] and ReCore [16]) yielded the best results [19].

EvoCrash. This approach utilizes a single-objective genetic algorithm to generate a CRT from a given stack trace and a *target frame* (*i.e.*, a frame in the stack trace that its class will be used as the class under test). The CRT generated by EvoCrash throws the same stack trace as the given one up to the target frame. For example, by passing the stack trace in Listing 1.2 and target frame 3 to EvoCrash, it generates a test case reproducing the first three frames of this stack trace (*i.e.*, thrown stack trace is identical from line 0 to 3).

EvoCrash uses a fitness function, called *WeightedSum*, to evaluate the candidate test cases. *WeightedSum* is the sum scalarization of three components: (i) the **target line coverage** (d_s), which measures the distance between the execution trace and the *target line* (*i.e.*, the line number pointed to by the target frame) using *approach level* and *branch distance*; (ii) the **exception type coverage** (d_e), determining whether the type of the triggered exception is the same as the given one; and (iii) the **stack trace similarity** (d_{tr}), which indicates whether the stack trace triggered by the generated test contains all frames (from the most in-depth frame up to the target frame) in the given stack trace.

Definition 1 (***WeightedSum*** *[19]***).** *For a given test case execution t, the WeightedSum (ws) is defined as follows:*

$$ws(t) = \begin{cases} 3 \times d_s(t) + 2 \times max(d_e) + max(d_{tr}) & \textit{if line not reached} \\ 3 \times min(d_s) + 2 \times d_e(t) + max(d_{tr}) & \textit{if line reached} \\ 3 \times min(d_s) + 2 \times min(d_e) + d_{tr}(t) & \textit{if exception thrown} \end{cases} \quad (1)$$

Where $d_s(t) \in [0,1]$ indicates how far t is from reaching the target line and is computed using the normalized approach level and branch distance: $d_s(t) = \|approachLevel_s(t) + \|branchDistance_s(t)\|\|$. Also, $d_e(t) \in \{0,1\}$ shows if the type of the exception thrown by t is the same as the given stack trace (0) or not (1). Finally, $d_{tr}(t) \in [0, 1]$ measures the stack trace similarity between the given stack trace and the one thrown by t. $max(f)$ and $min(f)$ denote the maximum and minimum possible values for a function f, respectively. In this fitness function, $d_e(t)$ and $d_{tr}(t)$ are only considered in the satisfaction of two *constraints*: (i) *exception type coverage* is relevant only when we reach the target line and (ii) *stack trace similarity* is important only when we both reach the target line and throw the same type of exception.

As an example, when applying EvoCrash on the stack trace from Listing 1.2 with the target frame 3, *WeightedSum* first checks if the test cases generated by the search process reach the statement pointed to by the target frame (line 413 in class `BaseClass` in this case). Then, it checks if the generated test can throw a `ClassCastException` or not. Finally, after fulfilling the first two constraints, it checks the similarity of frames in the stack trace thrown by the generated test case against the given stack trace in Listing 1.2.

EvoCrash uses **guided** initialization, mutation and single-point crossover operators to ensure that the target method (*i.e.,* the method appeared in the target frame) is always called by the different tests during the evolution process.

According to a recent study, EvoCrash outperforms other non-search-based crash reproduction approaches in terms of *effectiveness in crash reproduction* and *efficiency* [19]. This study also shows the helpfulness of tests generated by EvoCrash for developers during debugging.

In this paper, we assess the impact of *BBC* as the secondary objective in the EvoCrash search process.

ReCore. This approach utilizes a genetic algorithm guided by a single fitness function, which has been defined according to the core dump and the stack trace produced by the system when the crash happened. To be more precise, this fitness function is a sum scalarization of three sub-functions: (i) **TestStackTraceDistance**, which guides the search process according to the given stack trace; (ii) **ExceptionPenalty**, which indicates whether the same type of exception as the given one is thrown or not (identical to ExceptionCoverage in EvoCrash); and (iii) **StackDumpDistance**, which guides the search process by the given core dump.

Definition 2 (***TestStackTraceDistance*** *[16]*). *For a given test case execution t, the TestStackTraceDistance (STD) is defined as follows:*

$$STD(R, t) = |R| - lcp - (1 - StatementDistance(s)) \tag{2}$$

Where $|R|$ is the number of frames in the given stack trace. And lcp is the longest common prefix frames between the given stack trace and the stack trace thrown by t. Concretely, $|R| - lcp$ is the number of frames not covered by t. Moreover, $StatementDistance(s)$ is calculated using the sum of the approach level and the normalized branch distance to reach the statement s, which is pointed to by the first (the utmost) uncovered frame by t: $StatementDistance(s) = approachLevel_s(t) + \|branchDistance_s(t)\|$.

Since using runtime data (such as core dumps) can cause significant overhead [5] and leads to privacy issues [13], the performance of ReCore in crash reproduction was not compared with EvoCrash in prior studies [19]. Although, two out of three fitness functions in ReCore use only the given stack trace to guide the search process. Hence, this paper only considers *TestStackTraceDistance* + *ExceptionPenalty* (called *STDistance* hereafter).

As an example, when applying ReCore with *STDistance* on the stack trace in Listing 1.2 with target frame 3, first, *STDistance* determines if the generated test covers the statement at frame 3 (line 413 in class `BaseClass`). Then, it checks the coverage of frame 2 (line 615 in class `PropertyClass`). After covering the first two frames by the generated test case, it checks the coverage of the statement pointed to by the deepest frame (line 45 in class `BaseStringProperty`). For measuring the coverage of each of these statements, *STDistance* uses the approach level and branch distance. After covering all of the frames, this fitness function checks if the generated test throws `ClassCastException` in the deepest frame.

In this study, we perform an empirical evaluation to assess the performance of crash reproduction using *STDistance* with and without *BBC* as the secondary objective in terms of *effectiveness in crash reproduction* and *efficiency*.

3 Basic Block Coverage

3.1 Motivating Example

During the search process, the fitness of a test case is evaluated using a fitness function, either *WeightedSum* or *STDistance* Since the search-based crash reproduction techniques model this task to a minimization problem, the generated test cases with lower fitness values have a higher chance of being selected and evolved to generate the next generation. One of the main components of these fitness functions is the coverage of specific statements pointed by the given stack trace. The distance of the test case from the target statement is calculated using the approach level and branch distance heuristics. As we have discussed in Sect. 2.1, the approach level and branch distance cannot guide the search process if the execution stops because of implicit branches in the middle of basic blocks (*e.g.*, a thrown `NullPointerException` during the execution of a basic block). As a consequence, these fitness functions may return the same fitness value for two tests, although the tests do not cover the same statements in the block of code where the implicit branching happens.

For instance, assume that the search process for reproducing the crash in Listing 1.2 generates two test cases T_1 and T_2. The first step for these test cases is to cover frame 3 in the stack trace (line `413` in BaseClass). However, T_1 stops the execution at line `404` due to a `NullPointerException` thrown in method `getName`, and T_2 throws a `NullPointerException` at line `405` because it passes a null value input argument to `map`. Even though T_2 covers more lines, the combination of approach level and branch distance returns the same fitness value for both of these test cases: approach level is 2 (nodes `407-408` and `410`)

Listing 1.3. *BBC* secondary objective computation algorithm

```
1  input: test T1, test T2, String method, int line
2  output: int
3  begin
4      FCB1 ← fullyCoveredBlocks(T1,method,line);
5      FCB2 ← fullyCoveredBlocks(T2,method,line);
6      SCB1 ← semiCoveredBlocks(T1,method,line);
7      SCB2 ← semiCoveredBlocks(T2,method,line);
8
9      if (FCB1 ⊂ FCB2 ∧ SCB1 ⊂ SCB2) ∨ (FCB2 ⊂ FCB1 ∧ SCB2 ⊂ SCB1):
10         return size(FCB2 ∪ SCB2) - size(FCB1 ∪ SCB1)
11     else if FCB1 = FCB2 ∧ SCB1 = SCB2:
12         closestBlock ← closestSemiCoveredBlocks(SCB1, method, line);
13         coveredLines1 ← getCoveredLines(T1,closestBlock);
14         coveredLines2 ← getCoveredLines(T2,closestBlock);
15         return size(coveredLines2) - size(coveredLines1);
16     else:
17         return 0;
18 end
```

and branch distance is measured according to the last predicate. This is because these two heuristics assume that each basic block is atomic, and by covering line 404, it means that lines 405 and 406 are covered, as well.

3.2 Secondary Objective

The goal of the Basic Block Coverage (*BBC*) secondary objective is to prioritize the test cases with the same fitness value according to their coverage within the basic blocks between the closest covered control dependency and the target statement. At each iteration of the search algorithm, test cases with the same fitness value are compared with each other using *BBC*. Listing 1.3 presents the pseudo-code of the *BBC* calculation. Inputs of this algorithm are two test cases T_1 and T_2, which both have the same fitness value (calculated either using *WeightedSum* or *STDistance*), as well as line number and method name of the target statement. This algorithm compares the coverage of basic blocks on the path between the entry point of the CFG of the given method and the basic block that contains the target statement (called *effective blocks* hereafter) achieved by T_1 and T_2. If *BBC* determines there is no preference between these two test cases, it returns 0. Also, it returns a value < 0 if T_1 has higher coverage compared to T_2, and vice versa. A higher absolute value of the returned integer indicates a bigger distance between the given test cases.

In the first step, *BBC* detects the effective blocks fully covered by each given test case (*i.e.*, the test covers all of the statements in the block) and saves them in two sets called $\texttt{FCB}_1$ and $\texttt{FCB}_2$ (lines 4 and 5 in Listing 1.3). Then, it detects the effective blocks semi-covered by each test case (*i.e.*, blocks where the test covers the first line but not the last line) and stores them in $\texttt{SCB}_1$ and $\texttt{SCB}_2$ (lines 6 and 7). The semi-covered blocks indicate the presence of implicit branches. Next, *BBC* checks if both fully and semi-covered blocks of one of the tests are subsets of the blocks covered by the other test (line 9). In this case, the test case that covers the most basic blocks is the winner. Hence, *BBC* returns the number of blocks only covered by the winner test case (line 10). If *BBC* determines T_2 wins over T_1, the returned value will be positive, and vice versa.

If none of the test cases subsumes the coverage of the other one, *BBC* checks if the blocks covered by T_1 and T_2 are identical (line 11). If this is the case, *BBC* checks if one of the tests has a higher line coverage for the semi-covered blocks closest to the target statement (lines 12 to 15). If this is the case, *BBC* will return the number of lines in this block covered only by the winning test case. If the lines covered are the same for T_1 and T_2 (*i.e.*, `coveredLines1` and `coveredLines2` have the same size), there is no difference between these two test cases and *BBC* returns value 0 (line 15). Finally, if each of the given tests has a unique covered block in the given method (*i.e.*, the tests cover different paths in the method), *BBC* cannot determine the winner and returns 0 (lines 16 and 17) because we do not know which path leads to the crash reproduction.

Example. When giving two tests with the same fitness value (calculated by the primary objective) T_1 and T_2 from our motivation example to *BBC* with target method `fromMap` and line number 413 (according to the frame 3 of Listing 1.2), this algorithm compares their fully and semi-covered blocks with each other. In this example both T_1 and T_2 cover the same basic blocks: the fully covered block is `403` and the semi-covered block is `404-406`. So, *BBC* checks the number of lines covered by T_1 and T_2 in block `404-406`. Since T_1 stopped its execution at line 404, the number of lines covered by this test is 1. In contrast, T_2 managed to execute two lines (404 and 405). Hence, *BBC* returns $size(coveredLines2) - size(coveredLines1) = 1$. The positive return value indicates that T_2 is closer to the target statement and therefore, it should have higher chance to be selected for the next generation.

Branchless Methods. *BBC* can also be helpful for branchless methods. Since there are no control dependent nodes in branchless methods, approach level and branch distance cannot guide the search process in these cases. For instance, methods from frames 1 and 2 in Listing 1.2 are branchless. So, we expect that *BBC* can help the current heuristics to guide the search process toward covering the most in-depth statement.

4 Empirical Evaluation

To assess the impact of *BBC* on search-based crash reproduction, we perform an empirical evaluation to answer the following research questions:

RQ$_1$: *What is the impact of BBC on crash reproduction in terms of effectiveness in crash reproduction ratio?*

RQ$_2$: *What is the impact of BBC on the efficiency of crash reproduction?* *In these two RQs we want to evaluate the effect of BBC* on the existing fitness functions, namely *STDistance* and *WeightedSum*, from two perspectives: effectiveness on crash reproduction ratio and efficiency.

4.1 Setup

Implementation. Since ReCore and EvoCrash are not openly available, we implement *BBC* in Botsing, an extensible, well-tested, and open-source search-based crash reproduction framework already implementing the *WeightedSum* fitness function and the guided initialization, mutation, and crossover operators. We also implement *STDistance* (ReCore fitness function) in this tool. Botsing relies on EvoSuite [9], an open-source search-based tool for unit test generation, for code instrumentation and test case generation by using `evosuite-client` as a dependency. We also implement the *STDistance* fitness function used as baseline in this paper.

Crash Selection. We select crashes from JCrashPack [17], a benchmark containing hard-to-reproduce Java crashes. We apply the two fitness functions with and without using *BBC* as a secondary objective to 124 crashes, which have also been used in a recent study [8]. These crashes stem from six open-source projects: JFreeChart, Commons-lang, Commons-math, Mockito, Joda-time, and XWiki. For each crash, we apply each configuration on each frame of the crash stack traces. We repeat each execution 30 times to take randomness into account, for a total number of 114,120 independent executions. We run the evaluation on two servers with 40 CPU-cores, 128 GB memory, and 6 TB hard drive.

Parameter Settings. We run each search process with five minutes budget and set the population size to 50 individuals, as suggested by previous studies on search-based test generation [14]. Moreover, as recommended in prior studies on search-based crash reproduction [19], we use the *guided mutation* with a probability $p_m = 1/n$ (n = length of the generated test case), and the *guided crossover* with a probability $p_c = 0.8$ to evolve test cases. We do note that prior studies do not investigate the sensitivity of the crash reproduction to these probabilities. Tuning these parameters should be undertaken as future works.

4.2 Data Analysis

To evaluate the crash reproduction ratio (*i.e.*, the ratio of success in crash reproduction in 30 rounds of runs) of different assessed configurations ($\mathbf{RQ}_1$), we follow the same procedure as previous studies [8,18]: for each crash C, we detect the highest frame that can be reproduced by at least one of the configurations (r_{max}). We examine the crash reproduction ratio of each configuration for crash C targeting frame r_{max}. Since crash reproduction data has a dichotomic distribution (*i.e.*, an algorithm reproduces a crash C from its r_{max} or not), we use the Odds Ratio (OR) to measure the impact of each algorithm in crash reproduction ratio. A value $OR > 0$ in a comparison between a pair of factors (A, B) indicates that the application of factor A increases the crash reproduction ratio, while $OR < 0$ indicates the opposite. Also, a value of $OR = 0$ indicates that both of the factors have the same performance. We apply Fisher's exact test, with $\alpha = 0.01$ for the Type I error, to assess the significance of results.

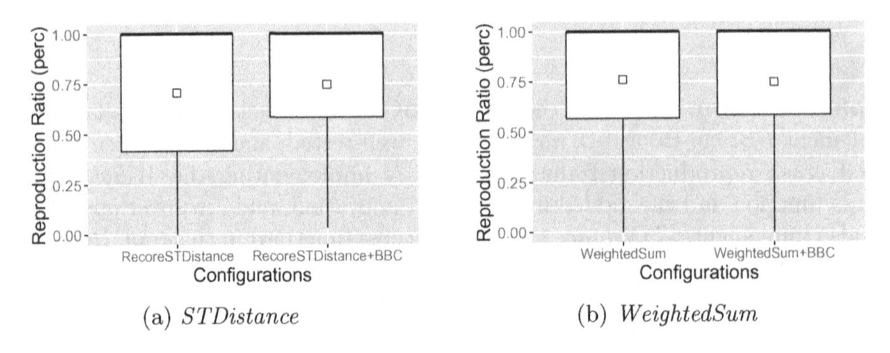

(a) *STDistance* (b) *WeightedSum*

Fig. 2. Crash reproduction ratio (out of 30 executions) of fitness functions with and without *BBC*. (□) denotes the arithmetic mean and the bold line (—) is the median.

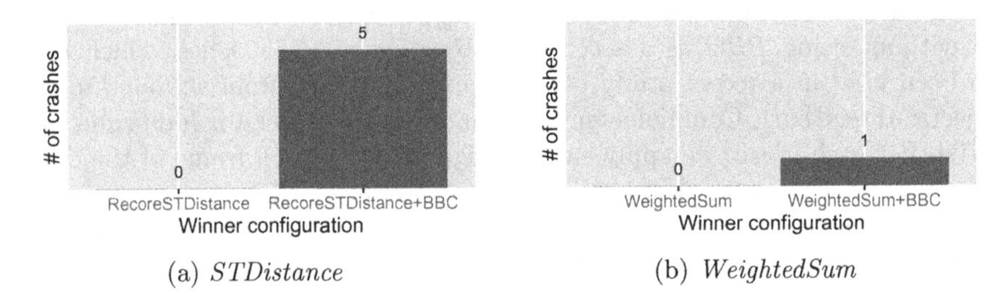

(a) *STDistance* (b) *WeightedSum*

Fig. 3. Pairwise comparison of impact of *BBC* on each fitness function in terms of crash reproduction ratio with a statistical significance <0.01.

To evaluate the efficiency of different configurations (**RQ**$_2$), we analyze the time spent by each configuration on generating a crash reproducing test case. We do note that the extra pre-analysis and basic block coverage in *BBC* is considered in the spent time. Since measuring efficiency is only possible for the reproduced crashes, we compare the efficiency of algorithms on the crashes that are reproduced at least once by one of the algorithms. In executions that an algorithm failed to reproduce a crash, we assume that it reached the maximum allowed budget (5 min).

In this study, we use the Vargha-Delaney $\hat{A}_{12}$ statistic [20] to examine the effect size of differences between using and not using *BBC* for efficiency. For a pair of factors (A, B) a value of $\hat{A}_{12} > 0.5$ indicates that A reproduces the target crash in a longer time, while a value of $\hat{A}_{12} < 0.5$ shows the opposite. Also, $\hat{A}_{12} = 0.5$ means that there is no difference between the factors. In addition, to assess the significance of effect sizes ($\hat{A}_{12}$), we utilize the non-parametric Wilcoxon Rank Sum test, with $\alpha = 0.01$ for the Type I error.

A replication package of this study has been uploaded to Zenodo [6].

5 Results

Crash Reproduction Effectiveness ($\mathbf{RQ_1}$). Figure 2 presents the crash reproduction ratio of the search processes guided by *STDistance* (Fig. 2a) and *WeightedSum* (Fig. 2), with and without *BBC* as a secondary objective. This figure shows that the crash reproduction ratio of *WeightedSum* improves slightly when using *BBC*. However, on average, the crash reproduction ratio achieved by *STDistance* + *BBC* is 4% better than *STDistance* without *BBC*. Also, the lower quartile of crash reproduction ratio using *STDistance* has been improved by about 30% by utilizing *BBC*.

Figure 3 depicts the number of crashes, for which *BBC* has a significant impact on the effectiveness of crash reproduction guided by *STDistance* (Fig. 3a) and *WeightedSum* (Fig. 3b). *BBC* significantly improves the crash reproduction ratio in 5 and 1 crashes for fitness functions *STDistance* and *WeightedSum*, respectively. Importantly, the application of this secondary objective does not have any significant negative effect on crash reproduction. Also, *BBC* helps *STDistance* and *WeightedSum* to reproduce 6 and 1 new crashes, respectively (in at least one out of 30 runs), that could not be reproduced without *BBC*.

Summary. *BBC* slightly improves the crash reproduction ratio when using the *WeightedSum* fitness function. However, on average, *BBC* achieves a higher improvement when used as a secondary objective with the *STDistance* function.

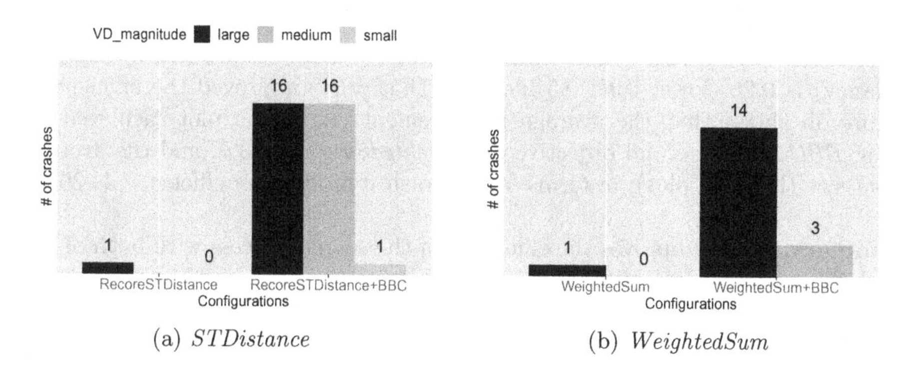

(a) *STDistance* (b) *WeightedSum*

Fig. 4. Pairwise comparison of impact of *BBC* on each fitness function in terms of efficiency with a small, medium, and large effect size $\hat{A}_{12} < 0.5$ and a statistical significance <0.01.

Crash Reproduction Efficiency ($\mathbf{RQ_2}$). Figure 4 illustrates the number of crashes, in which *BBC* significantly affects the time consumed by the crash reproduction search process. As Fig. 4b shows, *BBC* significantly improves the speed of crash reproduction guided by *WeightedSum* in 17 crashes (13.7% of cases), while it lost efficiency in the reproduction of only one crash. In cases that *BBC* significantly improves the efficiency of *WeightedSum*, on average, the efficiency is improved for about 40%. Moreover, Fig. 4a shows that *BBC* has a higher

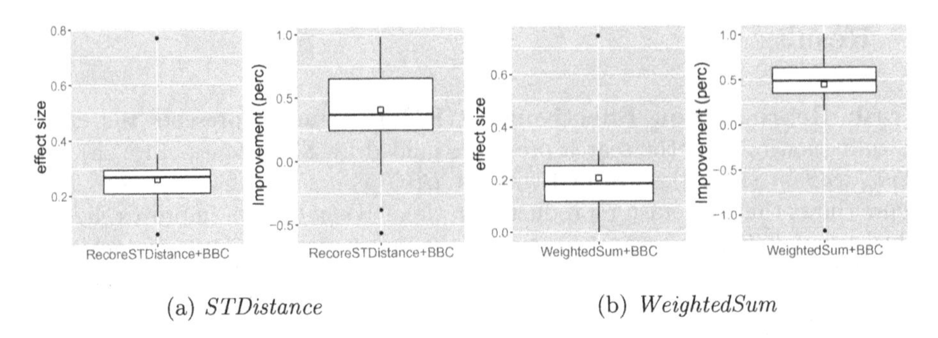

(a) *STDistance* (b) *WeightedSum*

Fig. 5. The effect size and the average improvement achieved by *BBC* on each of the fitness functions in cases that *BBC* makes a significant difference in terms of efficiency.

positive impact on the efficiency of the search process guided by *STDistance* It significantly reduces the time consumed by the search process in 33 crashes (26.6% of cases), while it had an adverse impact on the reproduction efficiency of only one crash. In cases that *BBC* significantly improves the efficiency of *STDistance*, on average, the efficiency is improved for about 53%.

Figure 5 depicts the average improvements in the efficiency and effect sizes for crashes where the difference in the consumed budget when using *BBC* or not was significant. According to the right-side plot in Fig. 5a, *BBC* reduces the time consumed by the search process guided by *STDistance* up to 98% (being 40.6% on average). Also, the left-side plot indicates that the average effect size of differences between *STDistance* and *STDistance* + *BBC* (calculated by Vargha-Delaney) is 0.26 (lower than 0.5 indicates that *BBC* improved the efficiency). Figure 5b shows that the average improvement (right-side plot) achieved by using *BBC* as the second objective of *WeightedSum* is 44.3%, and the average effect size (left-side plot), in terms of the crash reproduction efficiency, is 20.5.

Summary. *BBC* improves the efficiency of the search process with both of the crash reproduction fitness functions.

6 Discussion

Generally, using *BBC* as secondary objective leads to a better crash reproduction ratio and higher efficiency in search-based crash reproduction. This improvement is achieved thanks to the additional ability to guide the search process when facing implicit branches during the search. Combining *BBC* with *STDistance* shows an important improvement compared to the combination of *BBC* with *WeightedSum*. This result was expected, since only one (out of three) component in *WeightedSum* is allocated to line coverage, and thereby most parts of the fitness function do not use the approach level and branch distance heuristics. In contrast, *STDistance* uses the approach level and branch distance to cover each of the frames in the given stack trace incrementally.

Our results show that *BBC* helps the crash reproduction process to reproduce new crashes. For instance, the crash that we used in this study (XWIKI-13377) can be reproduced only by *STDistance* + *BBC*. Considering our results, we believe that the usage of approach level and branch distance can be improved in other areas of search-based test generation (*e.g.*, unit testing) by taking the *implicit branches* into account. However, it can be expensive to apply this secondary objective in cases where the search process tries to cover multiple paths. Assessing the impact of *BBC* on other search-based test generation techniques is part of our future research agenda.

Threats to Validity. We cannot guarantee that our implementation of BOTSING is bug-free. However, we mitigated this threat by testing our tool and manually examining some samples of the results. We cannot ensure that our results are generalizable to all crashes. However, we used an earlier established benchmark for crash reproduction containing 124 hard-to-reproduce crashes provoked by real bugs in a variety of open-source applications. Moreover, by following the guidelines of the related literature [3], we executed each configuration 30 times to take the randomness of the search process into account. Finally, we provide BOTSING as an open-source tool. Also, the data and the processing scripts used to present the results are available as a replication package on Zenodo [6].

7 Conclusion and Future Work

Approach level and branch distance are two well-known heuristics, widely used by search-based test generation approaches to guide the search process towards covering target statements and branches. These heuristics measure the distance of a generated tests from covering the target using the coverage of control dependencies. However, these two heuristics do not consider implicit branches. For instance, if a test throws an exception during the execution of a non-branch statement, approach level and branch distance cannot guide the search process to tackle this exception. In this paper, we introduced a secondary objective called *BBC* to address this issue. To assess *BBC*, we used it for search-based crash reproduction due to the high chance of implicit branch occurrence and the limited number of basic blocks that should be covered. Our results show that *BBC* helps *STDistance* and *WeightedSum* to reproduce 6 and 1 new crashes, respectively. Also, *BBC* significantly improves the efficiency in 26.6% and 13.7% of the crashes using *STDistance* and *WeightedSum*, respectively.

In our future work, we will investigate the application of *BBC* for other search-based test generation techniques (such as unit and integration).

Acknowledgements. The authors would like to thank Carolin Brandt for her valuable feedback on the paper. This research was partially funded by the EU Horizon 2020 ICT-10-2016-RIA "STAMP" project (No. 731529).

References

1. Allen, F.E.: Control flow analysis, vol. 5, pp. 1–19. ACM, New York (1970). https://doi.org/10.1145/390013.808479
2. Arcuri, A.: RESTful API automated test case generation with Evomaster. ACM Trans. Softw. Eng. Methodol. (TOSEM) **28**(1), 1–37 (2019)
3. Arcuri, A., Briand, L.: A Hitchhiker's guide to statistical tests for assessing randomized algorithms in software engineering. Softw. Test. Verif. Reliab. **24**(3), 219–250 (2014). https://doi.org/10.1002/stvr.1486
4. Borba, P., Cavalcanti, A., Sampaio, A., Woodcook, J. (eds.): PSSE 2007. LNCS, vol. 6153. Springer, Heidelberg (2010). https://doi.org/10.1007/978-3-642-14335-9
5. Chen, N., Kim, S.: STAR: stack trace based automatic crash reproduction via symbolic execution. IEEE Trans. Softw. Eng. **41**(2), 198–220 (2015)
6. Derakhshanfar, P., Devroey, X.: Replication package of basic block coverage for search-based crash reproduction. https://doi.org/10.5281/zenodo.3953519
7. Derakhshanfar, P., Devroey, X., Panichella, A., Zaidman, A., van Deursen, A.: Towards integration-level test case generation using call site information. arXiv preprint arXiv:2001.04221 (2020)
8. Derakhshanfar, P., Devroey, X., Perrouin, G., Zaidman, A., van Deursen, A.: Search-based crash reproduction using behavioural model seeding. Softw. Test. Verif. Reliab. (2020). https://doi.org/10.1002/stvr.1733
9. Fraser, G., Arcuri, A.: Evosuite: automatic test suite generation for object-oriented software. In: Proceedings of the 19th ACM SIGSOFT Symposium and the 13th European Conference on Foundations of software Engineering, pp. 416–419 (2011)
10. Fraser, G., Arcuri, A.: Whole test suite generation. IEEE Trans. Softw. Eng. **39**(2), 276–291 (2012)
11. Fraser, G., Arcuri, A.: 1600 faults in 100 projects: automatically finding faults while achieving high coverage with Evosuite. Empirical Software Eng. **20**(3), 611–639 (2015)
12. McMinn, P.: Search-based software test data generation: a survey. Software Test. Verif. Reliab. **14**(2), 105–156 (2004)
13. Nayrolles, M., Hamou-Lhadj, A., Tahar, S., Larsson, A.: Jcharming: a bug reproduction approach using crash traces and directed model checking. In: International Conference on Software Analysis, Evolution, and Reengineering (SANER), pp. 101–110. IEEE (2015)
14. Panichella, A., Kifetew, F.M., Tonella, P.: Automated test case generation as a many-objective optimisation problem with dynamic selection of the targets. IEEE Trans. Software Eng. **44**(2), 122–158 (2018)
15. Rojas, J.M., Campos, J., Vivanti, M., Fraser, G., Arcuri, A.: Combining multiple coverage criteria in search-based unit test generation. In: Barros, M., Labiche, Y. (eds.) SSBSE 2015. LNCS, vol. 9275, pp. 93–108. Springer, Cham (2015). https://doi.org/10.1007/978-3-319-22183-0_7
16. Rößler, J., Zeller, A., Fraser, G., Zamfir, C., Candea, G.: Reconstructing core dumps. In: Proceedings of the International Conference on Software Testing, Verification and Validation (ICST), pp. 114–123. IEEE (2013)
17. Soltani, M., Derakhshanfar, P., Devroey, X., van Deursen, A.: A benchmark-based evaluation of search-based crash reproduction. Empirical Software Eng. **25**(1), 96–138 (2020). https://doi.org/10.1007/s10664-019-09762-1

18. Soltani, M., Derakhshanfar, P., Panichella, A., Devroey, X., Zaidman, A., van Deursen, A.: Single-objective versus multi-objectivized optimization for evolutionary crash reproduction. In: Colanzi, T.E., McMinn, P. (eds.) SSBSE 2018. LNCS, vol. 11036, pp. 325–340. Springer, Cham (2018). https://doi.org/10.1007/978-3-319-99241-9_18
19. Soltani, M., Panichella, A., Van Deursen, A.: Search-based crash reproduction and its impact on debugging. IEEE Trans. Software Eng. 1 (2018)
20. Vargha, A., Delaney, H.D.: A critique and improvement of the CL common language effect size statistics of McGraw and Wong. J. Educ. Behav. Stat. **25**(2), 101–132 (2000)
21. Xuan, J., Xie, X., Monperrus, M.: Crash reproduction via test case mutation: let existing test cases help. In: Proceedings of the Joint Meeting on Foundations of Software Engineering (ESEC/FSE), pp. 910–913. ACM (2015)
22. Zeller, A.: Why programs fail, second edition: a guide to systematic debugging, 2nd edn. Morgan Kaufmann Publishers Inc., San Francisco (2009)

Search-Based Testing for Scratch Programs

Adina Deiner, Christoph Frädrich, Gordon Fraser(✉), Sophia Geserer, and Niklas Zantner

University of Passau, Innstr. 33, 94032 Passau, Germany
gordon.fraser@uni-passau.de

Abstract. Block-based programming languages enable young learners to quickly implement fun programs and games. The Scratch programming environment is particularly successful at this, with more than 50 million registered users at the time of this writing. Although Scratch simplifies creating syntactically correct programs, learners and educators nevertheless frequently require feedback and support. Dynamic program analysis could enable automation of this support, but the test suites necessary for dynamic analysis do not usually exist for Scratch programs. It is, however, possible to cast test generation for Scratch as a search problem. In this paper, we introduce an approach for automatically generating test suites for Scratch programs using grammatical evolution. The use of grammatical evolution clearly separates the search encoding from framework-specific implementation details, and allows us to use advanced test acceleration techniques. We implemented our approach as an extension of the Whisker test framework. Evaluation on sample Scratch programs demonstrates the potential of the approach.

Keywords: Search-based testing · Block-based programming · Scratch

1 Introduction

Visual, block-based programming languages are a popular means to introduce young learners to programming. Programs are created by visually arranging high-level program statements, thus allowing learners to quickly and easily create engaging programs and fun games. There are many different programming environments built on this idea, and the Scratch [15] programming environment is one of the most popular of these, with more than 50 million registered users at the time of this writing[1]. While the visual representation ensures that statements can only be assembled in syntactically valid ways, achieving desired functionality can nevertheless be challenging: Testing, debugging, and fixing programs can challenge learners, as well as educators who may aim to support or

[1] https://scratch.mit.edu/statistics/, last accessed 9.6.2020.

Authors listed in alphabetical order.

A. Aleti and A. Panichella (Eds.): SSBSE 2020, LNCS 12420, pp. 58–72, 2020.
https://doi.org/10.1007/978-3-030-59762-7_5

assess them. In regular programming, this support is often provided by dynamic analysis: Given a test suite, one can check the runtime properties of a program, determine whether functionality is satisfied, and locate possible causes of failures. However, test suites for SCRATCH programs do not typically exist.

The WHISKER testing framework [22] has been introduced as a means to automate testing of SCRATCH programs. WHISKER-tests interact with SCRATCH programs through the user interface, for example by providing key-presses and mouse-clicks as inputs. In WHISKER, these inputs have to be scripted by a tester, or can be generated randomly. While SCRATCH programs are often trivially simple, this is not always the case, thus challenging the test generator. We therefore cast the problem of generating WHISKER tests as search problem.

We use many-objective optimisation to evolve sets of tests that cover as many as possible program statements. We use grammatical evolution, where search is applied to an integer-list representation using traditional search operators, and the integers are decoded to UI events using a dynamically generated input grammar for the SCRATCH program under test. Fitness evaluation requires test execution, which is challenging because (1) SCRATCH programs are UI-centric and often encode timed behaviour, thus making test execution slow, and because (2) SCRATCH programs are interpreted by the SCRATCH virtual machine (VM) based on a custom internal representation, thus making program instrumentation difficult. Our implementation overcomes these challenges by integrating an accelerated, headless execution framework to speed up test generation.

In detail, the contributions of this paper are as follows:

- We cast SCRATCH testing as a many-objective search problem using grammatical evolution (Sect. 3.1) and many-objective search (Sect. 3.2).
- We define coverage-based fitness functions (Sect. 3.3) and provide efficient means to evaluate fitness (Sect. 3.4).
- We illustrate the problem domain and potential of the approach using example SCRATCH programs (Sect. 4).

2 Background

2.1 SCRATCH Programs

A SCRATCH program consists of the *stage*, which represents the application window and background image, and a collection of *sprites* that are rendered as different images on top of the stage. The stage and each of the sprites contain a number of *scripts* S that define the program logic. Scripts are created by visually arranging *blocks* that correspond to syntactical elements of the language, such as control-flow structures or expressions. A *script* $s = (L, X, G, l_0) \in S$ is a tuple that represents a control-flow automaton, with the set of *control locations* L, the set of *data locations* X, the *control transition relation* $G \subseteq L \times Ops \times L$ with possible program operations Ops, and the *initial control location* l_0. Executing a SCRATCH program results in the creation of a collection of concurrent *processes* P, each process $p \in P$ being an instance of a script s. A *concrete state* $c = \langle p_1, \ldots, p_n \rangle \in C$ of a Scratch program is modelled as a list of concrete process states, which map a concrete value to each data location $x \in X_p$. Scratch programs are controlled by the user, using mouse, keyboard, or microphone. That is, a program can react to mouse movement, mouse button presses, keyboard key presses, sound levels, or entering answers to "ask"-blocks. Typically, the first statement of a script s is an event handler block (*hat block*) that links the execution of the script to the occurrence of an event (user events or internally triggered events, such as broadcasts or clone events).

2.2 The Whisker Test Framework

Whisker [22] is an automated testing framework for Scratch programs. A Whisker test consists of a *test harness*, which takes the role of stimulating the Scratch program under test with inputs, and a set of Scratch *observers*, which encode properties that should be checked on the program under test. To execute tests, the Whisker virtual machine wraps the Scratch virtual machine and its step-function. Before each execution step, the test harness is used to produce inputs that are sent as messages/events to the Scratch program under test, and after each execution of step the Scratch observers check whether the resulting state satisfies the given properties. Each invocation of step executes the processes concurrently, and it is possible that several control-flow transitions are taken in one step. Whisker supports *static test harnesses*, where the system is stimulated with inputs encoded in JavaScript, or *dynamic test harnesses*, where the system is stimulated with randomly determined sequences of inputs. Although random inputs are often sufficient to fully cover simple programs, previous work has shown [22] that programs are not always fully covered. Therefore, the aim of this paper is to use metaheuristic search to automatically generate static test harnesses, i.e., test suites that reach all statements of a program under test.

3 Search-Based Testing for Scratch

3.1 Encoding Scratch Tests Using Grammatical Evolution

Grammatical evolution [19] (GE) describes a form of Genetic Algorithms (GAs) where the mapping from genotype to phenotype is performed using a problem-specific grammar $G = \langle T, N, P, n_s \rangle$: T is a set of terminals, which are the items that will appear in the resulting phenotype; N are non-terminals, which are intermediate elements associated with the production rules $P : N \rightarrow (N \cup T)*$. The element $n_s \in N$ is the start symbol, which is used at the beginning of the mapping process.

The genotype is typically represented simply as a list of integers (*codons*). The mapping of a list of codons to the phenotype creates a derivation of the grammar as follows: Beginning with the first production of starting symbol n_s of the grammar, for each non-terminal x on the right hand side of the production we choose the rth production rule out of all n rules available for x. Given a codon c and n productions for non-terminal x, the number r of the production rule to choose is determined as follows:

$$r = c \; mod \; n$$

Each time a production rule is selected, the decoding moves on to the next codon of the genotype. If the end of the genotype has been reached and there are non-terminals left, then usually the selection of codons starts over at the beginning of the genotype.

In order to instantiate GE, we need to define a grammar that represents these tests. A test case is a sequence of user inputs (UI events). Thus, the starting production for a test case of length n is given by the following:

$$testcase ::= input_1 \; input_2 \; \ldots \; input_n$$

User inputs can be events sent via mouse, keyboard, or microphone. Therefore, terminals in the test grammar will denote concrete mouse, keyboard, or sound events, which may be parameterised. The following grammar thus defines possible inputs:

$$input ::= \mathsf{determineEvents}(S, C)$$

The function determineEvents(S, C) returns a list of all events that the SCRATCH program consisting of scripts S with a given concrete state C supports. In particular, we support the following events:

- KeyPress: One KeyPress event is created for each key for which an event handler exists in S.
- KeyDown: One KeyDown event is created for each key for which a key-sensing block exists in S.
- ClickSprite: One ClickSprite event is created for each sprite that contains a click-event handler. Furthermore, for each such sprite we create an additional ClickSprite event for each clone that exists of the sprite.
- ClickStage: If the stage has a click event handler, then this event is created.
- TypeText: If S uses the *answer* block and sensing is active at state C, then we create one TypeText event for each concrete string contained in the program.
- MouseDown: If S contains a sensing mouse button block, this event is added, which toggles the state of the mouse button.
- MouseMove: If S contains a sensing mouse position block, this event is added. The x and y location of the move are determined by the two following codons in the genotype (i.e., MouseMove implicitly contains a production with two further non-terminals for the coordinates).
- Sound: If S contains event handlers that check the loudness, then for each handler one such event is created, parameterised with the volume checked in the event handler.
- Wait: We always create a Wait event with the default step duration. In addition, one Wait event is created for each distinct delay value in S (e.g., parameters of wait, say, think, or glide).

SCRATCH programs usually contain main scripts triggered by the Greenflag event. We trigger the Greenflag event at the beginning of each test, and therefore do not include it in the grammar. Although the grammar for tests is quite simple, GE offers large implementation benefits as it cleanly separates search operators from the phenotype. For example, tests can easily be extended by modifying the grammar, without requiring any modifications of the search operators.

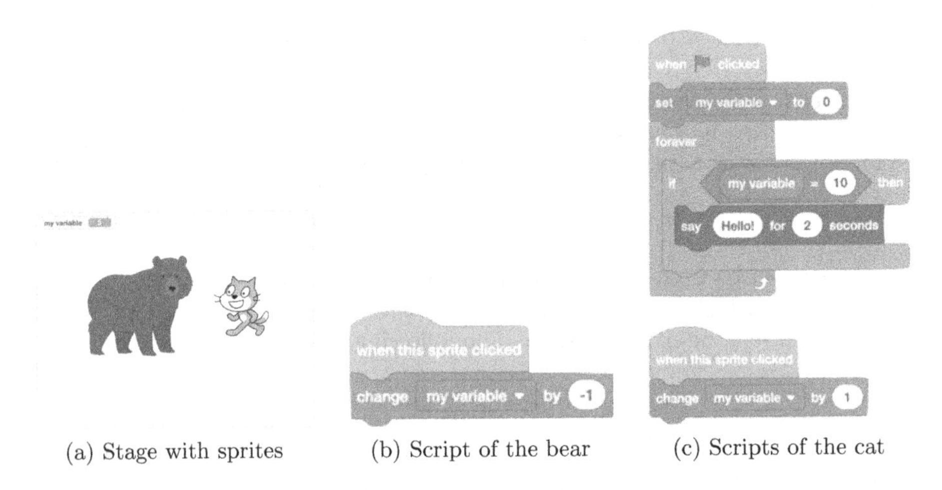

(a) Stage with sprites (b) Script of the bear (c) Scripts of the cat

Fig. 1. Example SCRATCH program: two sprites controlling the value of a shared variable with *click*-event handlers.

As an example of this encoding, consider the SCRATCH project shown in Fig. 1, which contains two sprites, a *cat* and a *bear*, and a variable `my_variable`. The cat's script with the green flag hat-block can be seen as the main function of the program, as SCRATCH programs are started by clicking a green flag in the user interface. When the green flag is clicked, the script initialises the variable `my_variable` with value 0, and then runs a forever-loop. In the forever-loop there is an if-condition which checks if the value of variable `my_variable` is 10; if so, the cat will say "Hello" for two seconds. Both, bear and cat, have event handlers that are triggered if the user performs a mouse click on either of the sprites. When the user clicks on the *bear* then `my_variable` is decremented by 1, if the user clicks on the *cat* then `my_variable` is incremented by 1.

The available events in this example program are independent of the concrete program state, as there are always only the two event handlers for clicking the two sprites. Thus, assuming a test of length 10, the following grammar describes the possible tests for this program:

$$testcase ::= input_1\ input_2\ \ldots\ input_{10}$$
$$input ::= \text{ClickSprite bear} \mid \text{ClickSprite cat} \mid \text{Wait default} \mid \text{Wait 2s}$$

Here, the Wait 2s event is based on the 2s delay in the *say*-block. Consider the following example chromosome in integer encoding:

$$T = \langle 4\;3\;5\;2\;2\;1\;4\;6\;3\;8 \rangle$$

The decoding would start with symbol *testcase* and codon 4. Since there are only 4 productions for *input*, the decoding to a test case looks as follows:

$$\begin{aligned} 4 \; mod \; 4 = 0 &\longrightarrow \text{ClickSprite bear} \\ 3 \; mod \; 4 = 3 &\longrightarrow \text{Wait 2s} \\ 5 \; mod \; 4 = 1 &\longrightarrow \text{ClickSprite cat} \\ &\;\;\cdots \\ 8 \; mod \; 4 = 0 &\longrightarrow \text{ClickSprite bear} \end{aligned}$$

3.2 Search Operators and Algorithm

Since the optimal number of events in a test case is problem specific and cannot be known ahead of time, we use a variable length encoding. This also provides an opportunity for the search to minimise the length of the tests. To generate a random individual for the initial population we select a random length n in the range $[1..max]$, where max is a predetermined parameter representing the maximum number of events in a test case. Then, we generate n random codons, each of which is selected from the range from 0 to 480; the value 480 is the width of the stage in pixels and thus the largest possible parameter any of our supported events can take. The mutation operator can probabilistically (1) replace codons with random codons, (2) insert new codons, and (3) delete codons, each with a certain probability dependent on the length. The crossover operator splits the two parent chromosomes into two parts at a randomly selected relative point (i.e., $[0..1]$). Then the codons on the right side of that splitting point are swapped between the chromosomes. The mutation and crossover operators are based on prior work on variable size search [6] with the aim of avoiding test length bloat [7].

The goal of the optimisation is to produce a set of test cases such that each program statement is covered, thus there is one objective function for each program statement. We therefore use the many-objective sorting algorithm (MOSA) [20], which overcomes the scalability problems of traditional many-objective algorithms. The initial random population evolves toward better populations through subsequent generations until a stopping condition is reached. In each generation an offspring population of the same size is created by selecting test cases from the parent population and modifying them using crossover and mutation. For this, rank selection is used, which gives better test cases a higher probability of being selected.

During evolution, the test cases in the parent and offspring populations are classified into different fronts. At first, for each uncovered statement, the shortest test case, that is closest to covering the statement, is computed and added to

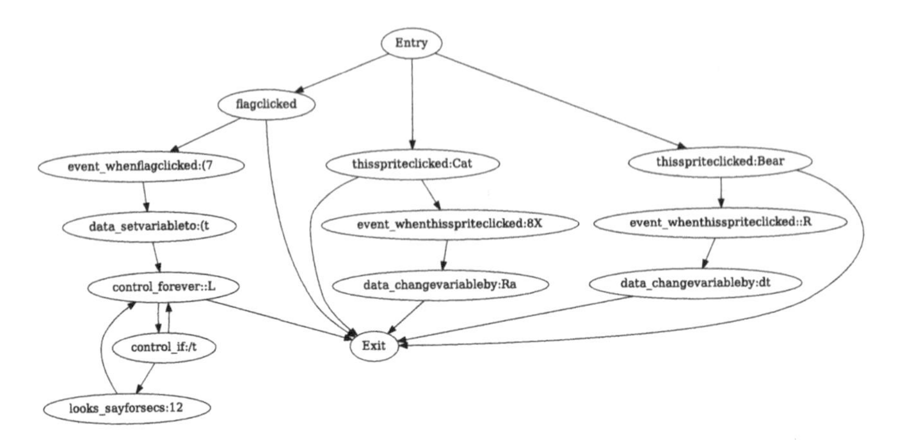

Fig. 2. Control flow graph created for the example program from Fig. 1.

$front_0$. The remaining test cases are sorted according to Pareto dominance: A test case x dominates another test case y, if x is better or equal than y for all uncovered statements and better than y for at least one uncovered statement. All test cases not dominating each other are assigned the same front. The obtained Pareto fronts $front_1$, ..., $front_n$ are sorted in descending order by dominance, meaning test cases in lower fronts dominate test cases in higher fronts. After ranking all test cases a new parent population is formed by adding the test cases in $front_0$ gradually followed by the subsequent fronts until the population size is reached. If the front is too big to add all test cases, a test case with a greater distance to other test cases is preferred in order to promote diversity. With this approach the search focuses towards the uncovered statements, because the best test cases for these statements are likely to survive. Furthermore, there exists an archive, which stores the shortest covering test case for each covered statement. The test cases in the archive are updated in each generation and form the test suite at the end of the algorithm.

3.3 Fitness Function

As basic coverage criterion we consider statement coverage, such that for each statement in a program under test we derive a separate fitness function. The fitness function for a given target statement is encoded using the traditional combination of approach level [26] and branch distance [13]. Given a test t and target statement s, the fitness function is defined as:

$$f(t, s) = approachLevel(t, s) + \alpha(branchDistance(t, s))$$

where α denotes a normalisation function in the range [0..1] [3].

A given target SCRATCH program consists of a number of scripts S (see Sect. 2); for each script $s = (L, X, G, l_0) \in S$ we derive the control flow graph (CFG), defined as $CFG = (L \cup \{entry, exit\}, G)$, i.e., a directed graph consisting of nodes L as well as dedicated *entry* and *exit* nodes, and edges G. We combine these intraprocedural CFGs to an interprocedural super-CFG as follows:

- For each event handler, we add an artificial node with edges to the event handler (*hat block*) as well as the *exit* node, effectively turning event nodes into branching nodes in the CFG. We further add an edge from *entry* to this artificial node for event handlers of user inputs.
- For each *broadcast* statement, we add an edge from the broadcast to all scripts that start with a matching receive event handler block.
- For each *create clone* statement, we add an edge from the *create clone* block to all scripts that start with a matching *When I start as clone* event handler block for the corresponding sprite.
- For each *procedure call* statement, we add an edge from the call to the start block of the procedure (custom block), and a return edge from the end of the procedure to the successor node in the calling script.

Figure 2 shows the interprocedural CFG for the program in Fig. 1. This CFG contains three artificial event nodes (*thisspriteclicked:Cat*, *thisspriteclicked:Bear* or *flagclicked*), each of which effectively is a branching statement depending on whether the event occurs. These branches turn the occurrence of events into control dependencies of the statements in the event handler code.

In order to measure the fitness, we instrument program executions to produce traces of the branching statements executed. Given a trace, the control dependence graph is used to calculate the approach level for a given target node. For each branch, the execution trace further contains information about the minimum branch distances (for evaluation to true and to false). For the fitness evaluation we then use the minimum branch distance of the branching node with the lowest approach level for our target node.

A particularly interesting aspect of SCRATCH programs is that predicates in the code often refer to the locations and interactions of the sprites on the stage, in particular to check whether a sprite touches another sprite. We instrument the corresponding *reporter block* such that it produces not just a binary true/false result, but an actual distance measurement. In case the sprites are touching the branch distance for the true evaluation is 0 by definition; if they are not touching we use the *distanceTo* function in SCRATCH to determine the distance between the sprites, and use that as the branch distance. Similarly, if the condition checks if a sprite is touching the edge of the stage, we can gather the position information and calculate the distance to all edges, and use the minimal distance as the branching distance. Further predicates (e.g., *touching colour*) can be approximated with branch distances similarly.

3.4 Headless Accelerated Test Execution

SCRATCH tests are executed by running a SCRATCH program and applying events encoded in a test to the SCRATCH VM. This kind of test mimics a normal execution of SCRATCH, which is as fast as the regular execution when a user runs the program. The result is that running tests can be very time consuming. We added two modifications to the SCRATCH VM to increase its execution speed, which in turn decreases the time to execute tests.

The SCRATCH VM updates its internal state and the UI representation with a given interval. The first modification introduces an acceleration factor which reduces the default update interval. By simply reducing the interval, the VM updates its state more often which leads to faster execution of regular blocks.

However, only increasing the number of state updates of the VM is not sufficient: Blocks that use time (e.g., waiting for x seconds) measure time not in state updates but in real time, and simply accelerating the state updates does not speed up the perceived time of blocks. A block that is waiting for 2 s will still wait 2 s even if we speed up that execution by a factor of 10.

To solve this problem we instrumented blocks that use time to speed up their waiting time according to the acceleration factor. The following example shows the waiting function of the *control_ wait* SCRATCH block. By default, the control block makes the application wait for the amount of seconds defined in `args.DURATION`:

```
1 wait (args, util) {
2   if (util.stackTimerNeedsInit()) {
3       const duration = Math.max(
4         0,
5         1000 * Cast.toNumber(args.DURATION)
6       );
7   ...
```

To reduce the waiting time of the block the `args.DURATION` has to be reduced by the acceleration factor. For example, to run the application with an acceleration factor of 10 instead of 1, the duration has to be divided by factor 10. Instrumentation of this function is achieved by dynamically replacing the function with a modified version as follows:

```
1 const accelerationFactor = 10;
2 const original = this.vm.runtime._primitives.control_wait;
3 const instrumented = (args, util) => {
4     const clone = {...args};
5     clone.DURATION = args.DURATION / accelerationFactor;
6     return original(clone, util);
7 };
8 this.vm.runtime._primitives.control_wait = instrumented;
```

As a third modification the SERVANT front-end was created for WHISKER. The SERVANT is a Node.js based command-line interface (CLI) which is based on *Puppeteer*[2], a *Headless Chrome Node.js API.* This new front-end makes it possible to use WHISKER in headless environments (i.e., without a graphical user interface), so that tests can be executed on computing clusters.

[2] https://pptr.dev/, last accessed 9.6.2020.

4 Case Study Examples

To illustrate the proposed approach, we use three example SCRATCH programs shown in Fig. 3. The Ping Pong game is a classical SCRATCH tutorial example, and represents the types of programs commonly created by early learners. It consists of 36 blocks in 4 scripts, and is controlled by five different key events. We modified the Ping Pong game to initialise all variables (missing variable initialisation is a common bug in SCRATCH [8]). The Fruit Catching game was used by Stahlbauer et al. [22] as part of their evaluation of WHISKER, and is taken from an educational context. It consists of 49 blocks in 4 scripts, and is controlled by the cursor keys. Green Your City is part of the popular CodeClub collection of example projects. It consists of 8 scripts and 52 blocks.

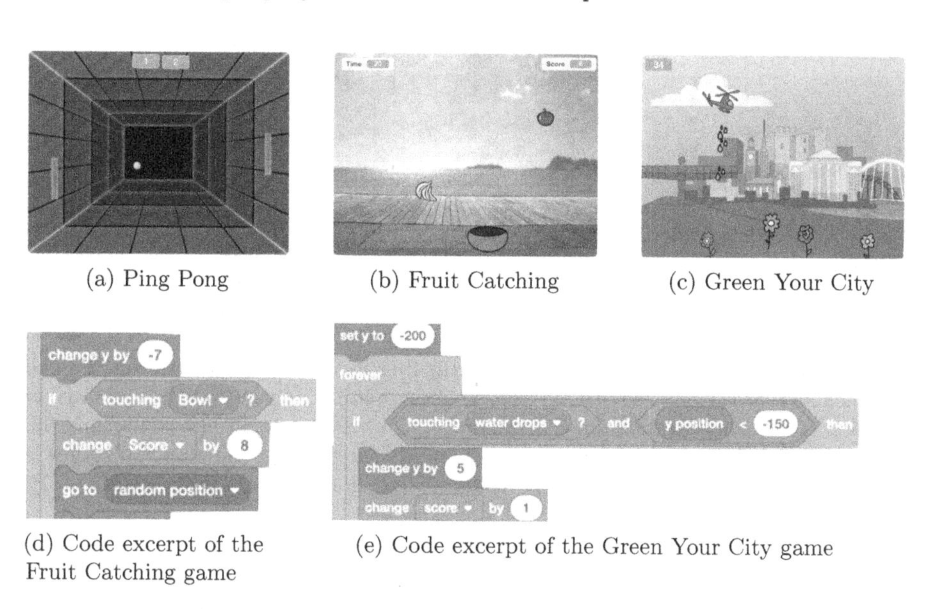

(a) Ping Pong (b) Fruit Catching (c) Green Your City

(d) Code excerpt of the Fruit Catching game

(e) Code excerpt of the Green Your City game

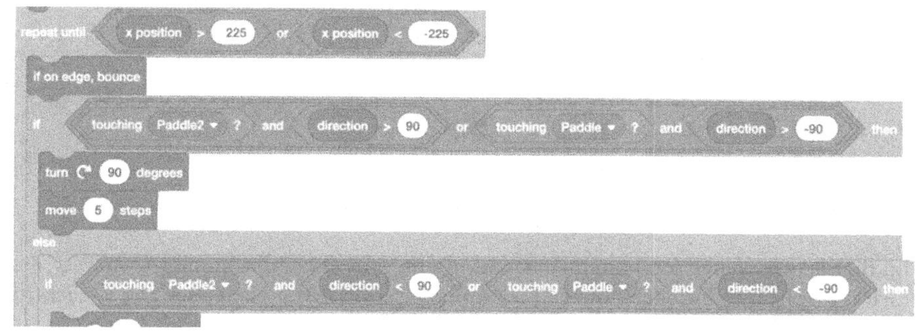

(f) Code excerpt of the Ping Pong game

Fig. 3. Case study programs and code examples.

(a) Branch distance 284

(b) Branch distance 188

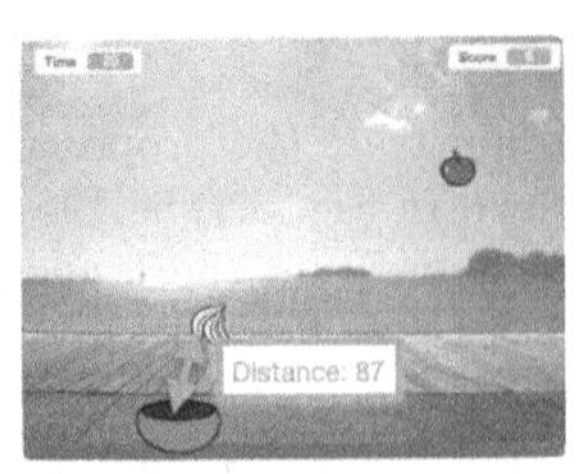

(c) Branch distance 87

Fig. 4. Fitness values for statements within the if-condition shown in Fig. 3d. The approach level is 0 in all cases, but the branch distance of the if-condition depends on the distance between the bowl and the banana.

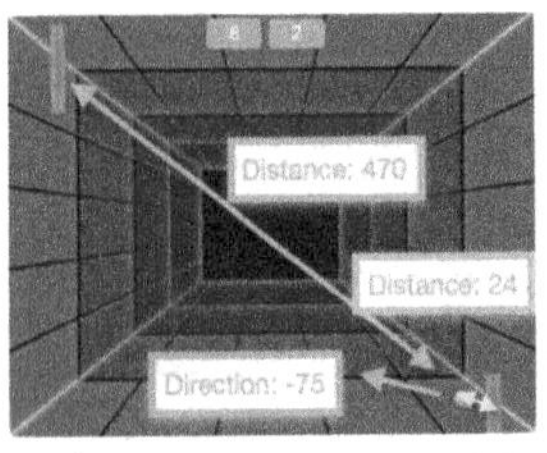

(a) Branch distance 190

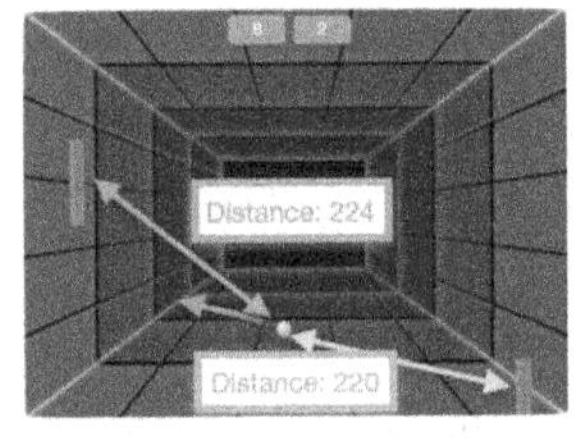

(b) Branch distance 224

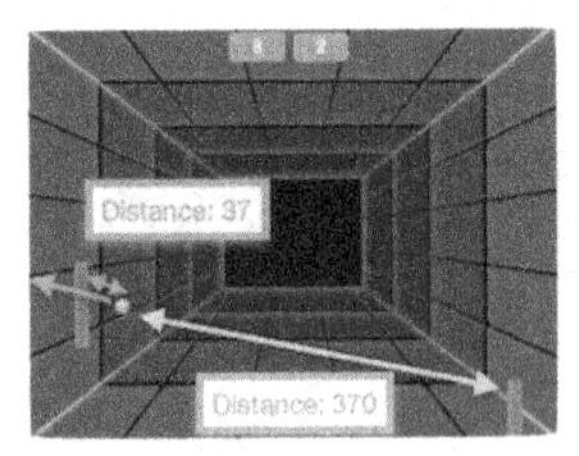

(c) Branch distance 37

Fig. 5. Fitness values for statements within the first if-condition shown in Fig. 3f. The approach level is 1 if $x > 225$ or $x < -225$, otherwise it is 0. Given that the ball faces in direction -75 (such that the branch distance for $-75 > 90$ is $90 - -75 + 1 = 166$, and the branch distance for $-75 > -90$ is 0) and is between the paddles, the branch distance is calculated as $min(distance(ball, right\_paddle) + 166, distance(ball, left\_paddle) + 0))$.

(a) Branch distance 136

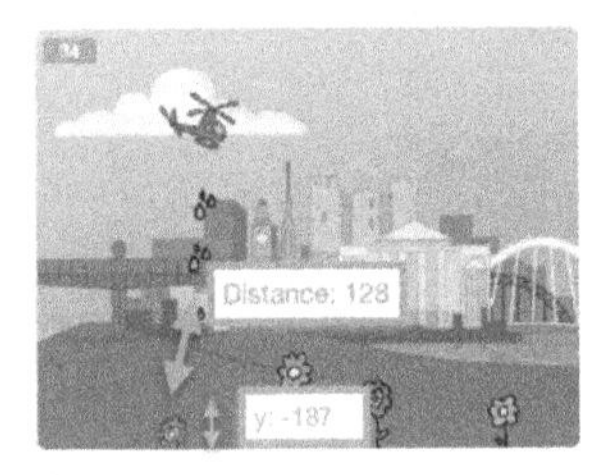

(b) Branch distance 128

(c) Branch distance 297

Fig. 6. Fitness values for statements within the first if-condition shown in Fig. 3e, which is duplicated for each flower. The approach level is 0 if the green flag is clicked; the branch distance is calculated based on the y-position and the distance to the closest water drop. Given that a flower needs to be below $y = -150$, the branch distance for the flower in Fig. 6a is $distance(flower, water) + (-129 - -150 + 1) = 136$. The flower in Fig. 6b is below $y = -150$, therefore the branch distance is $distance(flower, water) + 0 = 128$, and for Fig. 6c it is 297. (Color figure online)

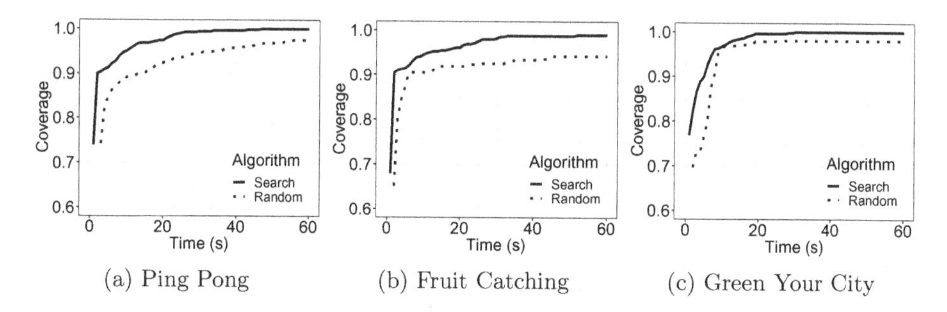

(a) Ping Pong (b) Fruit Catching (c) Green Your City

Fig. 7. Coverage over time, averaged over 30 runs.

In a typical SCRATCH program, many statements are trivially covered, but there can always be tricky code that poses challenges for automated testing. Figure 3d shows an excerpt of the code of the Fruit Catching game, taken from the banana sprite: In order to execute the statements in the then-block of the shown branch, the player-controlled bowl needs to touch the banana. The *touching* predicate is reflected by a branch distance that guides the test generator towards achieving this, as illustrated in Fig. 4 for different states of the game.

The if-condition of the Ping Pong game shown in Fig. 3f requires a branch distance calculation by applying the rules on nested logical expressions to combine the distances between the ball and the two paddles, as well as the required and current orientation of the ball. As suggested by the example states shown in Fig. 5, the fitness landscape induced by the complex expression is less convenient than in the simple *touching*-predicate shown in Fig. 4. Green Your City also contains logical expressions combining various distance measurements; in particular each of the flowers contains the code shown in Fig. 3e. Flowers grow vertically each time they touch a water drop, the example branch is covered once this has happened sufficiently often to let the flower grow above $y = -150$ and it touches a water drop again; Fig. 6 shows examples of this distance for different flowers.

To see the search in action, Fig. 7 shows the coverage over time, averaged over 30 runs of WHISKER using the original configuration [22] ("Random", i.e., 60 s, events every 250 ms, program reset every 10 s), and the search extension ("Search", 60 s runtime, population size 10, 250 ms default event duration, acceleration factor 5, crossover probability 0.8, initial length 2). The improvement of the search over random testing after 60 s is statistically significant with $p < 0.001$ in all three cases: For Ping Pong the Vargha-Delaney effect size $\hat{A}_{12} = 0.7$, for the Fruit Catching game $\hat{A}_{12} = 1.0$, and for Green Your City $\hat{A}_{12} = 1.0$. For Ping Pong (Fig. 7a) the search covers most statements quickly, but generally needs longer ($\approx$30 s) to cover the two branches discussed in example Fig. 5. Random testing sometimes hits the ball by accident, but generally needs to play much longer before that happens. Similarly, for the Fruit Catching game Fig. 7b shows, while again many statements are easy to cover, the search needs about 30 s to succeed in catching both types of fruit, while random takes substantially longer.

The results for Green Your City (Fig. 7c) are similar, in that the search requires around 30 s until all statements are covered, while random usually takes longer; the tricky branches are those requiring to hit flowers with water drops, as well as one statement that requires the helicopter to reach the bottom, and thus a longer play time than random testing used would be necessary. While these are promising initial results, a more thorough evaluation on a larger number of SCRATCH projects is planned as future work. The randomised and game-like nature of most SCRATCH projects can be a cause for test flakiness; however, this flakiness can be easily contained [22].

Note that an automatically generated coverage test suite is only one of the stepping stones towards providing automated feedback to learners: The generated tests tend to be short and their randomised nature potentially makes them difficult to understand, and so we do not expect that the tests would be shown *directly* to learners. The tests are rather meant as input to further dynamic analysis tools, for example in order to serve as test harness for WHISKER [22], where manually written SCRATCH observers would provide insight into which aspects of the functionality are correctly implemented in a concrete learner's implementation of a predefined programming challenge. We envision that the concrete feedback to learners will be provided in terms of textual hints or code suggestions [21,25], and the generated tests are a prerequisite for achieving this.

5 Related Work

The rising popularity of block-based programming languages like SCRATCH creates an increased demand for automated analysis to support learners. It has been shown that learners tend to adopt bad habits [17] and create "smelly" code [1,11,24]; these code smells have been shown to have a negative impact on understanding [10]. To counter this, automated tools can help to identify and overcome such problems. For example, the DR. SCRATCH [18] website points out a small number of code smells to learners, using the HAIRBALL [4] static analysis tool, and similar smells are identified by QUALITY HOUND [23] and SAT [5]. LITTERBOX [8] can identify patterns of common bugs in SCRATCH programs. Besides pointing out possible mistakes, it is desirable to also identify helpful suggestions and feedback, such as what step to take next [21] or how to remove code smells [25]. The majority of existing approaches are based on static program analysis, and can therefore only provide limited reasoning about the actual program behaviour. The ITCH tool [12] translates a small subset of SCRATCH programs to Python programs (textual interactions via say/ask blocks) and then allows users to run tests on these programs. WHISKER [22] takes this approach a step further and, besides execution of automated tests directly in SCRATCH, also provides automated property-based testing.

We introduce search-based testing as a means to fully automate the generation of test suites for SCRATCH programs. These test suites are intended to be the input to dynamic analysis tools that can then use the dynamic information to produce hints and feedback. Our approach is based on evolutionary search, which is common for API-level test generation [6], but has also been applied to GUI testing [9,14,16]. The concept of grammatical evolution [19] has not been thoroughly explored in the context of test generation yet [2].

6 Conclusions

In this paper, we have introduced the idea to apply search-based testing for the problem of generating coverage-oriented test suites for SCRATCH programs. The use of Grammatical Evolution allows a clean separation between aspects of the meta-heuristic search, and the technical challenges posed by the testing environment. The specific graphical nature of SCRATCH programs provides opportunities for guidance beyond those common in regular programs. Our extension of the WHISKER test generator has demonstrated its potential on a number of example programs. However, there are remaining challenges to be addressed in future work, such as refined support for all user actions, better integration of seeding, consideration of program state (which, for example, is often encoded by the costumes/backdrops in use), and many others, permitting a larger scale evaluation. Furthermore, future work will be able to build on the test suites generated by our approach for further analysis and for generating actionable feedback to users. For example, we anticipate that an example application scenario will be that where a teacher produces a golden solution, then generates a test suite for it, and this test suite then serves for fault localisation or repair suggestions. To support this future work, our extensions to WHISKER are available as open source at:

https://github.com/se2p/whisker-main

Acknowledgements. This work is supported by EPSRC project EP/N023978/2 and DFG project FR 2955/3-1 "TENDER-BLOCK: Testing, Debugging, and Repairing Blocks-based Programs".

References

1. Aivaloglou, E., Hermans, F.: How kids code and how we know: an exploratory study on the scratch repository. In: Proceedings of ICER, pp. 53–61 (2016)
2. Anjum, M.S., Ryan, C.: Seeding grammars in grammatical evolution to improve search based software testing. In: Hu, T., Lourenço, N., Medvet, E., Divina, F. (eds.) EuroGP 2020. LNCS, vol. 12101, pp. 18–34. Springer, Cham (2020). https://doi.org/10.1007/978-3-030-44094-7_2
3. Arcuri, A.: It really does matter how you normalize the branch distance in search-based software testing. Softw. Test. Verif. Reliab. **23**(2), 119–147 (2013)
4. Boe, B., Hill, C., Len, M., Dreschler, G., Conrad, P., Franklin, D.: Hairball: lint-inspired static analysis of scratch projects. In: Proceedings of SIGCSE, pp. 215–220 (2013)
5. Chang, Z., Sun, Y., Wu, T.Y., Guizani, M.: Scratch analysis tool (SAT): a modern scratch project analysis tool based on ANTLR to assess computational thinking skills. In: Proceedings of IWCMC, pp. 950–955. IEEE (2018)

6. Fraser, G., Arcuri, A.: Whole test suite generation. IEEE Trans. Softw. Eng. **39**(2), 276–291 (2012)
7. Fraser, G., Arcuri, A.: Handling test length bloat. Softw. Test. Verif. Reliab. **23**(7), 553–582 (2013)
8. Frädrich, C., Obermüller, F., Körber, N., Heuer, U., Fraser, G.: Common bugs in scratch programs. In: Proceedings of ITICSE, pp. 89–95 (2020)
9. Gross, F., Fraser, G., Zeller, A.: Search-based system testing: high coverage, no false alarms. In: Proceedings of ISSTA, pp. 67–77 (2012)
10. Hermans, F., Aivaloglou, E.: Do code smells hamper novice programming? A controlled experiment on scratch programs. In: Proceedings of ICPC, pp. 1–10. IEEE (2016)
11. Hermans, F., Stolee, K.T., Hoepelman, D.: Smells in block-based programming languages. In: Proceedings of VL/HCC, pp. 68–72. IEEE (2016)
12. Johnson, D.E.: Itch: individual testing of computer homework for scratch assignments. In: Proceedings of SIGCSE, pp. 223–227 (2016)
13. Korel, B.: Automated software test data generation. IEEE Trans. Softw. Eng. (TSE) **16**, 870–879 (1990)
14. Mahmood, R., Mirzaei, N., Malek, S.: Evodroid: segmented evolutionary testing of android apps. In: Proceedings of ESEC/FSE, pp. 599–609 (2014)
15. Maloney, J., Resnick, M., Rusk, N., Silverman, B., Eastmond, E.: The scratch programming language and environment. TOCE **10**(4), 1–15 (2010)
16. Mao, K., Harman, M., Jia, Y.: Sapienz: multi-objective automated testing for android applications. In: Proceedings of ISSTA, pp. 94–105 (2016)
17. Meerbaum-Salant, O., Armoni, M., Ben-Ari, M.: Habits of programming in scratch. In: Proceedings of ITICSE, pp. 168–172 (2011)
18. Moreno-León, J., Robles, G.: Dr. scratch: a web tool to automatically evaluate scratch projects. In: Proceedings of WIPSCE, pp. 132–133 (2015)
19. O'Neill, M., Ryan, C.: Grammatical evolution. IEEE Trans. Evol. Comput. **5**(4), 349–358 (2001)
20. Panichella, A., Kifetew, F.M., Tonella, P.: Reformulating branch coverage as a many-objective optimization problem. In: Proceedings of ICST, pp. 1–10 (2015)
21. Price, T.W., Dong, Y., Lipovac, D.: iSnap: towards intelligent tutoring in novice programming environments. In: Proceedings of SIGCSE, pp. 483–488 (2017)
22. Stahlbauer, A., Kreis, M., Fraser, G.: Testing scratch programs automatically. In: Proceedings of ESEC/SIGSOFT FSE, pp. 165–175. ACM (2019)
23. Techapalokul, P., Tilevich, E.: Quality hound-an online code smell analyzer for scratch programs. In: Proceedings of VL/HCC, pp. 337–338. IEEE (2017)
24. Techapalokul, P., Tilevich, E.: Understanding recurring quality problems and their impact on code sharing in block-based software. In: Proceedings of VL/HCC, pp. 43–51. IEEE (2017)
25. Techapalokul, P., Tilevich, E.: Code quality improvement for all: Automated refactoring for scratch. In: Proceedings of VL/HCC, pp. 117–125. IEEE (2019)
26. Wegener, J., Baresel, A., Sthamer, H.: Evolutionary test environment for automatic structural testing. Inf. Softw. Tech. **43**(14), 841–854 (2001)

Solving Schedulability as a Search Space Problem with Decision Diagrams

Dimitri Racordon(✉), Aurélien Coet, Emmanouela Stachtiari, and Didier Buchs

Computer Science Department, Faculty of Science, University of Geneva, Geneva, Switzerland
{dimitri.racordon,aurelien.coet,emmanouela.stachtiari,didier.buchs}@unige.ch

Abstract. Real-time system design involves proving the schedulability of a set of tasks with hard timing and other constraints that should run on one or several cores. When those requirements are known at design time, it is possible to compute a fixed scheduling of tasks before deployment. This approach avoids the overhead induced by an online scheduler and allows the designer to verify the schedulability of the taskset design under normal and degraded conditions, such as core failures. In this context, we propose to solve the schedulability problem as a state space exploration problem. We represent the schedulings as partial functions that map each task to a core and a point in time. Partial functions can be efficiently encoded using a new variant of decision diagrams, called Map-Family Decision Diagrams (MFDDs). Our setting allows first to create the MFDD of all possible schedulings and then apply homomorphic operations directly on it, in order to obtain the schedulings that respect the constraints of the taskset.

Keywords: Search problems · Decision diagrams · Schedulability · Real-time systems · Resilient systems · Multi-core architectures

1 Introduction

Multi-core architectures have become ubiquitous, as an answer for the exponential growth in computer performance required by modern applications. This observation obviously applies to small-scale real-time and cyber-physical systems as well. Unlike more general applications, these systems often have to run tasks with hard deadlines, to interact with their hardware components. It follows that one of their essential requirements is to guarantee that they are able to perform all their tasks on time, by providing a scheduling that assigns each task to a specific core at a specific point in time. Such a scheduling can be built along with the execution of the system, according to some heuristics [11], or pre-computed statically to avoid the overhead induced by an online scheduler. An additional advantage of the latter approach is that it allows to study the system's performance under various scenarios, not only to make sure it is actually capable of

A. Aleti and A. Panichella (Eds.): SSBSE 2020, LNCS 12420, pp. 73–87, 2020.
https://doi.org/10.1007/978-3-030-59762-7_6

running its workload, but also to check whether it can be resilient to hardware faults (e.g. the failure of one of its cores).

Typically, each task has timing constraints, such as a *release time*, after which it can be executed, a *worst-case execution time*, which is the most pessimistic assumption for the time it takes to complete, and a *deadline* for its completion. Other non-timing constraints may exist and should also be taken into account in the scheduling. Common constraints include *precedence*, which indicates that a predecessor task has to be completed before a successor task starts. Schedulability analysis verifies that a *feasible* scheduling exists for a given taskset respecting all the timing, precedence, and other constraints. Roughly speaking, it suffices to check that all tasks will complete before their deadline.

Several analysis tools have been used to solve the multicore schedulability problem. For example, *utilization bound checks* [12] have been proposed for testing the schedulability of a taskset analytically. Though these checks are efficient, they inherently are pessimistic, often rejecting valid tasksets. Furthermore, they cannot handle multiple constraints over tasks. Other tools have been developed to analyze tasksets with complex constraints [14], and rely on *simulation* to check schedulability. However, simulation is known to cover only a sample of possible scenarios, thus it can lead to falsely feasible schedulings, which is not acceptable for critical real-time systems.

The aforementioned approaches in the literature target a specific variation of the problem, such as the existence of task parallelism or interference among tasks that cause delays. In this paper, we opt for a different, more generic approach, which relies on model checking. Unlike simulation, model checking explores the entire space of possible states. In our case, this translates into an enumeration of all possible schedulings, that we can filter to remove specific instances for which the constraints are not satisfied. In practice, such an approach is often intractable due to the *state-space explosion problem*. However, we mitigate this issue with decision diagrams, a data structure that encodes large sets of data into a memory-efficient representation by exploiting the similarities between each element. Our work is related to the technique proposed in [17]. The authors solve the schedulability problem as a state-space exploration using Data Decision Diagrams [6], which encode sets of variable assignments. We use a slightly different flavor of decision diagrams, called Map-Family Decision Diagrams (MFDDs), that encode sets of partial functions, and allow for a more direct translation of the problem. Each constraint is represented as a homomorphic operation that is applied directly on the encoded form, in a fashion reminiscent to Fourier filters.

This paper offers the following contributions:

1. An exhaustive search methodology for the multi-core schedulability problem, based on inductive homomorphisms, i.e. structure-preserving transformations, that compute all solutions at once.
2. A compact, human-readable representation of the solution set, which can be easily inspected during design and efficiently stored as a database of pre-computed schedulings in production.

3. A refinement of the seminal work proposed in [17] that fixes a flaw in the authors' method, offers a simpler way to define filters and supports additional constraints, such as transient core failures.

2 Related Work

In this section, we include works related to the offline computation of global multicore scheduling for time-triggered tasks. In particular, we discuss approaches that employ model checking, linear programming and decision diagrams.

Model checking has been a reliable tool for schedulability verification. Its inputs are a model of the system (e.g., tasks, scheduler) with finite reachable states and a set of properties that characterize the valid reachable states, e.g., states where no task has missed its deadline. Properties may be expressed as temporal logic formulae or be designated as *error states* in the model. An exhaustive search explores the reachable states of the system model and checks if properties hold. Violating states are returned as *counter-example*, which can be used to refine the model.

In [1], periodic tasksets are modeled as timed automata and schedulability holds if a certain state can be reached within an expected time window. In a similar approach [8], each task and its constraints are modeled as a timed automaton, on which it is verified that the task can meet its deadline. Interactions between tasks are modeled by composing their corresponding automata. Another work [18] for self-suspending tasks models each task as a set of segments, the end of which corresponds to a suspension point. Generally, model checking can model complex factors on schedulability, such as stochastic execution times. However, it is computationally expensive and therefore only able to handle small tasksets. Moreover, the generated solutions are independent to each other and can not be narrowed down efficiently to account for additional task constraints or be compactly stored. In [7], statistical model checking is proposed to reduce the undecidability of symbolic (i.e., more efficient but over-approximating) model checking for the scheduleability of tasks with uncertain response and blocking times. In this setting, the system is simulated for a finite number of runs to test the satisfaction probability of a given property. Tasksets that were found unschedulable by symbolic model checking can be probed for useful information, such as the probability of a certain violation, or bounds on blocking times. Other timed automata extensions, such as Priced Automata [2], have been suggested to efficiently identify subsets of feasible schedulings.

Other approaches to schedulability rely on linear programming. In [16], schedulability for multi/many-core architectures is studied for three different platform architectures using integer linear programming (ILP). Cache conscious real-time schedulability using ILP has also been the target in [15]. In their setting, they assume that all tasks are connected by the dependency relation, which reduces candidate solutions, compared to our setting that does not assume that. Linear programming uses heuristics algorithms to efficiently compute schedulings close to the optimal ones, but it does not return all schedulings.

Decision diagrams have been proposed for representing and solving schedulability In [10] the authors find an optimal schedule for tasks with arbitrary execution times using a breath firth search on a Binary Decision Diagram. However, their representation considers uniform cores, while our model can be also applied to non-uniform cores, e.g., when not all cores are suitable for every task. In another work [5], the authors proposed searching for the optimal single-core scheduling, using a Multi-Valued Decision Diagram that represents an overapproximated set of possible schedulings. As opposed to theirs, our method handles many cores and computes all feasible solutions.

3 Background

Decision diagrams were originally proposed as a data structure to represent and manipulate Boolean functions [4]. Since then, numerous variants have been developed that suit other domains (see [13] for a survey). Nonetheless, all of them share the same principle; they encode each element as a path in a finite directed acyclic graph (DAG), from its root to a terminal node (i.e. a node without successors). Non-terminal nodes are labeled with a variable, and arcs are labeled with the value assigned to this variable. Our approach uses one specific variant of decision diagrams, named MFDDs, which we developed to encode sets of partial functions $f : A \rightarrow B$, where A and B are any sets and $dom(f) \subseteq A$ is finite. Partial functions typically correspond to dictionaries or mappings in regular programming languages, and are well-suited to encode various kinds of data structures. For instance, a list can be seen as a partial function that maps numerical indices to the list's elements.

In a MFDD that encodes a set of partial functions $A \rightarrow B$, all non-terminal nodes are labeled with a value from A, while arcs are labeled with a value from $B \cup \{\epsilon\}$, where ϵ represents the absence of any value. A diagram may have up to two terminal nodes, labeled with $\top$ and $\bot$, representing the acceptance of a path from the root and its rejection, respectively. Hence, a function is encoded as a path from the root to the accepting terminal node.

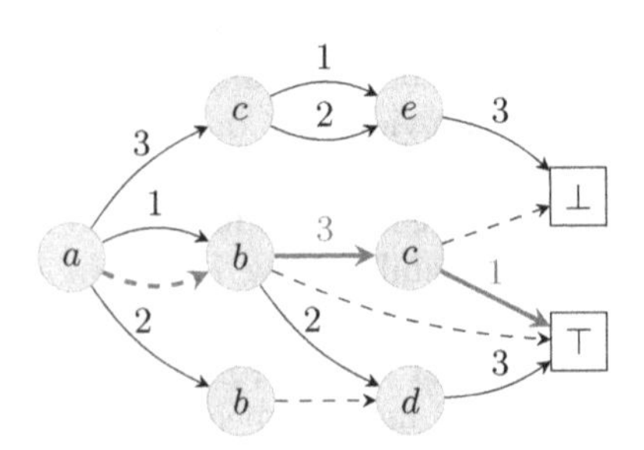

Fig. 1. Example of a (non-canonical) MFDD

Example 1. Figure 1 depicts a decision diagram that encodes a set of seven different partial functions from a domain $A = \{a, b, c, d, e\}$ to a codomain $B \subseteq \mathbb{N}$. Circles denote domain values, while squares denote terminal nodes. Dashed edges represent ϵ-arcs, while solid ones are labeled with a number. The path highlighted with thick, green arrows encodes a function f such that $dom(f) = \{b, c\}$ (i.e., it is undefined for any other value in A) and $f(b) = 3, f(c) = 1$.

Any path that leads to the rejecting terminal (i.e., $\bot$) is said *rejected* and denotes a function that is not present in the encoded set. If some domain value v is absent from a path, then it is assumed that the corresponding function is not defined for v. All functions that cannot be associated with any accepted path in the diagram are considered absent from the encoded set. It follows that a given set of functions can have multiple representations[1].

Definition 1 (Map-Family Decision Diagram). *Let A and B be a domain and a codomain set, respectively. The set of Map-Family Decision Diagrams $\mathbb{M}_{A,B}$ that encode families of partial functions $A \to B$ is inductively defined as the minimum set, such that:*

- *$\{\bot, \top\} \in \mathbb{M}_{A,B}$ are terminal nodes,*
- *$\langle a, s\rangle \in \mathbb{M}_{A,B}$ is a non-terminal node labeled with $a \in A$, whose successors are given by the partial function $s : B \cup \{\epsilon\} \to \mathbb{M}_{A,B}$.*

Efficient implementations of decision diagrams rely on representation uniqueness to share identical sub-graphs, in order to reduce the memory footprint and cache the result of each homomorphic operation. We provide a canonical representation of MFDDs by applying some constraints. Firstly, we require that domain A be associated with a total order, and that all successors of a node be either a terminal node or a node labeled with a greater value. This is the case in the MFDD of Fig. 1, assuming that the members in A are ordered lexicographically. Secondly, we require that all non-terminal nodes have at least one arc not labeled with ϵ. Nodes that do not satisfy this constraint can be safely removed, as they do not carry any information. For instance, node b at the bottom path of Fig. 1 is redundant. Finally, we require that all rejected paths be removed. Recall that functions which cannot be associated with an accepted path are assumed to be rejected (i.e. absent from the encoded set). Hence, rejected paths, like the two top paths of Fig. 1, do not add any information. Note that we cannot get rid of the rejecting terminal itself, as it is necessary to represent the empty set of functions. The canonical form of the MFDD from Fig. 1 is shown in Fig. 2a.

Definition 2 (Canonicity). *Let A be a domain set with a total order $< \subseteq A \times A$ and B be a codomain set. A MFDD $d \in \mathbb{M}_{A,B}$ is canonical if and only if:*

[1] Incidentally, as MFDDs are *finite* graphs, it follows that all encoded functions f have a finite domain $dom(f) \subseteq A$, represented by the non-terminal nodes along an accepting path, even if A is infinite.

- $d \in \{\bot, \top\}$ *is a terminal node, or*
- $d = \langle a, s\rangle$ *is a non-terminal node such that* $\exists b \in dom(s), b \neq \epsilon$ *and* $\forall b \in dom(s), s(b) \in \mathbb{M}_{\{x \in A | a < x\}, B} - \{\bot\}$.

Let $\langle a, s\rangle$ be a non-terminal node. The first canonicity constraint is enforced by $s(b) \in \mathbb{M}_{\{x \in A | a < x\}, B}$, which prescribes that successor nodes be labeled with greater domain values. The second constraint is enforced by $\exists b \in dom(s), b \neq \epsilon$. Finally, although the third constraint is not enforced explicitly, it is easy to show that requiring $s(b) \neq \bot$ inductively prevents rejected paths to be encoded.

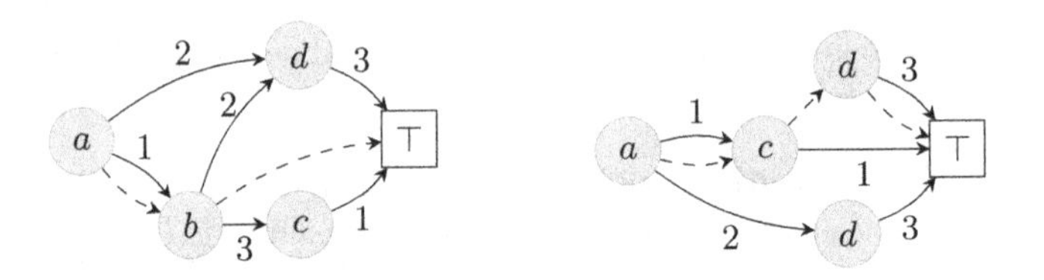

Fig. 2. Two examples of canonical MFDDs

Example 2. Consider the MFDD shown in Fig. 2a. The root node is a tuple $\langle a, s_a\rangle$, where s_a is a function such that $dom(s_a) = \{\epsilon, 1, 2\}$ and $s_a(\epsilon) = s_a(1) = \langle b, s_b\rangle$ and $s_a(2) = \langle d, s_d\rangle$. The function s_b is defined such that $dom(s_b) = \{\epsilon, 3\}$ and $s_b(\epsilon) = \top$ and $s_b(3) = \langle c, s_c\rangle$. Finally, s_c (resp. s_d) is defined such that $s_c(1) = \top$ (resp. $s_d(3) = \top$) and is undefined for any other value in $B \cup \{\epsilon\}$.

In addition to their compact representation, another advantage of MFDDs is that they can be manipulated by the means of homomorphisms. These operations can modify multiple elements at once, as any alteration of a prefix has a direct impact on all elements encoded by its suffixes.

Example 3. Suppose we were to remove b from the domain of all partial functions encoded by the MFDD in Fig. 2a. Rather than enumerating all seven instances to apply the filter, we could define a homomorphism that simply removes the node corresponding to b's bindings, and rewires its incoming arcs to existing or new nodes, as shown in Fig. 2b. In other words, a homomorphism can modify the domain of all encoded functions by rewriting the decision diagram.

A key property of MFDD homomorphisms is that they preserve set-theoretic operations, such as union and intersection. More formally, let $d_1, d_2 \in \mathbb{M}_{A,B}$ be two MFDDs, and Φ be a homomorphism, then $\Phi(d_1 \cup d_2) = \Phi(d_1) \cup \Phi(d_2)$. This allows homomorphisms to be rearranged for efficient computations.

4 Methodology

In the context of search based problems, MFDDs present the advantage that they can be used to both store and compute sets of solutions efficiently. Their compact representation is able to encode large sets with minimal memory footprint, while homomorphisms on map families allows the construction of solutions with a smaller computational overhead than traditional approaches.

There are two main approaches to use MFDDs in search based problems. The first consists of exploring the state space of a problem to build its solution set incrementally, augmenting it with new instances that satisfy the problem's constraints as they are found. This process ends when a fixpoint is reached or if the entire space has been visited. The second technique, more reminiscent of Answer Set Programming (ASP), proposes to start from a MFDD representing the entire state space of the problem before filtering out instances that do not satisfy the problem's constraints.

The *n-queens puzzle* is a simple example of a problem that can be solved with the second approach. The puzzle consists of finding all possible ways that n different queens can be placed on a $n \times n$ chessboard without being able to attack each other. More formally, let $Col_n = \{a, b, \dots\}$ denote a set of column identifiers for some $n \times n$ chessboard. Similarly, let $Row_n = \{1, 2, \dots, n\}$ denote row identifiers. Let $C_n = Col_n \times Row_n$ denote the set of coordinates on the board. Let $I \in \mathcal{P}(C_n)$ denote a set of coordinates at which queens are placed, and R denote a relation on coordinates which indicates whether a position can be reached from another by a queen, according to the rules of chess. $(d, 3) \in I$ indicates for example that a queen lies at row d and column 3 of the board, and $\langle (d, 3), (g, 6) \rangle \in R$, because the position $(g, 6)$ can be reached from $(d, 3)$ by a queen. A configuration I is said to be a solution to the n-queens puzzle if it contains exactly n coordinates, and if for all $a \in I$, there is no $b \in I$ such that $\langle a, b \rangle \in R$. Hence, the set of all solutions S is formally given as:

$$S = \{I \mid (\| I \| = n) \wedge (\forall a, b \in C_n, a \in I \wedge \langle a, b \rangle \in R \Rightarrow b \notin I\}$$

Using MFDDs, one can express the n-queens puzzle as the following algorithm:

1: $S_n \leftarrow \mathcal{P}(C_n)$
2: **for all** $a \in C_n$ **do**
3: $\quad S_n \leftarrow \{I \in S_n \mid a \in I \Rightarrow \nexists b \in I, \langle a, b \rangle \in R\}$
4: **end for**
5: $S_n \leftarrow \{I \in S_n \mid \| I \| = n\}$

At line 1, all possible configurations are computed, as the powerset of the coordinates C_n, and added to the set of candidate solutions S_n. Then, at line 3, the set of candidate solutions is refined by removing all configurations in which a queen can attack another. Finally, at line 5, all sets with cardinality other than n are filtered out, so as to keep in S_n only the actual solutions to the puzzle.

Implementing such an algorithm with MFDDs can be quite efficient, thanks to the use of homomoprhisms. As mentioned in the previous sections, they allow to perform filtering directly on the shared structure of a MFDD, therefore modifying large subsets of the encoded family at once. For instance, in the above algorithm, the actions performed on S_n operate on each map it contains in parallel, without the need to iterate over each one of them separately. In addition, MFDDs allow the maps in S_n to share nodes, so as to keep the overall representation compact. Although a 8×8 chessboard has roughly 10^{19} configurations, it takes only 2.5 KB to store in memory.

5 Multi-core Schedulability

The multi-core schedulability problem [17] can be viewed as an assignment problem that consists of verifying whether a set of tasks can be executed on a multi-core system, with respect to task-specific timing constraints and dependencies between the tasks. We focus on the most common type of precedence dependencies, where the predecessor must finish before the successor can start.

We define tasks T with associated triples $\langle r, c, d \rangle$, where r denotes their release time, c denotes their worst-case execution time and d denotes their (absolute) deadline. We assume that all tasks can potentially meet their deadline, i.e., $r + c \leq d$. A *task model* $M = \langle T, \mu, D \rangle$ is a DAG consisting of a set of tasks T as its vertices, a function $\mu : T \to \mathbb{N}^3$ defining tasks characteristics and a relation $D \subseteq T \times T$ describing *direct* dependencies between tasks as its edges.

Consider the task model depicted in Fig. 3, that features five tasks. Task t_0 can be scheduled at time 0, i.e., immediately when it is released, since it has no dependencies. Tasks t_2 and t_4 cannot be scheduled at time 0, even though they have no dependencies, because they are released at times 4 and 8, respectively. Task t_1 depends on both t_0 and t_2. Hence, it cannot start before they have completed, even though its release time is earlier. Finally, task t_3 depends on t_1, meaning that it also indirectly depends on t_0 and t_2 (i.e. t_1's dependencies).

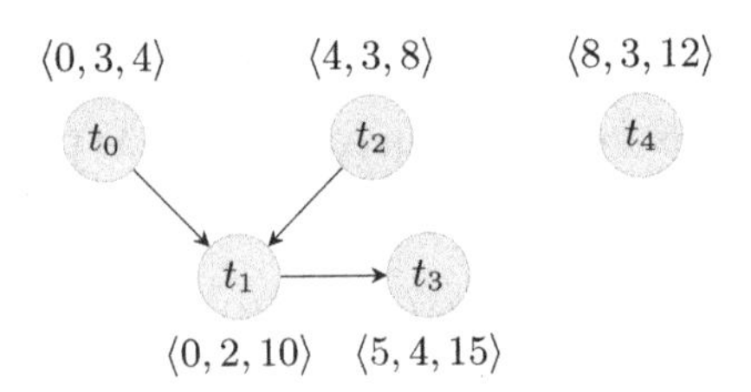

Fig. 3. Example of a task model consisting of 5 tasks.

Given a task model $M = \langle T, \mu, D \rangle$ and a set of cores C, a *scheduling* is a partial function $s : T \to C \times \mathbb{N}$, that assigns a task to a core at a specific time. Notice that this definition assumes that tasks cannot be preemptively suspended to execute another task on the same core, effectively meaning that all tasks are assumed to be executed from beginning to end. A scheduling is *feasible* if:

- all tasks are scheduled after their dependencies have finished;
- all tasks are scheduled after their release time; and
- the deadlines of all tasks are met.

Moreover, a scheduling is *consistent* if there are no tasks scheduled on the same core at the same time.

Figure 4 illustrates two examples of schedulings for the task model in Fig. 3. Both schedulings are consistent, since tasks do not overlap. However, only the left one is feasible, since the right one does not satisfy t_3's dependencies.

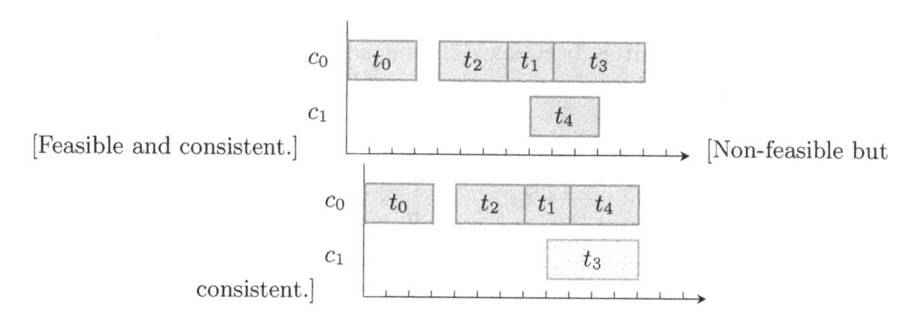

Fig. 4. Two examples of schedulings.

Definition 3 (Schedulability problem). *Given a task model and a set of cores, the schedulability problem consists of determining whether there exists at least one feasible and consistent scheduling.*

5.1 Schedulings as Decision Diagrams

Recall that MFDDs encode sets of partial applications. Since a scheduling is a partial function mapping each task to a core and a start time, encoding a set of schedulings with a MFDD is straightforward: tasks are represented by non-terminal nodes, whereas the start time and the core on which they are scheduled label the arcs. There are however two issues to address, which relate to the same limitation of decision diagrams. Because they are DAGs, extracting information from suffixes is usually challenging. Not only it initiates recursive explorations, which are costly on large diagrams, it also requires complex transformations to preserve the consistency of the prefix that leads to a particular node, when a parent is mutated. Such operations are usually avoided, in favor of homomorphisms that depend on values which are read on a prefix to a given node.

This limitation impacts the initial construction of the MFDD. Scheduling a task on a given core necessitates to know *when* the core is next available, which is an information that depends on the suffix of a given path. One way to tackle this issue is to lift the necessary information at the root of the MFDD. That way, we can first collect the next available time for a given core, before inserting a new mapping and schedule a new task on it.

The second issue relates to the handling of task dependencies. In order to decide if a scheduling is feasible, we need to determine whether each task is scheduled after all its dependencies. Consequently, the order in which tasks appear along a path in the MFDD must be chosen carefully, so that dependencies are laid out deeper. That way, we can first dive to the nodes representing a specific task, and then remove all suffixes for which its dependencies are either not scheduled or finish too late.

Figure 5 represents the two schedulings depicted in Fig. 4. The top path (resp. bottom path) corresponds to the left (resp. right) scheduling. The shared suffix shows how similarities between different schedulings can be exploited to compact the representation of a large set of possibilities.

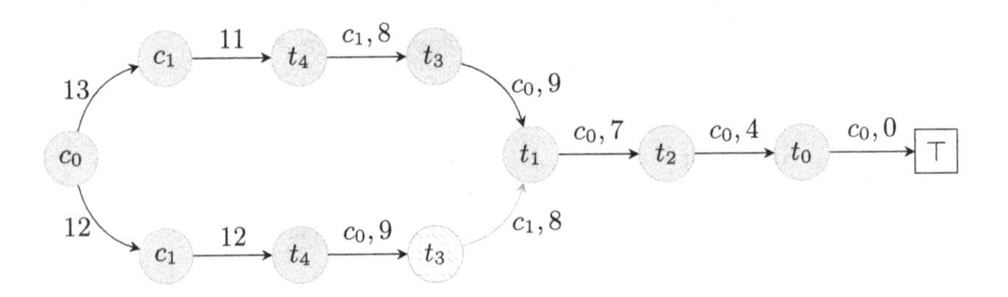

Fig. 5. Two schedulings represented as a MFDD.

Note that the order in which the tasks are laid out has a direct impact on sharing. Despite the constraint we mentioned earlier, namely that tasks should appear before their dependencies, there is a lot of freedom to pick this order. This is generally a difficult problem in the context of decision diagram optimizations [3, Chapter 7]. There are nonetheless some intuitions about what constitutes a good order. Indeed, sharing is most likely to occur on the bottom of the MFDD. Thus, tasks that are more loosely constrained, in terms of dependencies, release times and deadlines, should appear closer to the top. That way, more tightly constrained tasks, which will intuitively have less possible different assignments, will be found in the suffixes of more nodes.

The reader will notice that our encoding does not handle sporadic tasks. Such tasks are common in real-time systems, and corresponds to actions that must be executed periodically. As we chose to formalize schedulings in the form of partial functions $T \to C \times N$, representing sporadic tasks would require T to be an infinite set. However, our definition of a task model is sufficient to describe a time window, in which occurrences of sporadic tasks can be enumerated. As a result, any feasible and consistent scheduling can represent one time window in the unbounded behavior of the system, consisting of repeated time windows.

5.2 Computing Schedulings

We now describe the computation of the set of all possible schedulings by a MFDD of $\mathbb{M}_{T \cup C, \mathbb{N} \cup (C \times \mathbb{N})}$. In a nutshell, the method consists of iteratively scheduling one task, at one possible time slot, in all schedulings that have been computed so far. This produces a new subset of schedulings at each iteration, that is merged with the original set, until a fixed point is eventually reached. The process is guaranteed to terminate, as we can assume that all tasks have a finite deadline. This limitation is consistent with the idea of using task models as a way to represent slices of a system's behavior. It follows that there is a finite number of time slots at which the algorithm should attempt to schedule a task.

The following pseudo-code describes our algorithm:

1: $S \leftarrow enc(\{[c \mapsto 0 \mid c \in C]\})$
2: $S' \leftarrow \bot$
3: **while** $S \not\equiv S'$ **do**
4: $\quad S' \leftarrow S$
5: $\quad$ **for all** $\langle t, c \rangle \in T \times C$ **do**
6: $\quad\quad$ **for all** $t_r \leq i \leq t_d - t_c$ **do**
7: $\quad\quad\quad S \leftarrow S \cup (sch(t, c, i) \circ fltr(t))(S)$
8: $\quad\quad$ **end for**
9: $\quad$ **end for**
10: **end while**
11: $S \leftarrow check(S)$

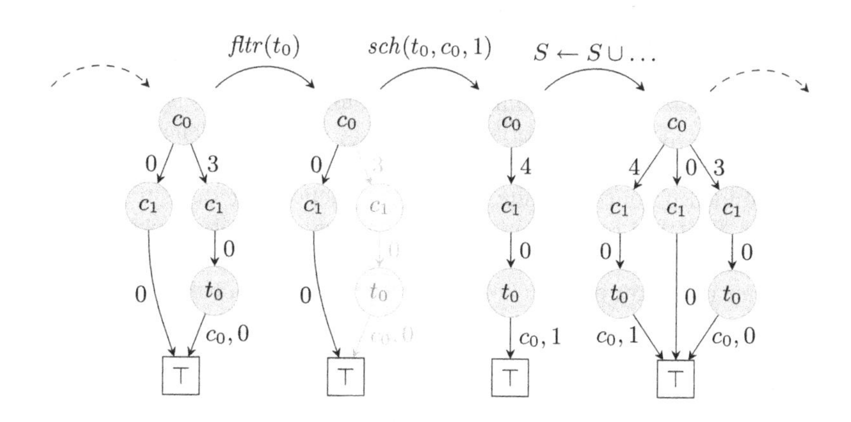

Fig. 6. One iteration of the scheduling construction algorithm's the inner loop

Line 1 creates a MFDD representing a singleton set which contains a function that maps all cores to 0 (i.e., their next available time). Then, most of the work is carried out at line 7, where a task's mapping (on a given core, at a specific time) is added to all schedulings. This process is illustrated for one single iteration in Fig. 6. Assume S to be the MFDD depicted on the left, which encodes the empty scheduling, as well as one mapping in which a task t_0 is being scheduled on core c_0. The first step consists of filtering out the mappings for which the task is already defined, with the *fltr* homomorphism, producing the MFDD in which the corresponding paths have been grayed out. The remaining mappings are then fed into the *sch* morphism, that actually inserts the task at the correct position in all paths. Finally, we compute the union of the resulting MFDD with S, effectively merging all new schedulings into the previous ones.

Note that task dependencies are not taken into consideration until line 11, where a third morphism *check* removes all non-feasible schedulings. This final step boils down to another filter that removes the paths that do not schedule all tasks, as well as those in which task constraints are not satisfied.

6 Experimental Results

We implemented our schedulability analysis technique with DDKit, a library to manipulate MFDDs. We tested it on several randomly generated task models of various sizes, for a 2-cores, a 3-cores and a 4-cores architecture. Our implementation closely follows the algorithm presented in Sect. 5.2. All tests were ran on a intel i9 at 2.3 GHz. Sources as well as the randomly generated datasets we used are available on GitHub[2], and distributed under the MIT license.

Test runs showed that our technique can quickly determine if a given model is not schedulable due to timing constraints alone. This is because schedulings are built incrementally, by adding a new unscheduled task at each pass. Paths in which a task is not schedulable due to timing constraints (i.e. its deadline cannot be met) are cut from the MFDD as soon as they are detected, actually reducing its size. As a result, subgraphs that were only reachable from such paths need no longer to be explored for scheduling the remaining tasks. This is akin to pruning in a classical backtracking algorithm. Filtering out dependency constraints is also efficient, and amounts to about only 5% of the total execution time. This can be explained by our encoding strategy. As dependent tasks tend to appear closer to the MFDD's root, filtering out a path in which its dependency is not scheduled removes a lot of subgraphs at once, and with them, a lot of possible solutions that no longer need to be explored. Furthermore, this mechanism deals gracefully with chains of dependencies.

Results are summarized in Fig. 7. As tasksets are generated randomly, computation times may vary from one task model to the other. Hence, we averaged all results on three different test cases, for each size of task model. We also removed models for which no schedulings can be found, as those are significantly faster to process. We measured the running time as well as the total memory consumed throughout the execution. Runs whose computation exceeded twelve hours were aborted, explaining missing results for both the 3-cores and 4-cores architectures.

As we can see, our algorithm scales much better with the number of tasks than with the number of cores. Although it took on average 4 min and a half

[2] https://github.com/kyouko-taiga/Schedulability.

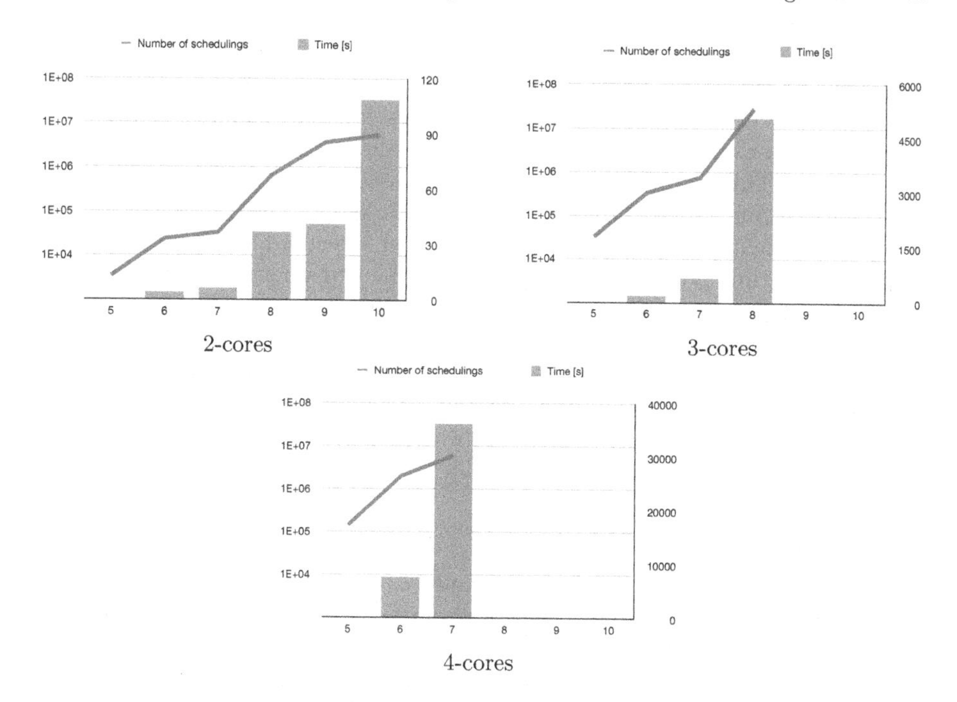

Fig. 7. Experimental results

to compute 150'000 solutions for a model consisting of 5 tasks on a 4-cores architecture, we were able to compute nearly 40 times more schedulings (i.e., roughly 5.5 millions) for a model of 10 tasks on a 2 core-architecture. This asymmetry can be explained by the way operations are optimized on decision diagrams. Recall that outgoing arcs of nodes that represent tasks are labeled with a pair denoting a core and a time. Increasing the number of cores increases the size of the domain from which these labels will be picked, which leads to reducing the possibility of sharing identical subgraphs. As a result, applying a homomorphism on the diagram is more expensive, as caching gets less efficient.

This scaling problem could be tackled by the means of *anonymization* [9]. Anonymization is an optimization technique for decision diagrams which aims to reduce the properties that make two configurations distinguishable, so as to leverage caching more aggressively. In our case, the specific core on which a task is scheduled could be forgotten, so that arcs would be labeled only by the start time. Obviously, this loss of information would make the encoded schedulings less precise, as we could only know *when* a task is scheduled, and not *where*. However, this would be sufficient to determine the number of solutions (if any) there exists to schedule a task model on a given architecture.

Our approach is a refinement of another similar work [17], which purposes the use of Data Decision Diagrams for computing sets of schedulings. Our technique brings a number of improvements. Firstly, we are able to compute all possible schedulings for a given time window, including those in which some cores may be idling (i.e., inactive in a period of time), which was suggested as future work in [17]. Hence, we are able to check for schedulability in the presence of transient errors, whereas the original method can only model permanent core failures. This refinement also solves an issue related to the handling of dependencies. Since their technique is not able to model idling, it is not able to schedule a task *after* its dependencies have been executed, on any core whose next available time precedes the completion of the dependency. Secondly, experimental results show that our approach is roughly twice as efficient in terms of computation time. We explain this disparity by the fact that we filter for dependency inconsistencies *after* having built the set of all possible schedulings, rather than after the addition of a new task. This alleviates the filtering effort and maintains a smaller MFDD during the task addition phase.

7 Conclusion

We presented a transformation of the multi-core schedulability problem as a state space exploration problem, in a style reminiscent to model checking. We showed how to build the set of all possible schedulings of a given taskset with Map-Family Decision Diagrams, a variant of decision diagrams that we designed to encode large sets of partial functions into a compact representation. We used homomorphic operations, in particular filters, to manipulate the set of schedulings and illustrated how to use these operations to analyze solution sets.

We envision future works along two main axes. The first one is to refine our encoding, so as to improve on the performance of our algorithm. A promising lead in this direction is to find better heuristic for task ordering, to maximize sharing. Another idea would be to exploit symmetries between schedulings, so as to prune the state space exploration. Finally, as mentioned in Sect. 6, anonymization could also be leveraged for specific analysis, for instance, to more quickly identify non-schedulable models [9]. The second axis for future works relates to the development of a framework to analyze scheduling sets. We already demonstrated that filtering homomorphisms can be efficiently leveraged to exclude schedulings based on dependency constraints. Other kind of constraints could be translated into filters as well, and applied on solution sets to check schedulability under more elaborate constraints. For instance, one could exclude schedulings in which two given tasks are not executed at the same time, on different cores, so as to account for possible communications.

References

1. Baro, J., Boniol, F., Cordovilla, M., Noulard, E., Pagetti, C.: Off-line (optimal) multiprocessor scheduling of dependent periodic tasks. In: Proceedings of the 27th Annual ACM Symposium on Applied Computing, pp. 1815–1820 (2012)
2. Behrmann, G., Larsen, K.G., Rasmussen, J.I.: Optimal scheduling using priced timed automata. SIGMETRICS Perform. Eval. Rev. **32**(4), 34–40 (2005)

3. Bergman, D., Ciré, A.A., van Hoeve, W., Hooker, J.N.: Decision Diagrams for Optimization. Artificial Intelligence: Foundations, Theory, and Algorithms. Springer, Heidelberg (2016). https://doi.org/10.1007/978-3-319-42849-9
4. Bryant, R.E.: Graph-based algorithms for boolean function manipulation. IEEE Trans. Comput. **35**(8), 677–691 (1986)
5. Cire, A.A., van Hoeve, W.J.: Multivalued decision diagrams for sequencing problems. Oper. Res. **61**(6), 1411–1428 (2013)
6. Couvreur, J.-M., Encrenaz, E., Paviot-Adet, E., Poitrenaud, D., Wacrenier, P.-A.: Data decision diagrams for petri net analysis. In: Esparza, J., Lakos, C. (eds.) ICATPN 2002. LNCS, vol. 2360, pp. 101–120. Springer, Heidelberg (2002). https://doi.org/10.1007/3-540-48068-4_8
7. David, A., Larsen, K.G., Legay, A., Mikucionis, M.: Schedulability of Herschel revisited using statistical model checking. Int. J. Softw. Tools Technol. Transfer **17**(2), 187–199 (2015). https://doi.org/10.1007/s10009-014-0331-4
8. Guan, N., Gu, Z., Deng, Q., Gao, S., Yu, G.: Exact schedulability analysis for static-priority global multiprocessor scheduling using model-checking. In: Obermaisser, R., Nah, Y., Puschner, P., Rammig, F.J. (eds.) SEUS 2007. LNCS, vol. 4761, pp. 263–272. Springer, Heidelberg (2007). https://doi.org/10.1007/978-3-540-75664-4_26
9. Hong, S., Kordon, F., Paviot-Adet, E., Evangelista, S.: Computing a hierarchical static order for decision diagram-based representation from P/T nets. Trans. Petri Nets Other Models Concurr. **5**, 121–140 (2012)
10. Jensen, A.R., Lauritzen, L.B., Laursen, O.: Optimal task graph scheduling with binary decision diagrams (2004)
11. Korousic-Seljak, B.: Task scheduling policies for real-time systems. Microprocess. Microsyst. **18**(9), 501–511 (1994)
12. Lehoczky, J.P., Sha, L., Ding, Y.: The rate monotonic scheduling algorithm: exact characterization and average case behavior. In: Real-Time Systems Symposium, pp. 166–171. IEEE Computer Society (1989)
13. Linard, A., Paviot-Adet, E., Kordon, F., Buchs, D., Charron, S.: polydd: towards a framework generalizing decision diagrams. In: Gomes, L., Khomenko, V., Fernandes, J.M. (eds.) International Conference on Application of Concurrency to System Design, pp. 124–133. IEEE Computer Society, New York (2010)
14. Mahadevan, S., Storgaard, M., Madsen, J., Virk, K.: ARTS: a system-level framework for modeling MPSoC components and analysis of their causality. In: International Symposium on Modeling, pp. 480–483. IEEE Computer Society (2005)
15. Nguyen, V.A., Hardy, D., Puaut, I.: Cache-conscious offline real-time task scheduling for multi-core processors. In: 29th Euromicro Conference on Real-time Systems (ECRTS 2017). Schloss Dagstuhl-Leibniz-Zentrum fuer Informatik (2017)
16. Puffitsch, W., Noulard, E., Pagetti, C.: Off-line mapping of multi-rate dependent task sets to many-core platforms. Real-Time Syst. **51**(5), 526–565 (2015). https://doi.org/10.1007/s11241-015-9232-1
17. Racordon, D., Buchs, D.: Verifying multi-core schedulability with data decision diagrams. In: Crnkovic, I., Troubitsyna, E. (eds.) SERENE 2016. LNCS, vol. 9823, pp. 45–61. Springer, Cham (2016). https://doi.org/10.1007/978-3-319-45892-2_4
18. Yalcinkaya, B., Nasri, M., Brandenburg, B.B.: An exact schedulability test for non-preemptive self-suspending real-time tasks. In: Teich, J., Fummi, F. (eds.) Design, Automation & Test in Europe, pp. 1228–1233. IEEE (2019)

Transforming Interactive Multi-objective Metamodel/Model Co-evolution into Mono-objective Search via Designer's Preferences Extraction

Wael Kessentini(✉) and Vahid Alizadeh(✉)

College of Computing and Digital Media, DePaul University, 243 South Wabash Avenue Chicago, Chicago, IL 60604, USA
{wkessent,alizadeh}@depaul.edu

Abstract. The simultaneous evolution of metamodels and models is called the meta-models/models co-evolution problem. While some Interactive/automated metamodel/model co-evolution techniques have been proposed using multi-objective search, designers still need to explore a large number of possible revised models. In this paper, we propose an approach to convert multi-objective search into a mono-objective one after interacting with the designer to identify a set of model changes based on his/her preferences. The first step consists of using a multi-objective search to generate different possible model edit operations by finding a trade-off between three objectives. Then, the designer may give feedback on some proposed solutions. The extracted preferences are used to transform the multi-objective search into a mono-objective one by generating an evaluation function based on the weights for the existing fitness functions that are automatically computed from the feedback. Thus, the designer will just interact with only one solution generated by the mono-objective search. We evaluated our approach on a set of metamodel/model co-evolution case studies and compared it to existing fully automated and interactive meta-model/model co-evolution techniques. The results show that the mono-objective search after the interaction with the users significantly improved the co-evolution changes for several widely used metamodels.

Keywords: Model co-evolution · Interactive multi-objective search

1 Introduction

Similar to the source code of large systems, the modeling languages (i.e., metamodels) are subject to evolution due to changing requirements and technological constraints requiring existing models to be adapted [1,2]. Thus, a set of changes must be applied to the initial model versions to fix the inconsistencies with the new metamodel version. This process is called metamodel/model co-evolution [1,3].

A. Aleti and A. Panichella (Eds.): SSBSE 2020, LNCS 12420, pp. 88–104, 2020.
https://doi.org/10.1007/978-3-030-59762-7_7

Several co-evolution studies are proposed where most of them are providing either a manual or semi-automated support based on pre-defined templates of evolution scenarios [4–8]. In addition to being pre-defined, these templates are specific to the artifact/models to co-evolve and the metamodel. Few fully automated co-evolution studies tried to find an entire edit operations sequence that revises models in accordance with the new metamodel version [3,9,10]. However, several transformations require interactions with the user especially when new elements are added to the new metamodel they are hard to full-automate. Recently, an approach has been proposed to interactively evaluate the co-evolved models using search-based software engineering [11]. The designers can provide feedback about the co-evolved models and introduce manual changes to some of the edit operations that revise the model. However, this interactive process can be expensive, and tedious since designers must evaluate every recommended set of edit operations and adapt them to the targeted design, especially in large models where the number of possible co-evolution strategies can grow exponentially. Besides, it is challenging to define upfront weights to some edit operation since the designer needs to have a look at the generated solutions to express his preferences.

In this paper, we propose an approach that takes advantage of existing Metamodel/Model co-evolution works. Thus, we propose a way to convert multi-objective search into a mono-objective one after few interactions with the developer. The first step consists of using a multi-objective search, based on the evolutionary algorithm NSGA-II [12], to generate a diverse set of model migration strategies by finding a trade-off between three objectives of reducing the number of edit operations, the dissimilarity with the initial models to reduce the information loss and the number of inconsistencies between the models and new metamodel. Then, the designer may give some feedback on the generated solutions to express his/her preferences by selecting relevant ones. The extracted preferences from the designer, via an analysis of the location of these solutions in the objective space, are used to transform the multi-objective search into a mono-objective one by generating an evaluation function based on the weights that are automatically calculated from the selected solutions. Therefore, the designer will interact in the next iterations with only one co-evolution solution generated by the mono-objective search.

Our approach is taking the advantages of mono-objective search, multi-objective search, and interactive computational intelligence. Multi-objective algorithms are powerful in diversifying solutions and finding trade-offs between many objectives but generate many solutions as an output. The interactive algorithm is useful in terms of extracting designers' knowledge and preferences. Mono-objective algorithms are the best in terms of optimization power once the evaluation function is well-defined and generate only one solution as an output. We selected 16 active developers to manually evaluate the effectiveness of our tool on four well-known metamodel/model co-evolution case studies. The results show that the participants found their desired revised models faster and more accurate than the current state of the art. A supplementary appendix materials can be found in the following link [13].

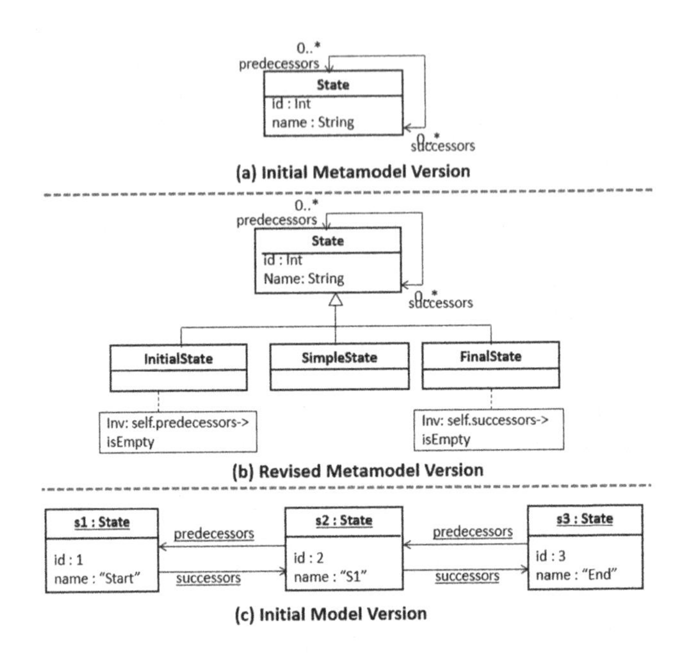

Fig. 1. A simplified metamodel evolution example

2 Background and Motivations

Figure 1 shows an example of a simplified metamodel evolution, based on the simple state machine language and an initial model conform to it. The metamodel evolution comprises three steps: extract sub-classes for *State* class resulting in *InitialState*, *SimpleState*, and *FinalState*, make class *State* abstract, and refine the cardinalities of the predecessor/successor references for the subclasses. This evolution results in the fact that, besides other constraints violations, the constraint which is shown in Listing 1.1 is violated when considering the initial model of Fig. 1c and its conformance to the new metamodel of Fig. 1b.

Listing 1.1. Type/Object relationship formalized as OCL constraint

```
context M!Object
  inv typeExists: MM!Class.allInstances() ->
    exists(c|c.name = self.type and not c.isAbstract)
```

To re-establish conformance for the given example, assume for now that only two operations on models are used in this context. Non-conforming objects may either be retyped (reclassified as instances of the concrete classes) or deleted.

Thus, the potential solution space for retyping or deleting non-conforming elements contains $(c+1)^O$ solutions (with c = number of candidate classes + 1 for deletion, o = number of non-conforming objects). This means, in our given example, we would end up with 64 possible co-evolutions.

Several co-evolution studies proposed to revise models after metamodels evolution from manual to fully automated approaches [2]. Recently, few automated/interactive tools [9–11] used search-based software engineering to generate revised models. The proposed tools refine an initial model instantiated from the previous metamodel version to make it as conform as possible to the new metamodel version by finding the best compromise between three objectives, namely minimizing (i) the non-conformities with new metamodel version (ii) the changes to existing models, and (iii) the dissimilarities between the initial and revised models. During the process, the designer may provide some feedback on the proposed solutions in order to improve them in the next iterations. The output is several equally good solutions (edit operations that revise the model) presented to designers to select the appropriate one based on his/her preferences.

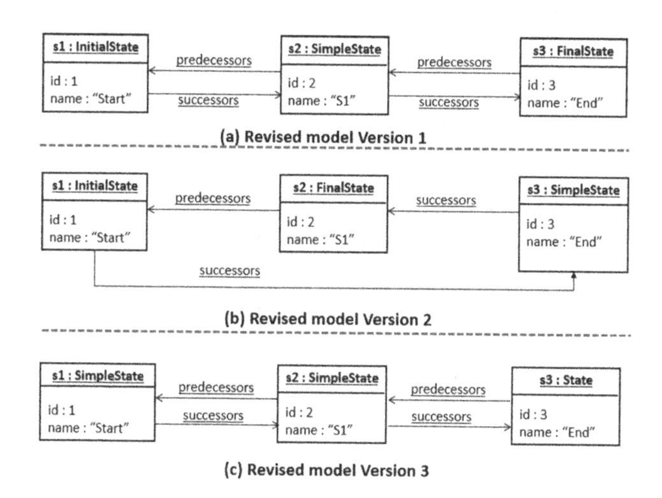

Fig. 2. Tentative revised models

Figure 2 shows modified models after applying a set of edit operations extracted from the output of an existing tool [11]. This figure shows that there may be several possible solutions where the user has to decide which one to select in the search space based on the preferences.

3 Approach Overview

Our approach includes three main phases. The first phase is the multi-objective algorithm, NSGA-II, executed for several iterations to generate a set of non-dominated co-evolution solutions called Pareto-optimal solutions [12], defined as a set of edit operations applied to the initial model, balancing the three objectives of minimizing the number of suggested edit operations, the deviation with the initial model, and the number conformance errors with the revised metamodel.

The output of the first component can be a large number of possible solutions. Thus, it is essential to provide the designers with additional support for interacting with this set of solutions. In the second phase, the user can interact with the tool at the solution level, by accepting or rejecting or modifying suggested edit operation(s) and s/he can also give a score to a selected solution between −1 and 1 (the highest is the better). Finally, we extract the preferences automatically and use them to transform the multi-objective problem into a mono-objective one by generating weights for each of the three objectives based on the selected solutions' locations in the objective space. The output of the mono-objective search is a single solution fitting the user's expectations and preferences; then the designer can interact with that solution if needed and continue the execution of the mono-objective algorithm until selecting a final co-evolution solution. In the following, we will explain, in detail, the phases of our approach.

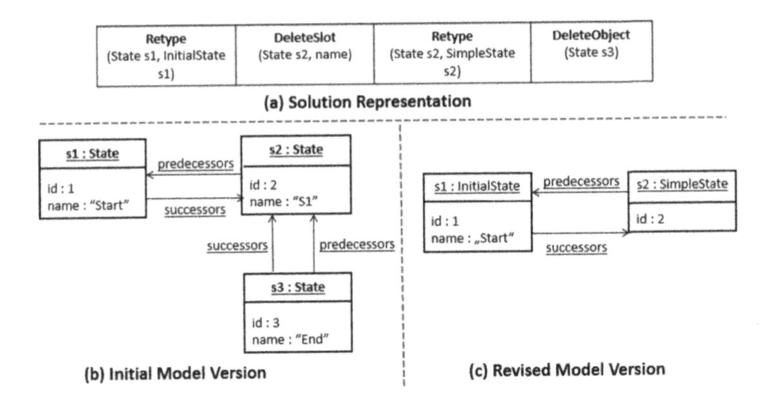

Fig. 3. Solution representation.

3.1 Phase 1: Multi-objective Metamodel/Model Co-evolution

Solution Representation. A co-evolution solution consists of a sequence of n edit operations to revise the initial model. The vector-based representation is used to define the edit operations sequence. Each vector's dimension has an operation, and its index in the vector indicates the order in which it will be applied. Consequently, vectors representing different solutions may have different sizes, i.e., the number of edit operations.

Table 1 shows the possible edit operations that can be applied to model elements. The instances of classes are called objects, instances of features are called slots, and instances of references are called links. These operations are inspired by the catalog of operators for metamodel/model co-evolution presented in [14]. The catalog includes both metamodel and model changes. Thus, we selected from it all the edit operations that can be applied to the model level since we are not changing the metamodels in this paper. Figure 3 represents a solution that can be applied to the initial model of our motivating example described in Sect. 2.

Table 1. Model edit operations.

Operations	Element	Description
Create/delete	Object, link, slot	Add/remove an element in the initial model.
Retype	Object	Replace an element by another equivalent element having a different type.
Merge	Object, link, slot	Merge several model elements of the same type into a single element.
Split	Object, link, slot	Split a model element into several elements of the same type.
Move	Link, slot	Move an element from an object to another.

Fitness functions. The investigated co-evolution problem involves searching for the best sequence of edit operations to apply among the set of possible ones. A good solution s is a sequence of edit operations to apply to an initial model with the objectives of minimizing the number of non-conformities $f_1(s) = nvc(s)$ with the new metamodel version, the number of changes $f_2(s) = nbOp(s)$ applied to the initial model, and the inconsistency $f_3(s) = dis(s)$ between the initial and the evolved models such as the loss of information.

The first fitness function $nvc(s)$ counts the number of violated constraints w.r.t. the evolved metamodel after applying a sequence s of edit operations. We apply, first, the sequence of edit operations (solution) on the initial model, then we load the evolved model on the target metamodel to measure the number of conformance errors based on the number of violated constraints. We consider three types of constraints, as described in [15]: related to model objects, i.e., model element (denoted by O.*), related to objects' values (V.*), and related to objects' links (L.*). We use the implementation of these constraints in our experiments inspired by Schoenboeck et al. [3] and Richters et al. [15] with slight adaptations. The constraints are hard-coded in the implementation of the algorithm, and most of them are from the EMF conformance verification constraints that already exist in EMF. The full list constraints can be found in this link [13].

For the second fitness function, which aims to minimize the changes to the initial models, we simply count the number of edit operations, $nbOp(s)$ of a solution s (size of s). The third fitness function dis(s) measures the difference between the model elements in the initial and revised model. As the type of a model element may change because of a change in the metamodel, we cannot rely on elements' types. Alternatively, we use the identifiers to assess whether the information was added or deleted when editing a model. In this case, the renamed or extracted model elements will be considered different than the initial model element. Thus, we considered the assumption that two model elements could be syntactically similar if they use a similar vocabulary. Thus, we calculated for the textual similarity based on the Cosine similarity [16]. In the first step, we tokenize the names of initial and revised model elements. The textual and context similarity between elements are grouped together to create a new class, which is an essential factor in evaluating the revised model's cohesion. The initial and revised models are represented as vectors of terms in n-dimensional space where n is the number of terms in all considered models. For each model, a weight is assigned to each dimension (representing a specific term) of the representative vector that corresponds to the term frequency score (TF) in the model. The similarity among initial and revised model elements is measured by the cosine of the angle between its representative vectors as a normalized projection of one vector over the other. This function will compare each of the initial model elements and all the elements of the revised model to find the best matching.

3.2 Phase 2: Interaction and Preference Extraction

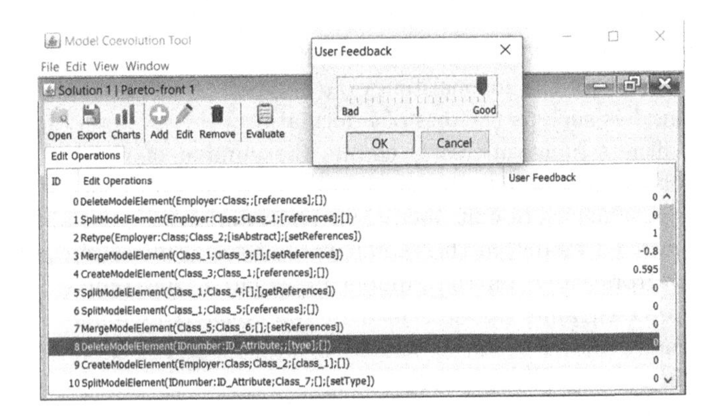

Fig. 4. User interactions with the solutions of the multi-objective search

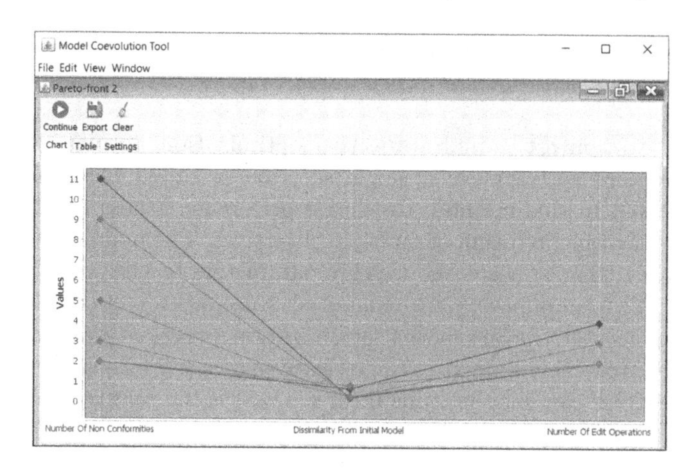

Fig. 5. The output of phase 2 (interactions with the user). (Color figure online)

The main goal of this step is to enable the designer's interactions with the solutions generated by the first phase of the multi-objective search. The interaction can be performed at the solution and edit operation levels depending on the user's desire. The feedback is quantified to a continuous score in the range of $[-1, 1]$. The user can evaluate a solution by modifying its edit operations (edit, add, delete, re-order) or just rate the whole solution, as shown in Fig. 4. After the designer's interaction, solution scores ($Score_{s_i}$) are computed as the average score of edit operations in a solution. The solutions with the highest score are considered as the region of interest. It indicates the preferred objectives and edit operations. The line chart of Pareto-front solutions after interactions is shown in Fig. 5. The color of each line indicates user preferences (green as preferred solutions versus red as non-preferred solutions.

3.3 Phase 3: Preference-Based Mono-objective Co-evolution

One of the main contributions of this paper is the ability to convert a multi-objective algorithm into a mono-objective one after interacting with the designer to extract his/her preferences. Mono-objective algorithms are known to be the best in terms of optimization but require that the fitness function should be well-defined based on the decision maker's preferences. The Multi-objective Evolutionary Algorithm used in Phase 1 might not provide high-quality solutions in the region of interest of the developer because of the high dimensionality nature of the problem and the need to find trade-offs. Therefore, it is important to consider the user preferences extracted in Phase 2.

The goal of this phase is to use the preferences extracted from the designer after the multi-objective optimization to transform the problem into a single objective optimization problem by aggregating objectives according to the user's preferences. This transformation gives the decision maker a single solution. Consequently, our proposed approach is a combination of all three categories of preference-based search where the preferences are expressed after the first evolutionary process, then they are incorporated to guide the single-objective optimization.

One way to convert a multi-objective optimization problem to a mono-objective problem and achieve a single solution is called the Weighted Sum Method (WSM). In this method, the single preference fitness function is computed as a linear weighted sum of multiple objectives. The main drawback of the WSM method is that it needs the weights parameters to be given. Fortunately, in our case, those parameters are computed automatically from the decision-maker preferences of the interactive optimization process (preferred solutions based on the interaction scores) in the objectives space. Thus, the weight of one or more objectives can get the value 0 (or almost) if the selected solution(s) by the developer penalized them while favoring other objectives. Also, the WSM is not computationally expensive, unlike the other scalarization methods.

In order to solve the converted mono-objective problem, we adopted a standard Genetic Algorithm (GA). To adapt the GA algorithm to our co-evolution problem, we use the same solution representation and fitness functions as reported in phase 1. The importance (weights) of the objectives are based on the preferred solutions by the user with an interaction score higher than 0.5. The obtained single fitness function is employed to evaluate the solutions in the execution of adapted GA. Thus, the weight of each objective is calculated as the average of the objective values of the preferred solution(s) by the user. We note that all the objectives of the multi-objective search are normalized using the min-max function.

Instead of generating the initial population randomly, we acquire the user preferred solutions as the elite set of solutions from which the search process is initiated. Thus, we do not generate solutions randomly for the mono-objective GA, but we take the preferred solutions as the initial population, so we do not lose the knowledge extracted from the developer. If the number of preferred solutions is low, then we apply the mutation operator to generate more solutions. The solutions are evaluated via the preference function aggregated from multiple objectives. When the stopping condition is satisfied, the single optimal solution is recommended to the user. Similar to Phase 1, the user can interact with this solution via editing/adding/removing the edit operations.

4 Evaluation

4.1 Research Questions and Experimental Setup

- **RQ1: Search validation.** How does our approach perform compared to random search (RS)?
- **RQ2: Benefits.** To what extent can our approach make relevant recommendations for designers compared to existing metamodel/model co-evolution techniques including multi-objective search and an existing deterministic method?

- **RQ3: The relevance of designers' preferences extraction.** To what extent can our approach reduce the interaction effort comparing to existing metamodel/model co-evolution techniques?

Studied Metamodels and Models. To answer the research questions, we considered the evolution of GMF covering a period of two years and the UML Class Diagram metamodel evolution from [17,18]. These case studies are interesting scenarios since they represent real metamodel evolutions, used in an empirical study [19] and studied in other contributions [20–22]. For GMF, we chose to analyze the extensive evolution of three Ecore metamodels. We considered the evolution from GMF's release 1.0 over 2.0 to release 2.1, covering a period of two years. For achieving a broad data basis, we analyzed the revisions of three metamodels, namely the Graphical Definition Metamodel (GMF Graph for short), the Generator Metamodel (GMF Gen for short), and the Mappings Metamodel (GMF Map for short). Therefore, the respective metamodel versions had to be extracted from GMF's version control system and, subsequently, manually analyzed. We created different scenarios based on the number of changes introduced at the metamodel level from the different metamodel releases of GMF and UML. We merged the releases that did not include extensive changes, and we generated two evolution scenarios per metamodel type.

The different models and metamodels can be classified as small-sized through medium-sized to large-sized. In our experiments, we have a total of 7 different co-evolution scenarios where each scenario included eight different models to evolve for the GMF case-studies. The percentage of changes between the different releases is estimated based on the number of modified metamodel elements divided by the size of the metamodel. The created models for our experiments are ensuring metamodels coverage. Furthermore, we used an existing set of 10 generated models for the case of UML metamodel class diagram evolution from the deterministic work of [17,18]; thus, we were not involved in the selection of models and metamodel changes. To ensure a fair comparison with Wimmer et al. [17], we only compared both approaches to the existing UML dataset. Table 2 describes the statistics related to the collected data.

Table 2. Statistics related to the collected data of the investigated cases.

Metamodels				Models		
Release	#of elements	#of changes	%of changes	#of models	#of model elements (Min,Max)	#of expected edit operations (Min, Max)
GMF Gen 1.41 to 1.90	From 885 to 1120	347	31%	8	389, 744	39, 70
GMF Gen 1.90 to 1.248	From 1120 to 1216	362	27%	8	433, 686	66, 83
GMF Map 1.45 to 1.52	From 382 to 413	62	15%	8	203, 394	46, 69
GMF Map 1.52 to 1.58	From 413 to 428	10	1.8%	8	347, 402	57, 81
GMF Graph 1.25 to 1.29	From 278 to 279	14	5%	8	142, 283	34, 55
GMF Graph 1.25 to 1.33	From 279 to 281	42	14%	8	149, 301	29, 43
UML CD [17]	From 23 to 29	8	8%	10	28, 49	11, 23

Evaluation Metrics. The quality of our results was measured by two methods: automatic correctness (AC) and manual correctness (MC). Automatic correctness consists of comparing the proposed edit operations to the reference ones, operation by operation using precision (AC-PR), and recall (AC-RE). For an operation sequence corresponding to a given solution, precision indicates the proportion of correct edit operations (w.r.t. the baseline sequence) in a solution. The recall is the proportion of correctly identified edit operations among the set of all expected operations. Both values range from 0 to 1, with higher values indicating good solutions. AC method has the advantage of being automatic and objective. However, since different edit operation combinations exist that describe the same evolution (different edit operations but same target model), AC could reject a good solution because it yields different edit operations from reference ones. To account for those situations, we used MC measured by the designers. It consists of the number of relevant edit operations identified by the designer over the total number of edit operations in the selected solutions. In addition, we report the number of interactions (NI) required on the Pareto front comparing to the one required once the mono-objective search is executed. This evaluation will help to understand if we efficiently extracted the developer preferences after the Pareto-front interactions. We decided to limit the comparison to only the interactive multi-objective work of Kessentini et al. [11] since it is the only approach offering interaction with the user, and it will help us understand the real impact of the knowledge extraction and mono-objective features (not supported by existing studies) on the recommendation and interaction effort. We also report the computation time (T) for the different evolution scenarios to estimate the effort required to obtain the best co-evolution solutions.

Study Participants. Our study involved 16 master students in Software Engineering. All the participants are volunteers and familiar with model-driven engineering and co-evolution/refactoring since they are part of a graduate course on Software Testing & Quality Assurance and most of them participated in similar experiments in the past, either as part of a research project or during graduate courses. Furthermore, 12 out of the 16 students are working as full-time or part-time developers in the software industry. Participants were first asked to fill out a pre-study questionnaire containing five questions. The questionnaire helped to collect background information such as their role within the company, their modeling experience, and their familiarity with model-driven engineering and co-evolution/refactoring. In addition, all the participants attended two lectures about model transformations and evolution, and passed six tests to evaluate their performance in evaluate and suggest model evolution solutions. We formed 4 groups, each composed of 4 participants. The groups were formed based on the pre-study questionnaire and the test results to ensure that all the groups have almost the same average skill level. We divided the participants into groups according to the studied metamodels, the techniques to be tested, and the developers' experience. The participants were asked to co-evolve the different models manually and evaluate the results of the different approaches based on a counter-balanced design [23].

Statistical Tests. Our experimental study is performed based on 30 independent simulation runs, and the obtained results by the alternative approaches are compared using the Wilcoxon rank-sum test [24] with a 95% confidence level. Roughly speaking, this test verifies the null hypothesis H0 that the observed differences between the alternative results were obtained by chance or if they are statistically significant (alternative hypothesis H1). The p-value of the Wilcoxon test corresponds to the probability of rejecting the null hypothesis H0 while it is true (type I error). A p-value that is less than or equal to 0.05) means that we reject H0 and accept H1. A p-value that is less than or equal to $\alpha(\leq 0.05)$ means that we accept H1, and we reject H0. However, a p-value that is strictly greater than $\alpha(> 0.05)$ means the opposite. In this way, we could decide whether the superior performance of NSGA-II to one of each of the others (or the opposite) is statistically significant or just a random result.

Parameter Settings. The stopping criterion was set to 100,000 evaluations for all search algorithms to ensure fairness of comparison (without counting the number of interactions since it is part of the users' decision to reach the best solution based on his/her preferences). The mono-objective search was limited to 10,000 evaluations after the interactions with the user. The other parameters' values are as follows for both the multi-objective and mono-objective algorithms: crossover probability = 0.4; mutation probability = 0.7, where the probability of gene modification is 0.5. Each parameter has been uniformly discrete in some intervals. Values from each interval have been tested for our application. Finally, we pick the best values for all parameters. Hence, a reasonable set of parameters values have been experimented.

4.2 Results

Results for RQ1: Search Validation. Figure 6 confirms that using NSGA-II produces results by far better (and statistically significant) than just randomly exploring a comparable number of solutions based on the three different metrics of precision, recall, and manual correctness on all the different evolution case studies. NSGA-II has precision (AC-PR and MC) and recall (AC-RE) more than twice higher than the ones of random search, as shown in Fig. 6 (~96% vs. ~42%). The difference in execution time in favor of random search (Table 3), due to the crossover and mutation operators, is largely compensated by the quality of the obtained results.

RS is not efficient in generating good co-evolution solutions using all the above metrics in all the experiments. Thus, an intelligent algorithm is required to find good trade-offs to propose efficient solutions. We conclude that there is empirical evidence that our multi-objective formulation surpasses the performance of RS search; thus, our formulation is adequate, and the use of metaheuristic search is justified (this answers RQ1).

Results for RQ2: Benefits. We report the results of the empirical evaluation in Fig. 6. The majority of the co-evolution solutions recommended by our approach were correct and validated by the participants on the different case

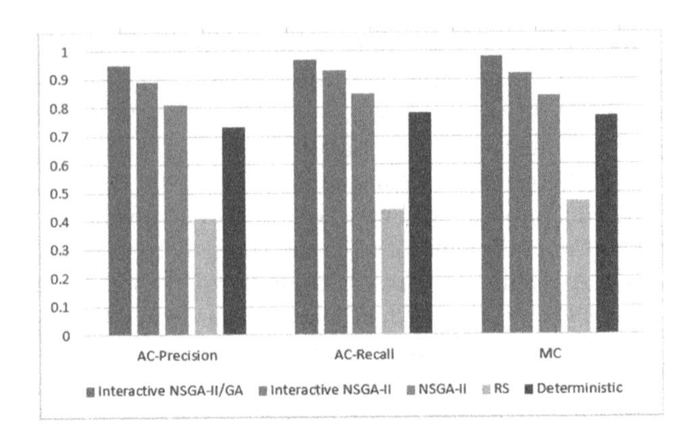

Fig. 6. The median evaluation scores on the four metamodel evolution scenarios with 95% confidence level ($\alpha = 5\%$)

Table 3. Median execution time, in minutes, on the different metamodel/model co-evolution scenarios and the number of interaction proposed by both interactive approaches

	Approaches						
	IMMO		I-NSGA-II		NSGA-II	RS	Deterministic
Metamodels	T	NI	T	NI	T	T	T
GMF Gen	44	18	71	32	38	30	31
GMF Map	30	16	55	25	28	22	17
GMF Graph	25	24	83	35	21	18	14
Class Diagram	20	8	39	5	16	14	12

studies. On average, for all of our four studied metamodels/models, our approach was able to recommend 96% of generated edit operations correctly. The remaining approaches have an average of 89% and 81%, respectively, for the interactive multi-objective approach [11] and the fully automated multi-objective approach [10]. Both of the interactive tools outperformed fully-automated ones, which shows the importance of integrating the human in the loop when co-evolving models. The deterministic approach defines generic rules for a set of possible metamodel changes that are applied to the co-evolved models. Figure 6 shows that our approach clearly outperforms, on average, the deterministic technique based on all measures: precision, recall, and manual correctness.

Results for RQ3: The relevance of Designers' Preferences Extraction. Table 3 summarizes the time, in minutes, and the number of interaction (for the interactive approaches) with the participants to find the most relevant solutions using our tool (IMMO), the interactive approach (I-NSGA-II) [11], the automated approach [10], Random search and the deterministic approach [17]. All the participants spent less time finding the most relevant model edit operations on the different metamodels than I-NSGA-II. For instance, the average time is reduced from 71 min to just 44 min for the case of GMF Gen. The time includes the execution of IMMO and the different phases of interaction until the designer is satisfied with a specific solution. It is clear as well that the time reduction is not correlated with the number of interaction. For instance, the deviation between IMMO and I-NSGA-II for GMF Graph in terms of the number of interactions is limited to 9 (24 vs. 35), but the time reduction is 58 min. The time includes the execution of the multi-objective and mono-objective search (if any) and the different phases of interaction until the designer is satisfied with a specific solution. The drop of the execution time is mainly explained by the fast execution of the mono-objective search and the reduced search space after the interactions with the designer.

It is clear that our approach reduced as well as the number of interaction comparing to I-NSGA-II. IMMO required much fewer designer interactions. For instance, only 16 interactions to modify, reject, and select solutions were observed on GMF Map using our approach, while 25 interactions were needed for I-NSGA-II. The reductions of the number of interactions are mainly due to the move from multi-objective to mono-objective search after one round of interactions since the designers will not deal anymore with a set of solutions in the front but only one.

5 Related Work

In one of the early works [25], the co-evolution of models is tackled by designing co-evolution transformations based on metamodel change types. In [7,8], the authors compute differences between two metamodel versions, which are then input to adapt models automatically. This is achieved by transforming the differences into a migration transformation with a so-called higher-order transformation, i.e., a transformation that takes/produces another transformation as input/output. In [26], the authors proposed an approach that compromises multiple steps for model co-evolution: change detection either by comparing between metamodels or by tracing and recording the changes applied to the old version of the metamodel. The second step is a classification of the changes in metamodel and their impact in its instances. Finally, an appropriate migration algorithm for model migration is determined. For initial model elements for which no transformation rule is found, a default copy transformation rule is applied. This algorithm has been realized in the model migration framework Epsilon Flock [27] and in the framework described in [28].

A comprehensive survey of interactive SBSE approaches can be found in [29]. The problems of contextualization to developer's regions of interest during the recommendation process have been treated in recent SBSE papers for the code refactoring problem [30–34]. Han et al. proposed in [32] an approach to enable the interactions with the user, then a Delta Table can select the next refactoring quickly to improve a specific objective without calculating a fitness function.

6 Conclusion

In this paper, we proposed a novel approach to extract designers' preferences to find good recommendations to co-evolve models. We combined the use of multi-objective search, mono-objective search, and user interaction in our approach. To evaluate the effectiveness of our tool, we conducted an evaluation with 16 participants who evaluated the tool and compared it with the state-of-the-art techniques. As part of our future work, we are planning to evaluate our approach on further metamodel evolution cases and a more extensive set of participants. We will also adapt our approach to address other problems requiring designers' interactions, such as model transformation rules.

References

1. Iovino, L., Pierantonio, A., Malavolta, I.: On the impact significance of metamodel evolution in MDE. J. Object Technol. (2012)
2. Hebig, R., Khelladi, D.E., Bendraou, R.: Approaches to co-evolution of metamodels and models: a survey. IEEE Trans. Softw. Eng. **43**(5), 396–414 (2017)
3. Schoenboeck, J., et al.: CARE: a constraint-based approach for re-establishing conformance-relationships. In: Proceedings of APCCM (2014)
4. Meyers, B., Wimmer, M., Cicchetti, A., Sprinkle, J.: A generic in-place transformation-based approach to structured model co-evolution. In: Proceedings of MPM Workshop (2010)
5. Meyers, B., Vangheluwe, H.: A framework for evolution of modelling languages. Sci. Comput. Program. **76**(12), 1223–1246 (2011)
6. Cicchetti, A., Ciccozzi, F., Leveque, T., Pierantonio, A.: On the concurrent versioning of metamodels and models: challenges and possible solutions. In: Proceedings IWMCP (2011)
7. Garcés, K., Jouault, F., Cointe, P., Bézivin, J.: Managing model adaptation by precise detection of metamodel changes. In: Paige, R.F., Hartman, A., Rensink, A. (eds.) ECMDA-FA 2009. LNCS, vol. 5562, pp. 34–49. Springer, Heidelberg (2009). https://doi.org/10.1007/978-3-642-02674-4_4
8. Cicchetti, A., Ruscio, D.D., Eramo, R., Pierantonio, A.: Automating co-evolution in model-driven engineering. In: Proceedings of EDOC (2008)
9. Kessentini, W., Sahraoui, H., Wimmer, M.: Automated metamodel/model co-evolution using a multi-objective optimization approach. In: Wąsowski, A., Lönn, H. (eds.) ECMFA 2016. LNCS, vol. 9764, pp. 138–155. Springer, Cham (2016). https://doi.org/10.1007/978-3-319-42061-5_9
10. Kessentini, W., Sahraoui, H.A., Wimmer, M.: Automated metamodel/model co-evolution: a search-based approach. Inf. Softw. Technol. **106**, 49–67 (2019)
11. Kessentini, W., Wimmer, M., Sahraoui, H.A.: Integrating the designer in-the-loop for metamodel/model co-evolution via interactive computational search. In: Wasowski, A., Paige, R.F., Haugen, Ø. (eds.) Proceedings of MODELS, pp. 101–111 (2018)
12. Deb, K., Agrawal, S., Pratap, A., Meyarivan, T.: A fast elitist non-dominated sorting genetic algorithm for multi-objective optimization: NSGA-II. In: Schoenauer, M., et al. (eds.) PPSN 2000. LNCS, vol. 1917, pp. 849–858. Springer, Heidelberg (2000). https://doi.org/10.1007/3-540-45356-3_83

13. Immo. https://sites.google.com/view/ssbse2020/
14. Herrmannsdoerfer, M., Vermolen, S.D., Wachsmuth, G.: An extensive catalog of operators for the coupled evolution of metamodels and models. In: Malloy, B., Staab, S., van den Brand, M. (eds.) SLE 2010. LNCS, vol. 6563, pp. 163–182. Springer, Heidelberg (2011). https://doi.org/10.1007/978-3-642-19440-5_10
15. Richters, M.: A precise approach to validating UML models and OCL constraints. Technical report (2001)
16. Muflikhah, L., Baharudin, B.: Document clustering using concept space and cosine similarity measurement. In: Proceedings of ICCTD (2009)
17. Wimmer, M., Kusel, A., Schoenboeck, J., Retschitzegger, W., Schwinger, W.: On using inplace transformations for model co-evolution. In: Proceedings of MtATL Workshop (2010)
18. Cicchetti, A., Ciccozzi, F., Leveque, T., Pierantonio, A.: On the concurrent versioning of metamodels and models: challenges and possible solutions. In: Proceedings of IWMCP (2011)
19. Herrmannsdoerfer, M., Ratiu, D., Wachsmuth, G.: Language evolution in practice: the history of GMF. In: van den Brand, M., Gašević, D., Gray, J. (eds.) SLE 2009. LNCS, vol. 5969, pp. 3–22. Springer, Heidelberg (2010). https://doi.org/10.1007/978-3-642-12107-4_3
20. Herrmannsdoerfer, M.: GMF: a model migration case for the transformation tool contest. In: Proceedings of TTC (2011)
21. Rose, L.M., et al.: Graph and model transformation tools for model migration - empirical results from the transformation tool contest. SoSym **13**(1), 323–359 (2014)
22. Di Ruscio, D., Lämmel, R., Pierantonio, A.: Automated co-evolution of GMF editor models. In: Malloy, B., Staab, S., van den Brand, M. (eds.) SLE 2010. LNCS, vol. 6563, pp. 143–162. Springer, Heidelberg (2011). https://doi.org/10.1007/978-3-642-19440-5_9
23. Pollatsek, A., Well, A.D.: On the use of counterbalanced designs in cognitive research: a suggestion for a better and more powerful analysis. J. Exp. Psychol. Learn. Mem. Cogn. **21**(3), 785 (1995)
24. Arcuri, A., Briand, L.: A practical guide for using statistical tests to assess randomized algorithms in software engineering. In: Proceedings of ICSE (2011)
25. Sprinkle, J., Karsai, G.: A domain-specific visual language for domain model evolution. J. Vis. Lang. Comput. **15**(3–4), 291–307 (2004)
26. Gruschko, B.: Towards synchronizing models with evolving metamodels. In: Proceedings of MoDSE Workshop (2007)
27. Rose, L.M., Kolovos, D.S., Paige, R.F., Polack, F.A.C.: Model migration with epsilon flock. In: Tratt, L., Gogolla, M. (eds.) ICMT 2010. LNCS, vol. 6142, pp. 184–198. Springer, Heidelberg (2010). https://doi.org/10.1007/978-3-642-13688-7_13
28. Narayanan, A., Levendovszky, T., Balasubramanian, D., Karsai, G.: Automatic domain model migration to manage metamodel evolution. In: Schürr, A., Selic, B. (eds.) MODELS 2009. LNCS, vol. 5795, pp. 706–711. Springer, Heidelberg (2009). https://doi.org/10.1007/978-3-642-04425-0_57
29. Ramirez, A., Romero, J.R., Simons, C.L.: A systematic review of interaction in search-based software engineering. IEEE Trans. Softw. Eng. **45**(8), 760–781 (2018)
30. Morales, R., Chicano, F., Khomh, F., Antoniol, G.: Efficient refactoring scheduling based on partial order reduction. J. Syst. Softw. **145**, 25–51 (2018)

31. Morales, R., Soh, Z., Khomh, F., Antoniol, G., Chicano, F.: On the use of developers' context for automatic refactoring of software anti-patterns. J. Syst. Softw. **128**, 236–251 (2017)
32. Han, A.R., Bae, D.H., Cha, S.: An efficient approach to identify multiple and independent move method refactoring candidates. IST **59**, 53–66 (2015)
33. Kessentini, W., Kessentini, M., Sahraoui, H., Bechikh, S., Ouni, A.: A cooperative parallel search-based software engineering approach for code-smells detection. TSE **40**(9), 841–861 (2014)
34. Alizadeh, V., Fehri, H., Kessentini, M.: Less is more: from multi-objective to mono-objective refactoring via developer's knowledge extraction. In: SCAM, pp. 181–192. IEEE (2019)

Evolutionary Grammar-Based Fuzzing

Martin Eberlein, Yannic Noller, Thomas Vogel(✉), and Lars Grunske

Software Engineering Group, Humboldt-Universität zu Berlin, Berlin, Germany
{eberlema,yannic.noller,thomas.vogel,grunske}@informatik.hu-berlin.de

Abstract. A *fuzzer* provides randomly generated inputs to a targeted software to expose erroneous behavior. To efficiently detect defects, generated inputs should conform to the structure of the input format and thus, grammars can be used to generate syntactically correct inputs. In this context, fuzzing can be guided by probabilities attached to competing rules in the grammar, leading to the idea of probabilistic grammar-based fuzzing. However, the optimal assignment of probabilities to individual grammar rules to effectively expose erroneous behavior for individual systems under test is an open research question. In this paper, we present EvoGFuzz, an *evo*lutionary *g*rammar-based *fuzz*ing approach to optimize the probabilities to generate test inputs that may be more likely to trigger exceptional behavior. The evaluation shows the effectiveness of EvoGFuzz in detecting defects compared to probabilistic grammar-based fuzzing (baseline). Applied to ten real-world applications with common input formats (JSON, JavaScript, or CSS3), the evaluation shows that EvoGFuzz achieved a significantly larger median line coverage for all subjects by up to 48% compared to the baseline. Moreover, EvoGFuzz managed to expose 11 unique defects, from which five have not been detected by the baseline.

Keywords: Grammar-based fuzzing · Probabilistic fuzzing · Software testing

1 Introduction

Software security vulnerabilities can be extremely costly [30]. Hunting down those issues has therefore been subject of intense research [10,25,31]. A typical example are internet browsers that combine a wide variety of interconnected components, using different interpreters and languages like JavaScript, Java, CSS, or JSON. This makes web browsers extremely prone to exploiting the growing set of embedded parsers and interpreters to launch malicious attacks. Hallaraker et al. [12] have shown that in particular the JavaScript interpreter, which is used to enhance the client-side display of web pages, is responsible for high-level security issues, allowing attackers to steal users' credentials and lure users into divulging sensitive information. Unfortunately, due to the steady increase in complexity, interpreters become increasingly hard to test and verify.

A. Aleti and A. Panichella (Eds.): SSBSE 2020, LNCS 12420, pp. 105–120, 2020.
https://doi.org/10.1007/978-3-030-59762-7_8

Fuzzing [8,22] has shown great success in finding vulnerabilities and erroneous behavior in a variety of different programs and software [34,35]. A *fuzzer* generates random input data and enhances or mutates them to trigger potential defects or software vulnerabilities. In general, fuzzing comes in various flavors: blackbox, whitebox, and greybox fuzzing [8]. While blackbox fuzzers have no knowledge about the internals of the application under test and apply random input generation, whitebox fuzzers can unleash the full power of program analysis techniques to use the retrieved context information to generate inputs. Greybox fuzzing strikes a balance between these two cases: it employs a light-weight instrumentation of the program to collect some feedback about the generated inputs, which is used to guide the mutation process. This approach reduces the overhead significantly and makes greybox fuzzing an extremely successful vulnerability detection technique [4]. Nevertheless, greybox fuzzers still struggle to create semantically and syntactically correct inputs [29]. The lack of the structural input awareness is considered to be the main limitation. Since greybox fuzzers usually apply mutations on the bit level representation of an input, it is hard to keep a high level, syntactically correct structure. Yet, to detect vulnerabilities deep inside programs, complex input files are needed.

Recently, Pavese et al. [28] presented an approach to generate test inputs using a grammar and a set of input seeds. By using the input seeds to obtain a probabilistic grammar, Pavese et al. generate similar inputs to the seeds, or by inverting probabilities of the grammar, generate dissimilar inputs. Similar input samples can be very useful, for instance, when learning from failure-inducing samples, while dissimilar inputs can be very useful for testing less common, and therefore less evaluated parts of a program. We pick up this general idea of generating inputs based on a probabilistic grammar and propose *evolutionary grammar-based fuzzing* (EvoGFuzz), which combines the technique with an evolutionary optimization approach to detect defects and unwanted behavior in parsers and interpreters. By using a probabilistic grammar, the fuzzer is able to generate syntactically correct inputs. Furthermore, our concept of an evolutionary process is able to generate *interesting* (i.e., failure-inducing inputs) and *complex* input files (e.g., nested loops in JavaScript or nested data structures in JSON). Utilizing the probabilistic grammar to generate new populations allows for good guiding properties. By selecting the most promising inputs of a population and by learning and evolving the probabilistic grammar accordingly, essentially favoring specific production rules from the previous population, this process allows the directed creation of inputs towards specific features. Additionally, EvoGFuzz aims to be language and grammar independent to appeal to a broader testing community.

To examine the effectiveness of our approach, we implemented EvoGFuzz as an extension of the tool by Pavese et al. [28] and conducted experiments on several subjects for three common input languages and their parsers: JSON, JavaScript, and CSS3. We compared EvoGFuzz with the original approach and observed that within the same resource budget our technique can significantly increase the program coverage. Moreover, EvoGFuzz has been able to trigger more exception types (EvoGFuzz 11 vs. the original approach 6).

In summary, this paper makes the following contributions:

- We propose an evolutionary grammar-based fuzzing approach (EvoGFuzz) that combines the concept of probabilistic grammars and evolutionary algorithms to generate test inputs that trigger defects and unwanted behavior.
- We implement EvoGFuzz as an extension of an probabilistic grammar-based fuzzer [28] and the ANTLR parser generator.
- We demonstrate the effectiveness of EvoGFuzz on ten real-world examples across multiple programming languages and input formats, and provide a comparison with the original approach [28].

2 Related Work

EvoGFuzz focuses on the generation of test inputs to reveal defects and unwanted behavior. Existing approaches in this area can be separated in *search-based*, *generative*, *learning-based*, and *combined* techniques [1,26].

Search-Based Input Generation. Test input generation can be formulated as a *search* problem to be solved by meta-heuristic search [14,38]. A simple way is to *randomly* generate inputs, as employed in the original work on fuzzing by Miller et al. [22]. More sophisticated random testing strategies are directed by *feedback* [27]. Evolutionary search applies fitness functions to select promising inputs, while the inputs are generated by mutating an initial population. Recent advances in *fuzzing* show the strength of such search-based techniques [5,19,34]. One of the most popular greybox fuzzers is AFL [34] that applies a genetic algorithm guided by coverage information. While these techniques can successfully generate error-revealing inputs, they miss required information about a program to generate syntactically and semantically correct inputs [29,33].

Generative Input Generation. Hanford [13] introduced grammar-based test generation with his *syntax machine*. Recent advances in *grammar-based fuzzing* pick up this idea and use a grammar to generate inputs that are syntactically correct [9,15]. The main focus of grammar-based fuzzers are parsers and compilers [15,37]. Having grammar production rules augmented with probabilities (aka *probabilistic grammars*) allows to generate inputs based on rule prioritization. Pavese et al. [28] employ this notion: they take an input grammar, augment it with probabilities and generate structured inputs that represent common or very uncommon inputs. In general, generative approaches require the input grammar or language specification, which might not always be a available or accurate enough. Therefore, Höschele and Zeller [16] proposed input grammar mining.

Learning-Based Input Generation. In addition to grammar mining, machine learning is increasingly applied for software testing [6,11,20]. Those techniques learn input structures from seed inputs and use them to generate new testing sequences. They target web browsers [11], compilers [6], and mobile apps [20].

Combined Techniques. Recently, a lot of research efforts focus on the combination of grammar-based and coverage-based fuzzing (CGF) with the goal to use the grammar to generate valid inputs but to use CGF to further explore the input space. Le et al. [18] propose a fuzzer that generates inputs for the worst-case analysis. While they leverage a grammar to generate seed inputs for a CGF, they continuously complement/refine the grammar. Atlidakis et al. [3] propose PYTHIA to test REST APIs. They learn a statistical model that includes the structure of specific API requests as well as their dependencies across each other. While PYTHIA's mutation strategies use this statistical model to generate valid inputs, coverage-guided feedback is used to select inputs that cover new behavior. Other fuzzing works aim to incorporate grammar knowledge within their mutation strategies [29,33]. Similarly to PYTHIA, we use seed inputs to generate an initial probabilistic grammar. However, with every iteration we retrieve new probabilities for the grammar while also mutating these probabilities, which enables evolution of the grammar and a broad exploration of the input space.

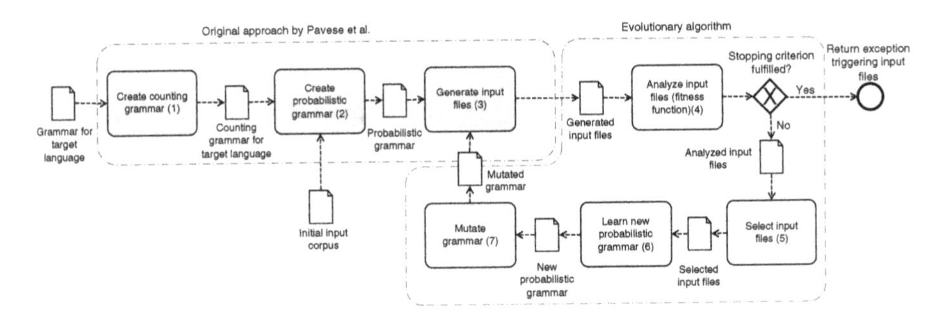

Fig. 1. Overview of EVOGFUZZ.

3 Evolutionary Grammar-Based Fuzzing (EvoGFuzz)

In this section, we will present EVOGFUZZ, a language-independent evolutionary grammar-based fuzzing approach to detect defects and unwanted behavior.

The key idea of EVOGFUZZ is to combine probabilistic grammar-based fuzzing and evolutionary computing. This combination aims for directing the probabilistic generation of test inputs toward "interesting" and "complex" inputs. The motivation is that "interesting" and "complex" inputs more likely reveal defects in a software under test (SUT). For this purpose, we extend an existing probabilistic grammar-based fuzzer [28] with an evolutionary algorithm (Fig. 1).

Similarly to the original fuzzer, EVOGFUZZ requires a correctly specified *grammar for the target language*, that is, the input language of the SUT. From this grammar, we create a so-called *counting grammar* (Activity 1 in Fig. 1) that describes the same language but allows us to measure how frequently individual choices of production rules are used when parsing input files of the given language. Thus, the counting grammar allows us to learn a *probabilistic grammar* from a sampled *initial corpus of inputs* (Activity 2). Particularly, we learn the probabilities for selecting choices of production rules in the grammar, which correspond to the relative numbers of using these choices when parsing the initial corpus. Consequently, we use the probabilistic grammar to generate more input files that resemble features of the initial corpus, that is, "more of the same" [28] is produced (Activity 3). Whereas this activity is the last step of the approach by Pavese et al. [28], it is the starting point of the evolutionary process in EvoGFuzz as it generates an initial population of test inputs. An individual of the population therefore corresponds to a single input file for the SUT.

The evolutionary algorithm of EvoGFuzz starts a new iteration with analyzing each individual using a fitness function that combines feedback and structural scores (Activity 4). By executing the SUT with an individual as input, the feedback score determines whether the individual triggers an exception. If so, this input is considered as an "interesting" input. The structural score quantifies the "complexity" of the individual. If the stopping criterion is fulfilled (e.g., a time budget has been completely used), the exception-triggering input files are returned. Otherwise, the "interesting" and most "complex" individuals are selected for evolution (Activity 5). The selected individuals are used to learn a *new probabilistic grammar*, particularly the probability distribution for the production rules similarly to Activity 2 that, however, used a sampled initial corpus of inputs (Activity 6). Thus, the new probabilistic grammar supports generating "more of the same" interesting and complex inputs. To mitigate a genetic drift toward specific features of the selected individuals, we mutate the new probabilistic grammar by altering the probabilities for randomly chosen production rules (Activity 7). Finally, using the mutated probabilistic grammar, we again generate new input files (Activity 8) starting the next evolutionary iteration.

Assuming that inputs similar to "interesting" and "complex" inputs more likely reveal defects in the SUT, EvoGFuzz guides the generation of inputs toward "interesting" and "complex" inputs by iteratively generating, evaluating, and selecting such inputs, and learning (updating) and mutating the probabilistic grammar. In contrast to typical evolutionary algorithms, EvoGFuzz does not directly evolve the individuals (test input files) by crossover or mutation but rather the probabilistic grammar whose probabilities are iteratively learned and mutated. In the following, we will discuss each activity of EvoGFuzz in detail.

3.1 Probabilistic Grammar-Based Fuzzing (Activities 1–3)

Pavese et al. [28] have proposed probabilistic grammar-based fuzzing to generate test inputs that are similar to a set of given inputs. Using a (context-free) grammar results in syntactically correct test inputs being generated. However, the production rules of the grammar are typically chosen randomly to generate inputs, which does not support influencing the features of the generated inputs. To mitigate this situation, Pavese et al. use a probabilistic grammar, in which probabilities are assigned to choices of production rules. The distribution of these probabilities are learned from a sample of inputs. Consequently, test inputs produced by a learned probabilistic grammar are therefore similar to the sampled inputs. Pavese et al. call this idea "more of the same" [28] because the produced inputs share the same features as the sampled inputs.

Technically, a given context-free grammar for the input language of the SUT is transformed to a so-called counting grammar by adding a variable to each choice of all production rules (Activity 1). Such a variable counts how often its associated choice of a production rule is executed when parsing a given sample of inputs. Knowing how often a production rule is executed in total during parsing, the probability distribution of all choices of the rule is determined. Thus, using the counting grammar to parse a sample of n_{sample} input files, the variables of the grammar are filled with values according to the executed production rules and their choices (Activity 2). This results in a probabilistic grammar, in which a probability is assigned to each choice of a production rule. Using this probabilistic grammar, we can generate new input files that resemble features of the sampled input files since both sets of input files have the same probability distribution for the production rules (Activity 3). Thus, EvoGFuzz uses the approach by Pavese et al. to initially learn a probabilistic grammar from n_{sample} input files (Activities 1 and 2), and to generate $|P|$ new input files for the (initial) population P (Activity 3), which starts an evolutionary iteration.

3.2 Evolutionary Algorithm (Activities 4–8)

The evolutionary algorithm of EvoGFuzz evolves a population of test input files by iteratively (i) analyzing the fitness of each individual, (ii) selecting the fittest individuals, (iii) learning a new probabilistic grammar based on the selected individuals, (iv) mutating the learned grammar, and (v) generating new individuals by the mutated grammar that form the population for the next iteration.

Analyze Individuals (Activity 4). Our goal is to evolve individuals toward "complex" and "interesting" test inputs as such inputs more likely detect defects and unwanted behavior. To achieve this goal, we use a fitness function that quantifies both aspects, the complexity and whether an input is of interest.

Concerning complexity, we focus on the structure of test input files assuming complex structures (e.g., nested loops in JavaScript) have a higher tendency to reveal uncommon behavior in the SUT (e.g., a JavaScript parser) than simple ones. However, we can make only few assumptions about the complexity of input files since EvoGFuzz is language independent and thus, it has no semantic knowledge about the language of test inputs besides the grammar. Thus, we can only rely on *generic* features of test input files and grammars to quantify the complexity of an input file. A straightforward and efficient metric to use would be the *length* of an input file in terms of the number of characters contained by the file. However, a fitness function maximizing file length would favor production rules that produce terminals being longer strings (e.g., "true" or "false" in JavaScript) over rules that produce more expansions through non-terminals to obtain complex structures (e.g., "if" branches or "for" loops). To mitigate this effect, we further use *the number of used expansions to derive an input file* because using more expansions to generate an input file makes the input file more complex. Thus, we build the *ratio* of the number of *expansions* to the *length* of an input file x to favor input files that were produced by more expansions and to punish lengthy input files that contain long strings of characters. Depending on the language of the input files, this ratio can be controlled by the parameter λ.

$$ratio(x) = \frac{expansions(x)}{\lambda \times length(x)} \tag{1}$$

Using this ratio, we score the structure of an input file x by multiplying the ratio and the number of expansions to put more weight on the expansions while a good ratio (>1) increases the score.

$$score_{structure}(x) = ratio(x) \times expansions(x) \tag{2}$$

Benefits of this score are its efficient computation and independence of the input language, although controlling λ allows accommodation of a specific language.

Concerning the "interesting" inputs, we rely on the feedback from executing the SUT with a concrete input x. Being interested in revealing defects in the SUT, we observe whether x triggers any exception during execution. If so, such an input will be assigned the maximum fitness and favored over all other non-exception triggering inputs. This results in a feedback score for an input file x:

$$score_{feedback}(x) = \begin{cases} \infty & \text{if } x \text{ triggers any exception} \\ 0 & \text{otherwise} \end{cases} \tag{3}$$

Moreover, EvoGFuzz keeps track of all exception-triggering inputs throughout all iterations as it returns these inputs at the end of the evolutionary search.

Finally, we follow the idea by Veggalam et al. [32] and combine the structural and feedback scores to a single-objective fitness function to be maximized:

$$fitness(x) = score_{feedback}(x) + score_{structure}(x) \tag{4}$$

Using this fitness function, all $|P|$ input files generated by the previous activity (Activity 3) are analyzed by executing them and computing their fitness.

Select Individuals (Activity 5). Based on the fitness of the $|P|$ input files, a strategy is needed to select the most promising files among them that will be used for further evolution. To balance selection pressure, EvoGFuzz uses elitism [7] and tournament selection [23]. By elitism, the top e_{rate}% of the $|P|$ input files ranked by fitness are selected. Additionally, the winners of n_{tour} tournaments of size s_{tour} are selected. The s_{tour} participants of each tournament are randomly chosen from the remaining $(100 - e_{rate})$% of the $|P|$ input files. In contrast to typical evolutionary algorithms, the selected individuals are not directly evolved by crossover or mutation, but they are used to learn a new probabilistic grammar.

Learn New Probabilistic Grammar (Activity 6). The selected input files are the most promising files of the population and they help in directing the further search toward "complex" and "interesting" inputs. Thus, these files are used to learn a new probabilistic grammar, particularly the probability distributions for all choices of production rules, by parsing them (cf. Activity 2 that learns a probabilistic grammar, however, from a given sample of input files). Thus, the learned probability distributions reflect features of the selected input files, and the corresponding probabilistic grammar can produce more input files that resemble these features. But beforehand, EvoGFuzz mutates the learned grammar.

Mutate Grammar (Activity 7). We mutate the learned probabilistic grammar to avoid a genetic drift [36] toward specific features of the selected individuals. With such a drift, the grammar would generate only input files with specific features exhibited by the selected individuals from which the grammar has been learned. Thus, it would neglect other potentially promising, yet unexplored features. Moreover, mutating the grammar maintains the diversity of input files being generated, which further could prevent the search from being stuck in local optima. In contrast to typical evolutionary algorithms, we do not mutate the individuals directly for two reasons. First, mutating an input file may result in a syntactically invalid file (i.e., the file does not conform to the given grammar). Second, a stochastic nature of the search is already achieved by using a *probabilistic* grammar to generate input files.

Therefore, we mutate the learned probabilistic grammar by altering the probabilities of individual production rules. The resulting mutated grammar produces syntactically valid input files whose features are similar to the selected individuals but that may also exhibit other unexplored features. For instance, a mutation could enable choices of production rules in the grammar that have not been used yet to generate input files because of being tagged so far with a probability of 0 that is now mutated to a value larger than 0. This increases the genetic variation.

For a single mutation of a probabilistic grammar, we choose a random production rule with n choices for expansions from the grammar. For each choice, we recalculate the probabilities p_i by selecting a random probability r_i from $(0, 1]$—we exclude 0 to enable all choices by assigning a probability larger than zero—and normalizing r_i with the sum of all of the n probabilities to ensure $\sum_{i=1}^{n} p_i = 1$ (i.e., the individual probabilities of all choices of a production rule sum up to 1). Thus, a probability p_i for one choice is calculated as follows:

$$p_i = \frac{r_i}{\sum_{j=1}^{n} r_j} \tag{5}$$

Finally, EvoGFuzz allows multiple of such mutations (n_{mut} many) of a probabilistic grammar in one iteration of search by performing each mutation independently from the other ones.

Generate Input Files (Activity 3). Using the learned and mutated grammar, EvoGFuzz generates $|P|$ new input files that resemble features of the recently selected input files but still diverge due to the grammar mutation. With the newly generated input files, the next iteration of the evolutionary process begins.

4 Evaluation

In this section, we evaluate the effectiveness of EvoGFuzz by performing experimentation on ten real-world applications.[1] We compare EvoGFuzz to a baseline being the original approach by Pavese et al. [28] (i.e., probabilistic grammar-based fuzzing), and ask the following research questions:

RQ1. Can evolutionary grammar-based fuzzing achieve a higher code coverage than the baseline?

RQ2. Can evolutionary grammar-based fuzzing trigger more exception types than the baseline?

4.1 Evaluation Setup

To answer the above research questions, we conducted an empirical study, in which we analyze the achieved line coverage and the triggered exception types. Line or code coverage [24] is a metric counting the unique lines of code of the targeted parser (i.e., the SUT) that have been executed during a test.

In order to examine the effectiveness of EvoGFuzz we evaluate our approach on the same test subjects that Pavese et al. have originally covered with their proposed probabilistic grammar-based fuzzing approach. These test subjects require three, in complexity varying input formats, namely *JSON*, *JavaScript*, and *CSS3*. ARGO, Genson, Gson, JSONJava, JsonToJava, MinimalJson, Pojo, and json-simple serve as the JSON parsers, whereas Rhino and cssValidator serve as the JavaScript and CSS parser, respectively. All parsers are widely used in browsers and web applications. A further description of all subjects along with their grammars can be found in the work of Pavese et al. [28]. All experiments have been performed on a virtual machine with Ubuntu 20.04 LTS featuring a Quad-Core 3 GHz Intel(R) CPU with 16 GB RAM.

4.2 Research Protocol

Giving both approaches the same starting conditions, we considered the same randomly selected input files from Pavese et al. to create the initial probabilistic grammar. The baseline uses this probabilistic grammar to generate "more of the same" inputs, whereas EvoGFuzz uses this grammar to generate the initial population followed by executing its evolutionary algorithm. In our evaluation, we observe the performance of both approaches for all subjects over a time frame of 10 min, that is, each approach runs for 10 min to test one subject.

For EvoGFuzz, a population consists of 100 individuals ($|P| = 100$) and one mutation of the grammar ($n_{mut} = 1$) is performed in each iteration of the search. The elitism rate e_{rate} is set to 5%, and for each generation ten tournaments of size ten were held ($n_{tour} = 10$ and $s_{tour} = 10$). In the fitness function, λ is set to 1.5 for JSON and 2.0 for JavaScript and CSS. Since the goal is to find

[1] Data and code artifacts are available here: https://doi.org/10.5281/zenodo.3961374.

exceptions, we configured the baseline to perform iterations of generating and executing 100 "more of the same" input files for 10 min. After 10 min the baseline and EvoGFuzz return all found exceptions and the exception-triggering test inputs. For each test subject and approach, we repeated these experiments 30 times.

4.3 Experimental Results

Figures 2(a) to (j) show the coverage results for the ten subjects. For each subject, we plot a chart showing the comparison of EvoGFuzz and the baseline with regard to the achieved line coverage. The vertical axis represents the achieved line coverage in percent, and the horizontal axis represents the time in seconds (up to 600 s = 10 min). The *median* runs for both approaches are highlighted, with all individual runs being displayed in the background.

RQ1 - Line coverage. To answer RQ1, we compare the line coverage achieved by both approaches. In particular, we investigate whether EvoGFuzz achieves at least the same percentage of line coverage than the baseline. Figs. 2(a) to (h) show the results for the JSON parsers, and Figs. 2(i) and (j) show the results for the JavaScript and CSS3 parser, respectively.

The results show that EvoGFuzz improves the coverage for all subjects and is able to increase the median line coverage for JSON by up to 18.43% (json-simple, Fig. 2(h)), for JavaScript by up to 47.93% (Rhino, Fig. 2(i)), and for CSS3 by up to 8.45% (cssValidator, Fig. 2(j)). These numbers are also listed in the column "Median increase" of Table 1.

The detailed investigation of Figs. 2(a) to (h) shows that for almost all JSON parsers both approaches eventually reach a plateau with regard to the achieved line coverage. The baseline reaches this plateau relatively early in the input generation process: there is no further improvement after only approximately 10 s. For EvoGFuzz, the point in time when reaching the plateau varies from parser to parser: between 10 s (Pojo, Fig. 2(g)) and 450 s (json-simple, Fig. 2(h)). In contrary, for Rhino (Fig. 2(i)) both approaches cannot achieve a plateau within 10 min as they are able to continuously increase the line coverage over the time.

Table 1 shows the accumulated coverage results for each subject and approach over all 30 repetitions. For both approaches, Table 1 shows the maximum and median line coverage, the standard deviation as well as the number of generated input files, along with the increase of the median line coverage of EvoGFuzz compared to the baseline. The improvement of the median line coverage ranges from 1.00% (Pojo) to 47.93% (Rhino). Additionally, the standard deviation (SD) values for the baseline in Table 1 indicate the existence of plateaus because all repetitions for each JSON parser show a very low (and often 0%) SD value.

To support the graphical evaluation, we do a statistical analysis to increase the confidence in our conclusions. As we consider independent samples and cannot make any assumption about the distribution of the results, we perform a non-parametric Mann-Whitney U test [2,21] to check whether the achieved median

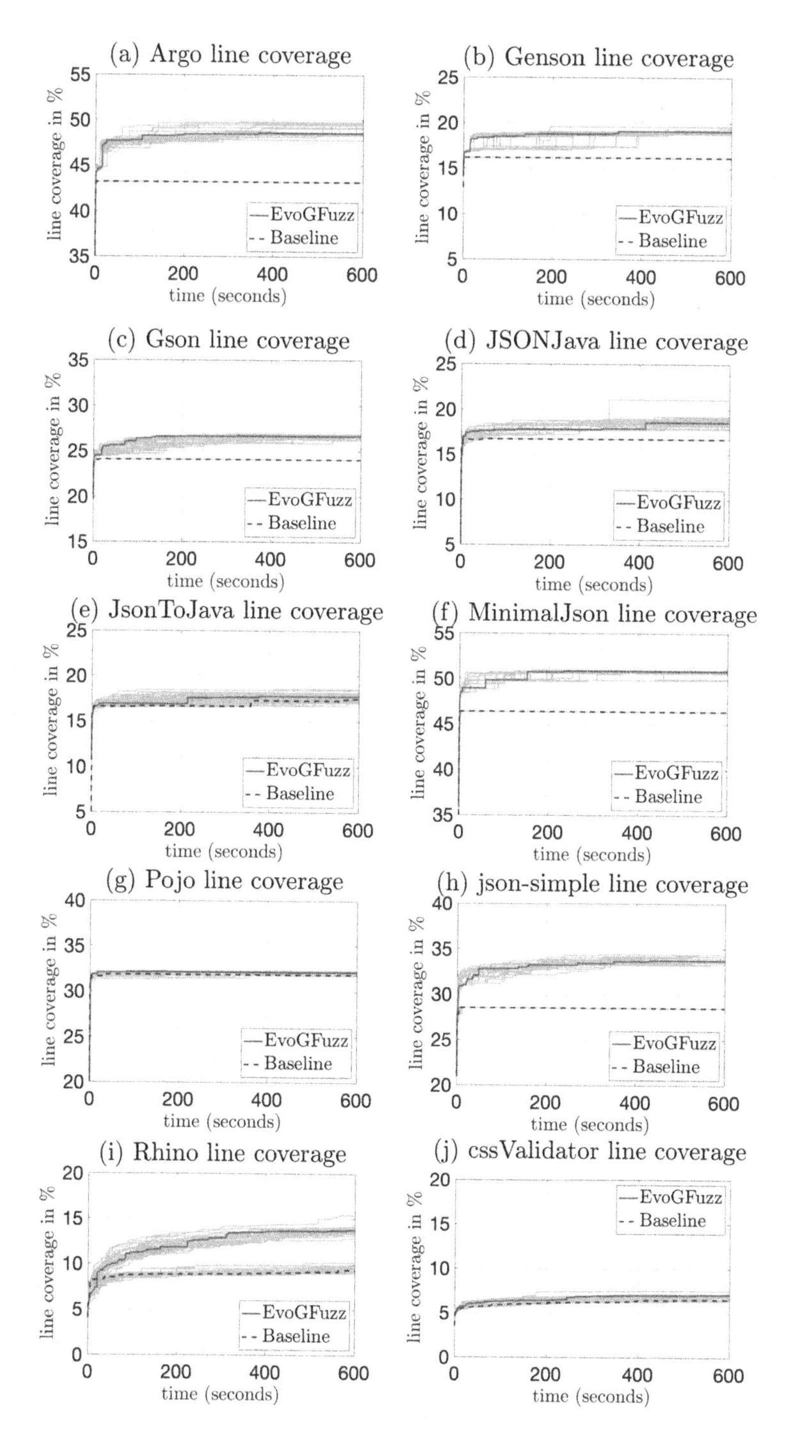

Fig. 2. Median and raw line coverage results for the ten subjects.

Table 1. Coverage results for each subject and approach over 30 repetitions.

Subject	LOC	EvoGFuzz				Baseline				Median	p-value
		Max	Median	SD	#files	Max	Median	SD	#files	Increase	
ARGO	8,265	49.78%	48.48%	0.60%	11,900	43.19%	43.19%	0%	13,900	12.25%	0.000062
Genson	18,780	19.65%	19.09%	0.19%	8,100	16.17%	16.17%	0%	9,000	18.06%	0.000063
Gson	25,172	26.92%	26.67%	0.15%	9,800	24.08%	24.08%	0%	11,200	10.76%	0.000064
JSONJava	3,742	21.09%	18.47%	0.59%	12,700	16.72%	16.72%	0%	15,000	10.47%	0.000064
JsonToJava	5,131	18.58%	17.90%	0.39%	11,400	17.58%	17.45%	0.09%	13,400	2.58%	0.020699
minimalJSON	6,350	51.06%	50.83%	0.26%	14,000	46.38%	46.38%	0%	16,600	9.59%	0.000055
Pojo	18,492	32.33%	32.17%	0.07%	10.600	31,88%	31.88%	0.02%	12,100	1.00%	0.000061
json-simple	2,432	34.44%	33.80%	0.33%	14,200	28.54%	28.54%	0%	16,700	18.43%	0.000059
Rhino	100,234	15.42%	13.95%	0.43%	3,200	10.20%	9.43%	0.28%	3,800	47.93%	0.000183
cssValidator	120,838	7.53%	7.06%	0.21%	1,000	6.62%	6.51%	0.06%	2,500	8.45%	0.000183

line coverage of both approaches differ significantly for each subject. This statistical analysis confirms that EvoGFuzz produces a significantly higher line coverage than the baseline for all subjects (cf. last column of Table 1).

The #files columns in Table 1 denote the average number of input files generated by one approach when testing one subject for 10 min. For all subjects, the baseline is able to generate on average more files than EvoGFuzz. These differences indicate the costs of the evolutionary algorithm in EvoGFuzz being eventually irrelevant due to the improved line coverage achieved by EvoGFuzz.

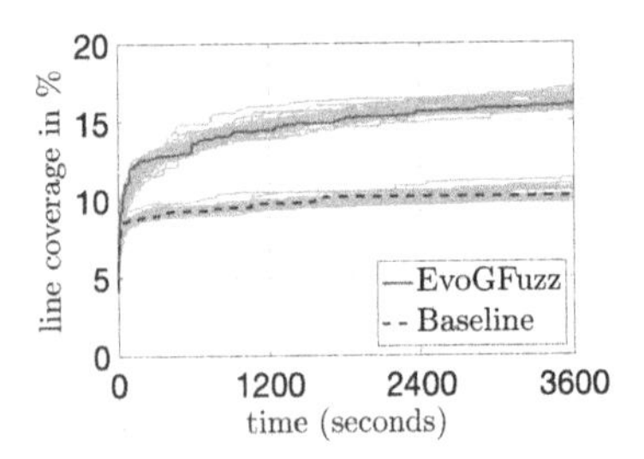

Fig. 3. Line coverage of Rhino.

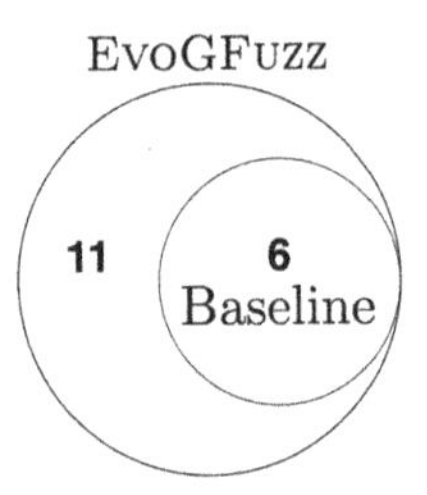

Fig. 4. Unique exceptions triggered by EvoGFuzz (11) and the baseline (6).

Since both approaches managed to continuously increase the line coverage for the Rhino parser (Fig. 2(i)), we conducted an additional experiment with the time frame set to one hour and again repeated the experiment 30 times. The results can be seen in Fig. 3. The chart shows that both approaches managed to further improve their (median) line coverage. EvoGFuzz was able to improve its previously achieved median line coverage of 13.95% to 16.10% with 18,500 generated input files, while the baseline improved from 9.43% to 10.23% with 22,700 generated input files, separating botch approaches even further.

Based on our evaluation, we conclude that EvoGFuzz is able to achieve a **significantly higher** line coverage than the baseline.

RQ2 - Exception Types. To answer RQ2, we compare the number of times a unique exception type has been triggered. Table 2 shows the thrown exception types per subject and input language. If neither approach was able to trigger an exception, the subject is not included in the table. For the Gson, JsonJava, simple-json, minimal-json, and cssValidator parsers no defects and exceptions have been found by both approaches. The ratios in the 4th and 5th column relate to the number of experiment repetitions in which EvoGFuzz and the baseline were able to trigger the corresponding exception type.

Table 2. Exception types that have been triggered by both approaches.

Input language	Subject	Exception types	EvoGFuzz	Baseline
JSON	ARGO	argo.saj.InvalidSyntaxException	**30**/30	0/30
	Genson	java.lang.NullPointerException	30/30	30/30
	jsonToJava	org.json.JSONException	30/30	30/30
	jsonToJava	java.lang.IllegalArgumentException	30/30	30/30
	jsonToJava	java.lang.ArrayIndexOutOfBoundsException	30/30	30/30
	jsonToJava	java.lang.NumberFormatException	**6**/30	0/30
	Pojo	java.lang.StringIndexOutOfBoundsException	30/30	30/30
	Pojo	java.lang.IllegalArgumentException	30/30	30/30
	Pojo	java.lang.NumberFormatException	**22**/30	0/30
JavaScript	Rhino	java.lang.IllegalStateException	**26**/30	0/30
	Rhino	java.util.concurrent.TimeoutException	**15**/30	0/30
CSS3	No exceptions triggered			
Total exception types			11	6

Table 2 and Fig. 4 show that during each experiment repetition, EvoGFuzz has been able to detect the same exception types than the baseline. Furthermore, EvoGFuzz was able to find five additional exception types that have not been triggered by the baseline. However, apart from the exception type *argo.saj.InvalidSyntaxException*, found in the ARGO parser, the other four exception types have not been identified by EvoGFuzz in all repetitions.

Overall, **11 different exception types** in five subjects have been found in our evaluation, incl. just two custom types (*org.json.JSONException* and *argo.saj.InvalidSyntaxException*). Out of these 11 exception types, **five** have **not** been triggered by the baseline. Figure 4 shows that **all six** exception types triggered by the baseline were also found by EvoGFuzz.

4.4 Threats to Validity

Internal Validity. The main threats to internal validity of fuzzing evaluations are caused by the random nature of fuzzing [2,17]. It requires a meticulous statistical assessment to make sure that observed behaviors are not randomly occurring. Therefore, we repeated all experiments 30 times and reported the descriptive statistics of our results. To match the evaluation of Pavese et al. [28], we used the same set of subjects and seed inputs. Furthermore, we automated the data collection and statistical evaluation. Finally, we did not tune the parameters of the baseline and EvoGFuzz to reduce the threat of overfitting to the given grammars and subjects. Only for the fitness function of EvoGFuzz, we determined appropriate λ values for the three input grammars by experiments.

External Validity. The main threat to external validity is the generalizability of the experimental results that are based on a limited number of input grammars and systems under test. However, similar to Pavese et al. [28], practically relevant input grammars with different complexities (small-sized grammars like JSON, and rather complex grammars like JavaScript and CSS) and widely used subjects (e.g., ARGO and Rhino) have been selected. As a result, we are confident that our approach will also work on other grammars and subjects.

5 Conclusion and Future Work

This paper presented EvoGFuzz, *evolutionary grammar-based fuzzing* that combines the technique by Pavese et al. [28] with evolutionary optimization to direct the generation of complex and interesting inputs by a fitness function. EvoGFuzz is able to generate structurally complex input files that trigger exceptions. The introduced mutation of grammars maintains genetic diversity and allows EvoGFuzz to discover features that have previously not been explored. Our experimental evaluation shows improved coverage compared to the original approach [28]. Additionally, EvoGFuzz is able to trigger more exception types undetected by the original approach. As future work, we want to investigate cases of having no precise grammar of the input space (cf. [18]) and using semantic knowledge of the input language to tune mutation operators. Finally, we want to compare EvoGFuzz with other state-of-the-art fuzzing techniques.

References

1. Anand, S., et al.: An orchestrated survey of methodologies for automated software test case generation. JSS **86**(8), 1978–2001 (2013)
2. Arcuri, A., Briand, L.: A Hitchhiker's guide to statistical tests for assessing randomized algorithms in software engineering. Softw. Test. Verif. Reliab. **24**(3), 219–250 (2014)
3. Atlidakis, V., Geambasu, R., Godefroid, P., Polishchuk, M., Ray, B.: Pythia: grammar-based fuzzing of REST APIs with coverage-guided feedback and learning-based mutations, pp. 1–12 (2020). http://arxiv.org/abs/2005.11498

4. Böhme, M., Pham, V.T., Nguyen, M.D., Roychoudhury, A.: Directed greybox fuzzing. In: Proceedings of the ACM SIGSAC Conference on Computer and Communications Security, CCS 2017, pp. 2329–2344. ACM (2017)
5. Böhme, M., Pham, V.T., Roychoudhury, A.: Coverage-based greybox fuzzing as markov chain. In: Proceedings of the ACM SIGSAC Conference on Computer and Communications Security, CCS 2016, pp. 1032–1043. ACM (2016)
6. Cummins, C., Petoumenos, P., Murray, A., Leather, H.: Compiler fuzzing through deep learning. In: Proceedings of the 27th ACM SIGSOFT International Symposium on Software Testing and Analysis, ISSTA 2018, pp. 95–105. ACM (2018)
7. Du, H., Wang, Z., Zhan, W., Guo, J.: Elitism and distance strategy for selection of evolutionary algorithms. IEEE Access **6**, 44531–44541 (2018)
8. Godefroid, P.: Fuzzing: hack, art, and science. Commun. ACM **63**(2), 70–76 (2020)
9. Godefroid, P., Kiezun, A., Levin, M.Y.: Grammar-based whitebox fuzzing. In: Proceedings of the 29th ACM SIGPLAN Conference on Programming Language Design and Implementation, PLDI 2008, pp. 206–215. ACM (2008)
10. Godefroid, P., Levin, M.Y., Molnar, D.: SAGE: whitebox fuzzing for security testing. Commun. ACM **55**(3), 40–44 (2012)
11. Godefroid, P., Peleg, H., Singh, R.: Learn&Fuzz: machine learning for input fuzzing. In: Proceedings of the 32nd International Conference on Automated Software Engineering, ASE 2017, pp. 50–59. IEEE (2017)
12. Hallaraker, O., Vigna, G.: Detecting malicious JavaScript code in Mozilla. In: Proceedings of the 10th IEEE International Conference on Engineering of Complex Computer Systems, ICECCS 2005, pp. 85–94. IEEE (2005)
13. Hanford, K.V.: Automatic generation of test cases. IBM Syst. J. **9**(4), 242–257 (1970)
14. Harman, M., McMinn, P., de Souza, J.T., Yoo, S.: Search based software engineering: techniques, taxonomy, tutorial. In: Meyer, B., Nordio, M. (eds.) LASER 2008-2010. LNCS, vol. 7007, pp. 1–59. Springer, Heidelberg (2012). https://doi.org/10.1007/978-3-642-25231-0_1
15. Holler, C., Herzig, K., Zeller, A.: Fuzzing with code fragments. In: Presented as part of the 21st USENIX Security Symposium, pp. 445–458. USENIX (2012)
16. Höschele, M., Zeller, A.: Mining input grammars with autogram. In: 39th International Conference on Software Engineering Companion, pp. 31–34. IEEE (2017)
17. Klees, G., Ruef, A., Cooper, B., Wei, S., Hicks, M.: Evaluating fuzz testing. In: Proceedings of the 2018 ACM SIGSAC Conference on Computer and Communications Security, CCS 2018, pp. 2123–2138. ACM (2018)
18. Le, X.B.D., Păsăreanu, C., Padhye, R., Lo, D., Visser, W., Sen, K.: Saffron: adaptive grammar-based fuzzing for worst-case analysis. SIGSOFT Softw. Eng. Notes **44**(4), 14 (2019)
19. Lemieux, C., Sen, K.: FairFuzz: a targeted mutation strategy for increasing greybox fuzz testing coverage. In: Proceedings of the 33rd ACM/IEEE International Conference on Automated Software Engineering, ASE, pp. 475–485. ACM (2018)
20. Liu, P., Zhang, X., Pistoia, M., Zheng, Y., Marques, M., Zeng, L.: Automatic text input generation for mobile testing. In: Proceedings of the 39th International Conference on Software Engineering, ICSE 2017, pp. 643–653. IEEE (2017)
21. Mann, H.B., Whitney, D.R.: On a test of whether one of two random variables is stochastically larger than the other. Ann. Math. Stat. **18**(1), 50–60 (1947)
22. Miller, B.P., Fredriksen, L., So, B.: An empirical study of the reliability of UNIX utilities. Commun. ACM **33**(12), 32–44 (1990)
23. Miller, B.L., Goldberg, D.E.: Genetic algorithms, tournament selection, and the effects of noise. Complex Syst. **9**, 193–212 (1995)

24. Miller, J.C., Maloney, C.J.: Systematic mistake analysis of digital computer programs. Commun. ACM **6**(2), 58–63 (1963)
25. Nilizadeh, S., Noller, Y., Păsăreanu, C.S.: Diffuzz: differential fuzzing for side-channel analysis. In: Proceedings of the 41st International Conference on Software Engineering, ICSE 2019, pp. 176–187. IEEE (2019)
26. Orso, A., Rothermel, G.: Software testing: a research travelogue (2000–2014). In: Future of Software Engineering, FOSE 2014, pp. 117–132. ACM (2014)
27. Pacheco, C., Ernst, M.D.: Randoop: feedback-directed random testing for Java. In: Proceedings of the 22nd Conference on Object-Oriented Programming Systems and Applications Companion, OOPSLA 2007, pp. 815–816. ACM (2007)
28. Pavese, E., Soremekun, E., Havrikov, N., Grunske, L., Zeller, A.: Inputs from hell: generating uncommon inputs from common samples. arXiv:1812.07525 [cs] (2018). http://arxiv.org/abs/1812.07525
29. Pham, V.T., Böhme, M., Santosa, A.E., Căciulescu, A.R., Roychoudhury, A.: Smart greybox fuzzing. IEEE Trans. Softw. Eng., 1–17 (2019). https://doi.org/10.1109/TSE.2019.2941681
30. Richardson, R.: CSI computer crime and security survey. Comput. Secur. Inst. **1**, 1–30 (2008)
31. Song, D., et al.: BitBlaze: a new approach to computer security via binary analysis. In: Sekar, R., Pujari, A.K. (eds.) ICISS 2008. LNCS, vol. 5352, pp. 1–25. Springer, Heidelberg (2008). https://doi.org/10.1007/978-3-540-89862-7_1
32. Veggalam, S., Rawat, S., Haller, I., Bos, H.: IFuzzer: an evolutionary interpreter fuzzer using genetic programming. In: Askoxylakis, I., Ioannidis, S., Katsikas, S., Meadows, C. (eds.) ESORICS 2016. LNCS, vol. 9878, pp. 581–601. Springer, Cham (2016). https://doi.org/10.1007/978-3-319-45744-4_29
33. Wang, J., Chen, B., Wei, L., Liu, Y.: Superion: grammar-aware greybox fuzzing. In: Proceedings of the 41st International Conference on Software Engineering, ICSE 2019, pp. 724–735. IEEE (2019)
34. Website: American Fuzzing Lop (AFL) (2018). http://lcamtuf.coredump.cx/afl/
35. Website: libFuzzer: a library for coverage-guided fuzz testing (2018). https://llvm.org/docs/LibFuzzer.html
36. Wright, S.: The evolution of dominance. Am. Nat. **63**(689), 556–561 (1929)
37. Yang, X., Chen, Y., Eide, E., Regehr, J.: Finding and understanding bugs in C compilers. SIGPLAN Not. **46**(6), 283–294 (2011)
38. Zeller, A., Gopinath, R., Böhme, M., Fraser, G., Holler, C.: The fuzzing book. In: The Fuzzing Book. Saarland University (2019). https://www.fuzzingbook.org/

Commonality-Driven Unit Test Generation

Björn Evers, Pouria Derakhshanfar, Xavier Devroey(✉), and Andy Zaidman

Delft University of Technology, Delft, The Netherlands
b.evers@student.tudelft.nl, {p.derakhshanfar, x.d.m.devroey,a.e.zaidman}@tudelft.nl

Abstract. Various search-based test generation techniques have been proposed to automate the generation of unit tests fulfilling different criteria (*e.g.*, line coverage, branch coverage, mutation score, *etc.*). Despite several advances made over the years, search-based unit test generation still suffers from a lack of guidance due to the limited amount of information available in the source code that, for instance, hampers the generation of complex objects. Previous studies introduced many strategies to address this issue, *e.g.*, dynamic symbolic execution or seeding, but do not take the internal execution of the methods into account. In this paper, we introduce a novel secondary objective called *commonality score*, measuring how close the execution path of a test case is from reproducing a *common* or *uncommon* execution pattern observed during the operation of the software. To assess the commonality score, we implemented it in EvoSuite and evaluated its application on 150 classes from JabRef, an open-source software for managing bibliography references. Our results are mixed. Our approach leads to test cases that indeed follow *common* or *uncommon* execution patterns. However, if the commonality score can have a positive impact on the structural coverage and mutation score of the generated test suites, it can also be detrimental in some cases.

Keywords: Search-based software testing · Automated unit testing · Common paths coverage · Secondary objective

1 Introduction

Search-based test case generation enables the automated generation of test cases according to predefined criteria. Among the different approaches, white-box search-based test case generation [13,15,29] relies on evolutionary algorithms to generate test cases from source code and maximize structural coverage [15,29] or mutation score [16]. Previous research has shown that automatically generated tests are effective for coverage and fault finding [3,23,28], can also find real faults [2], and are useful for debugging [5].

Electronic supplementary material The online version of this chapter (https://doi.org/10.1007/978-3-030-59762-7_9) contains supplementary material, which is available to authorized users.

A. Aleti and A. Panichella (Eds.): SSBSE 2020, LNCS 12420, pp. 121–136, 2020.
https://doi.org/10.1007/978-3-030-59762-7_9

Despite several advances, search-based unit test generation still faces many challenges. Among those are (i) the crafting of complex objects and values used during test generation [2], and (ii) the indirect coverage of encapsulated elements (*e.g.*, private methods and class attributes) through the invocation of specific paths in public methods [32]. Various approaches address those challenges by relying on *dynamic symbolic execution* to generate complex objects and values using constraint solvers [18,19,21,24]; *seeding* to identify objects and values from the application source and test code that are later reused during the search [31]; or class usages, learned from static analysis of the source code [8,17] and dynamic execution of the existing tests [8], and used to generate realistic objects.

However, if complex objects and values can indeed lead to an improvement in the coverage, it does not always succeed in covering all the elements of a class under test. For instance, if the indirect coverage of a private method requires specific executions paths in a public method, the current fitness functions will not be able to provide sufficient guidance to the search process [32].

In this paper, we hypothesize that common and uncommon execution paths, observed during the actual operation of the system, can lead to better guidance of the search process, and hence, better coverage. Complementing previous work on seeding [31], which is aimed at triggering different execution paths in the methods under test, we consider the *commonality* of those execution paths. For that, we approximate commonality using weights for the different code blocks, and define a secondary objective called the *commonality score*, denoting how close an execution path is from common or uncommon executions of the software.

We implemented the commonality score in EvoSuite [13] and evaluated it on 150 classes from JabRef, an open-source bibliography reference manager, for common and uncommon behaviors. We compare the commonality score (**RQ.1**), the structural coverage (**RQ.2**), and the fault-finding capabilities (**RQ.3**) of the thus generated tests to tests generated by the standard EvoSuite implementation. Our results are mixed but show that this secondary objective significantly improves the number of covered common paths in 32.6% of the classes. Although the average structural coverage remains stable, the commonality score significantly improves the line (resp. branch) coverage in three (resp. four) classes, but also negatively impacts the coverage for eight (resp. nine) classes. Finally, the commonality score impacts the number of killed mutants for 22 classes (11 positively and 11 negatively). Our implementation is openly available at https://github.com/STAMP-project/evosuite-ramp, and the replication package of our evaluation and data analysis have been uploaded to Zenodo [11,12].

2 Background and Related Work

2.1 Search-Based Unit Test Generation

Search-based unit test generation has been extensively investigated by prior studies [13,15,29]. These studies have confirmed that it achieves a high level of coverage [15,29], detects faults in real-world applications [2,16], and reduces the debugging costs [30]. Most search-based unit test generation approaches abstract the source code of a method to a control flow graph:

Definition 1 (Control Flow Graph (CFG) [1]**).** *A control flow graph for a method m is a directed graph $G = (B, E)$, where B is the set of* basic blocks *(i.e., sequences of statements that do not have any branch before the last statement), E is the set of* control flow edges *connecting the basic blocks.*

For instance, for the method with the pseudo-code presented in Fig. 1(a), the corresponding CFG for this method is depicted in Fig. 1(b).

Search-based software unit test generation approaches use meta-heuristics to *evolve* a set of test cases. These techniques start with generating an *initial population* of randomly produced test cases. The *fitness* of each individual in the population is evaluated using a fitness function, which is usually defined according to the coverage in the CFGs of the target class. Next, a subset of the fittest individuals is *selected* for evolution and leads to the generation of a new population. The evolving process contains three steps: (i) *crossover*, which randomly mixes two selected individuals to generate new offspring; (ii) *mutation*, which randomly changes, adds, or removes a statement in an individual; and (iii) *insertion*, which reinserts the modified individuals into the population for the next iteration of the algorithm. This process continues until either satisfactory individuals are found, or the time budget allocated for the search is consumed. Among the different approaches, EvoSuite [13] uses genetic algorithms to evolve Java test suites in order to cover a class under test.

MOSA [27] and *DynaMOSA* [29] are two new many-objectives genetic algorithms proposed for unit test generation. These algorithms consider test cases as individuals and incorporate separate fitness functions for separate coverage goals (*e.g.*, covering each branch in the CFGs will be an independent search objective). They use non-dominated fronts to generate test cases in the direction of multiple coverage goals in parallel, and thereby generate tests aiming to cover specific goals, while not letting the test generation be trapped for covering a single goal. Panichella *et al.* [27] show that *MOSA* outperforms the original EvoSuite approach in terms of structural coverage and mutation score.

In this paper, we use *MOSA* to automatically generate test cases according to the collected logs during the production phase. Future work includes the evaluation of our approach with other multi and many-objectives algorithms, like *DynaMOSA*.

2.2 Usage-Based Test Generation

The majority of search-based test generation techniques aim to achieve high coverage for various metrics (*e.g.*, line coverage, branch coverage, or more recently mutation coverage). Despite their considerable achievements, they do not consider the execution patterns observed in production use for automatic generation of unit tests. Hence, Wang *et al.* [37] investigated how developer-written tests and automatically generated tests represent typical execution patterns in production. Their study confirms that these tests are not a proper representation of real-world execution patterns.

The behavior of actual users may reveal faults, which are not detected by the existing test cases. For instance, a piece of code in the software under test that is not often used in practice may be left relatively untested because it is rarely exercised in production. A recent method from Wang *et al.* [38], based on symbolic execution, recreates users behaviors using log data from a system run in production, which has allowed to find the same faults in a system encountered by a user. This paper aims to expand upon generating tests based on the actual usage of a system at the unit level. In contrast to Wang *et al.* [38], where the aim is to replicate a full behavior executed by a user by using symbolic execution, we aim to guide the search process in a genetic algorithm towards executing common or uncommon behaviors. In the same vein as Wang et al. [38], log data is used to determine the execution counts of code branches.

Other approaches consider user feedback [20], or *usage models* of the application and *statistical testing* [9,22,34,35] to generate and prioritize test cases at the system level. A usage model consists in a state machine where transitions have been labelled with a probability of being executed. Unlike those approaches, we consider test case generation at the unit level.

3 Test Generation for Common and Uncommon Behaviors

Intuitively, *commonality* describes to what extent a test exercises code branches that are executed often during the normal operation of the system under test. If a test executes branches that are often (respectively rarely) executed in practice, it will have a high (respectively low) *commonality score*. The commonality score has a value between 0 and 1 and is computed based on an annotated control flow graph [1]:

Definition 2 (Annotated Control Flow Graph). *An annotated control flow graph is a directed graph $G = (B, E, \gamma)$, where $G = (B, E)$ is a control flow graph, and $\gamma : B \rightarrow \mathbb{R}$ is a labelling function giving for the basic blocks in B an* execution weight *denoting how often the block is executed during operations.*

The *execution weights* can be derived from the operation logs of the system, an instrumented version of the system (like in our evaluation), or assigned manually.

Let us define the *commonality score*. For a test case, its commonality score depends only on the branches it covers and on the highest and lowest execution weights in the class under test. Branches without execution weights are ignored and branches covered multiple times (*e.g.*, in a loop) are counted only once.

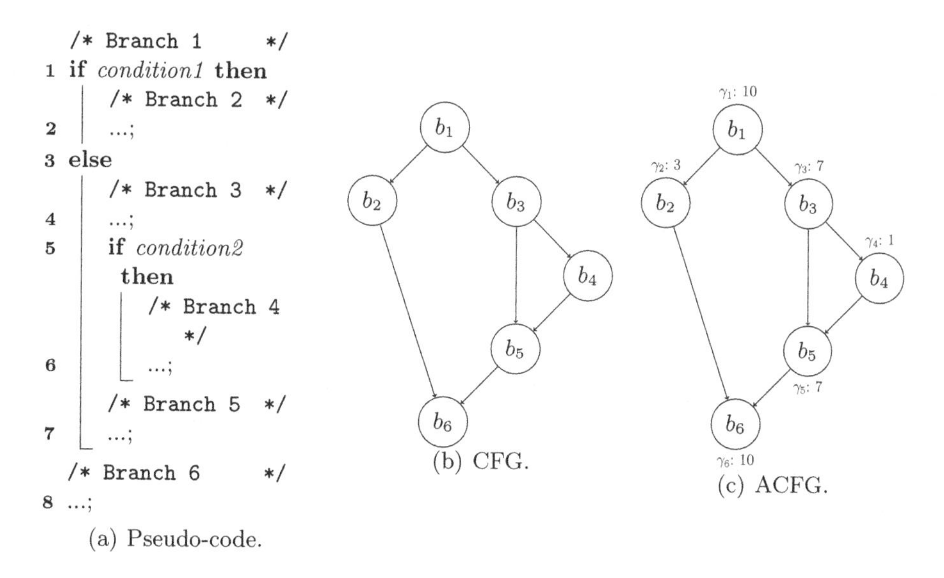

(a) Pseudo-code. (b) CFG. (c) ACFG.

Fig. 1. Example of pseudo-code and its corresponding annotated control flow graph. The γ_i indicate to the execution weight of the node.

Definition 3 (Commonality score). *For a test case t executing n basic blocks b_i labelled by a function γ, the highest execution weight in the class under test h, the lowest execution weight in the class under test l, the commonality score of t, denoted $c(t)$ is defined as:*

$$c(t) = \frac{\sum_{i=1}^{n} (\gamma b_i - l)}{n \times (h - l)}$$

The commonality score for a *test suite* s is defined as the average of the commonality scores of its test cases: $c(s) = \left(\sum_{t_i \in s} c(t_i)\right) / |s|$.

For instance, considering a class containing a single method with the pseudo-code presented in Fig. 1(a), the corresponding annotated control flow graph in Fig. 1(b), and a test suite containing three test cases t_1 covering (b_1, b_2, b_6), t_2 covering (b_1, b_3, b_5, b_6), and t_3 covering $(b_1, b_3, b_4, b_5, b_6)$. The commonality scores are $c(t_1) = ((10-1)) + (3-1) + (10-1)) / (3 \times (10-1)) = 20/27 \approx 0.741$, $c(t_2) = 5/6 \approx 0.833$, and $c(t_3) = 2/3 \approx 0.667$.

3.1 Commonality as a Secondary Objective

Secondary objectives are used to choose between different test cases in case of a tie in the main objectives. For instance, the default secondary objective used by MOSA [27] minimizes the test case length (*i.e.*, the number of statements) when two test cases satisfy the same main objectives (*e.g.*, cover the same branches). Using test case length minimization as a secondary objective addresses the *bloating effect* [33] by preventing the search process from always generating longer test cases. Since this is a desirable property, we combine the test case length minimization with the commonality of the test case using a weighted sum when comparing two test cases.

Definition 4 (Commonality secondary objective). *For two test cases t_1, t_2 with lengths l_1, l_2, the comparison between the two test cases is done using the following formula:*

$$common(t_1, t_2) = \frac{\left(\alpha\left(\frac{l_1}{l_2}\right) + \beta\left(\frac{1-c(t_1)}{1-c(t_2)}\right)\right)}{(\alpha + \beta)}$$

If $common(t_1, t_2) \leq 1$, then t_1 is kept, otherwise t_2 is kept.

Similarly, for the uncommonality between two test cases, we will have the following definition.

Definition 5 (Uncommonality secondary objective). *For two test cases t_1, t_2 with lengths l_1, l_2, the comparison between the two test cases is done using the following formula:*

$$uncommon(t_1, t_2) = \frac{\left(\alpha\left(\frac{l_1}{l_2}\right) + \beta\left(\frac{c(t_1)}{c(t_2)}\right)\right)}{(\alpha + \beta)}$$

If $uncommon(t_1, t_2) \leq 1$, then t_1 is kept, otherwise t_2 is kept.

In our evaluation, we use commonality and uncommonality with MOSA to answer our different research questions.

4 Empirical Evaluation

To assess the usage of commonality as a secondary objective for test case generation, we performed an empirical evaluation using 150 classes from JabRef[1], an open source bibliography reference manager, to answer the following research questions:

RQ.1. How does the *commonality score* of the generated tests compare when using the *common*, *uncommon*, and *default* secondary objectives?
RQ.2. How does the *line* and *branch coverage* of the generated tests compare when using the *common*, *uncommon*, and *default* secondary objectives?
RQ.3. How does the *mutation score* of the generated tests compare when using the *common*, *uncommon*, and *default* secondary objectives?

We implemented the secondary objectives from Sect. 3 in EvoSuite [13], a state-of-the-art white-box unit test generator tool for Java. Our implementation is openly available at https://github.com/STAMP-project/evosuite-ramp, and the replication package of our evaluation and data analysis have been uploaded to Zenodo [11,12].

[1] https://www.jabref.org.

4.1 Subject and Execution Weights

Collecting Execution Weights. For our evaluation, we choose JabRef (46 KLOC), an open-source Java bibliography reference manager with a graphical user interface working with BibTex files. To determine the execution weights of the different branches, we instrumented JabRef using Spoon[2] and added log statements producing a message with a unique identifier each time a branch is executed. These identifiers are then mapped to a source code location, identified by the *class name*, the *method name*, and the *line number*. Furthermore, the number of occurrences of the identifier in the log messages is established. We then asked five people (including the first author) to use our modified JabRef implementation to perform various tasks (adding a reference, updating a reference, removing a reference, *etc.*) and collected the produced logs. In an industrial context, operations logs can be analyses and traced back to the source code to identify the execution weights [39].

Classes Under Test. We sampled 150 classes. We excluded classes from the `org.jabref.gui` and `org.jabref.logic.importer.fileformat` packages as they respectively work with JavaFX and perform input-output operations. From the remaining classes and following the best practices of the search-based unit testing community [26], we selected 75 classes with the highest cyclomatic complexity, as classes with a higher cyclomatic complexity are harder to process for unit test generation tools and 38 classes with the largest number of lines of code. Additionally, we selected 37 classes that were executed the most by our modified JabRef implementation.

Configuration Parameters. We ran EvoSuite with the default coverage criteria (line, branch, exception, weak mutation, input, output, method, method without exceptions, and context branch) and three different secondary objectives: (i) *default*, minimizing the test case length, (ii) *commonality*, as described in Definition 4, and (iii) *uncommonality*, as described in Definition 5. We executed EvoSuite on each class under test 30 times with the MOSA algorithm [27] and a search budget of three minutes, offering a good compromise between runtime and coverage [15,28]. All other configuration parameters were left to their default value.

4.2 Data Analysis

For each of the 13,500 execution (150 classes $\times$ 30 repetitions $\times$ 3 configurations), we collected the commonality score and structural coverage information from EvoSuite. Additionally, we performed a mutation analysis of the generated test suites using Pit [7]. For 46 classes (out of 150), EvoSuite could not complete 30 executions using our different configurations. We excluded those classes to keep the comparison fair and performed our analysis on the 104 remaining classes.

[2] http://spoon.gforge.inria.fr/.

To compare the commonality score, the structural coverage, and the mutation score, we used the non-parametric Wilcoxon Rank Sum test, with $\alpha = 0.05$ for Type I error, and the Vargha-Delaney statistic $\widehat{A}_{12}$ [36] to evaluate the effect size between two configurations. An $\widehat{A}_{12}$ value lower than 0.5 for a pair of configurations (A,B) indicates that A increases the score or coverage compared to B, and a value higher than 0.5 indicates the opposite. The Vargha-Delaney magnitude measure also allows partitioning the results into three categories having large, medium, and small impact [36].

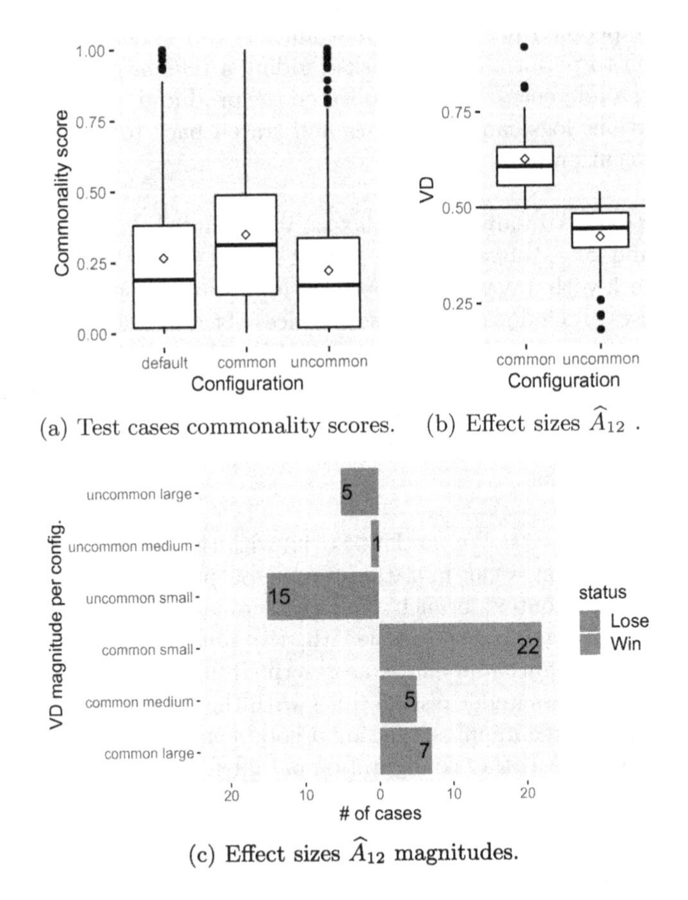

(a) Test cases commonality scores. (b) Effect sizes $\widehat{A}_{12}$.

(c) Effect sizes $\widehat{A}_{12}$ magnitudes.

Fig. 2. Test cases commonality values and comparison to *default*. Diamonds (◇) indicate mean values and horizontal bars (–) indicate median values.

5 Results

5.1 Commonality Score (RQ.1)

In this section we answer the question: *How does the commonality score of the generated tests compare when using the common, uncommon, and default secondary objectives?*

Figure 2 illustrates the impact of using *commonality* and *uncommonality*, as the secondary objective, on the commonality score of the generated test cases. Figure 2(a) shows that the average and median of the commonality score is improved by 8% and 12%, respectively, compared to *default* when using *commonality* as secondary objective. In parallel, using *uncommonality* as secondary objective reduces the commonality score by, on average, 5% (2.5% for median) compared to *default*. Moreover, Fig. 2(c) presents the number of cases (*i.e.*, classes used as the target class for unit testing), in which the application of *commonality* and *uncommonality* significantly (*p-value* < 0.05) changes the commonality score with effect size magnitude of large, medium, or small. As we can see in this figure, utilizing *commonality* always leads to a significant improvement in the commonality score (blue bars), and in contrast, using *uncommonality* always reduces this score (red bars). In total, *commonality* significantly improves the commonality score in 34 cases (32.6% of classes), and *uncommonality* significantly reduces this score in 21 classes (20.1% of cases). Figure 2(b) depicts the effect sizes of differences observed in these cases. Consistent with the previous figures, the average effect size ($\widehat{A}_{12}$) achieved by *commonality* is higher than 0.5 (*i.e.*, commonality score has been improved). However, this value is lower than 0.5 for *uncommonality*.

Summary. Using *commonality* as secondary objective in the EvoSuite search-based test case generation process leads to test cases that exhibit an improved commonality score. In parallel, the application of *uncommonality* leads to the reduction of the commonality score.

5.2 Structural Coverage (RQ.2)

In this section we provide an answer to the following research question: *How does the line and branch coverage of the generated tests compare when using the common, uncommon, and default secondary objectives?*

Figure 3 shows the line and branch coverage achieved by using *commonality* and *uncommonality* as secondary objectives compared to *default*. Figure 3(a) indicates that the average coverage is the same for all of the assessed configurations.

Looking at the comparison of the structural coverage values achieved by each secondary objective in each class, we can see that the line and branch coverage is significantly impacted by *commonality* and *uncommonality* in some cases. Figure 3(c) presents the number of cases that these secondary objectives significantly (*p-value* < 0.05) reduce ($\widehat{A}_{12} < 0.5$) or increase ($\widehat{A}_{12} > 0.5$) the line and branch coverage with effect size magnitude small, medium, or large. According to this figure, in general, utilizing *commonality* leads to a significant improvement for line and branch coverage in three and four classes, respectively. Nevertheless, this secondary objective reduced the line and branch coverage in eight and nine classes, respectively. A list of these classes is provided as the supplementary material to this submission.

Also, we can see a similar result for *uncommonality*: significant improvements in three and five classes and significant reductions in seven and nine cases for line and branch coverage. Since the number of cases in which *commonality* and *uncommonality* lead to a significantly lower structural coverage is higher than the number of cases in which we see a significant improvement in coverage, the average effect size of differences (Fig. 3(b)) is slightly less than 0.5 for both line (0.47 for both secondary objectives) and branch coverage (0.46 for both).

Summary. On average, using *commonality* or *uncommonality* does not impact the line and branch coverage. However, these two secondary objectives can significantly impact the structural coverage in specific cases.

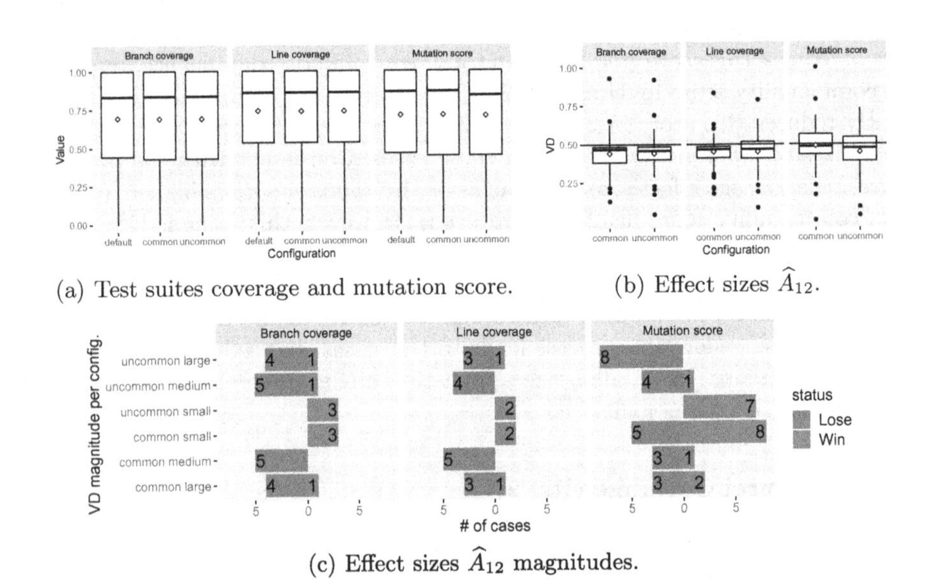

(a) Test suites coverage and mutation score.

(b) Effect sizes $\widehat{A}_{12}$.

(c) Effect sizes $\widehat{A}_{12}$ magnitudes.

Fig. 3. Test suites coverage and mutation score, and comparison to *default.* Diamonds (◇) indicate mean values and horizontal bars (–) indicate median values.

5.3 Mutation Analysis (RQ.3)

In the final research question we reflect on *How does the mutation score of the generated tests compare when using the common, uncommon, and default secondary objectives?*

Figure 3(a) depicts the mutation score achieved by using *commonality* and *uncommonality* compared to *default*. Like line and branch coverage, the average mutation scores achieved by these secondary objectives is similar to the one achieved by *default*. However, Fig. 3(c) shows that *commonality* and *uncommonality* can significantly (p-$value$ < 0.05) impact the mutation score achieved by unit test generation. The *commonality* secondary objective significantly increases the number of mutants killed for 11 classes but, at the same time, also decreases the mutation score in another 11 cases. Moreover, *uncommonality* significantly changes the mutation score in 20 cases (8 wins against 12 losses). Figure 3(b) shows the effect size of differences in these cases for both *commonality* and *uncommonality* secondary objectives. According to this Figure, the average $\widehat{A}_{12}$ estimations are 0.49 and 0.47. Since these values are lower than 0.5, on average, the difference achieved by these two secondary objectives is negative. However, the outliers in this figure show us that the effect sizes of *commonality* above 0.75 in some specific cases. Hence, the graphs in Fig. 3(c) indicate that using *commonality* and *uncommonality* can improve the mutation score in specific cases.

Summary. On average, using *commonality* or *uncommonality* does not have any effect on the mutation score achieved by the generated test suites. However, these two secondary objectives can significantly change the killed mutants in some cases.

6 Discussion

6.1 Execution Weights

In our evaluation, we collected execution weights using an instrumented version of JabRef distributed to five different users. As a result, a large number of log messages allowed us to have execution weights for many different classes. Such data collection is not realistic in an industrial setting as the collection and analysis of log data is challenging for large applications [4]. It is likely that the collected data will not cover the complete system but only a subset of its classes.

However, we believe that the development of scalable *software analytics* practices [25] represents an opportunity to include information from the software operations environment in various development activities, including testing [4,6]. For instance, Winter *et al.* [39] recently brought information about the number of times a log statement is executed to the developer's IDE to raise awareness about the load it represented during the operations of the system. Similar information can be collected for seeding [8,31] or annotating a control flow graph, like in our approach, and allow developers to generate new tests from their IDE.

Finally, our current approach considers execution weights individually to approximate usages, which allows us to compute a commonality score quickly. Different definitions of the commonality score based, for instance, on full and partial execution paths identified from the operations logs are possible. Those finer grained definitions of commonality would allow to take multiple executions of the same code blocks (like loops) into account, at the expense of a higher computational cost. Exploration and evaluation of such definitions are left for future work.

6.2 Impact on Mutation Analysis

After manual investigation of the generated tests[3], we saw that the classes on which *commonality* performs relatively well are often executed. The class on which the performance was especially good compared to *default* was an enum class (`org.jabref.logic.util.StandardFileType`). In a large majority of the cases (25 out of 30), the tests generated using the *commonality* secondary objective contain a method sequence that is not present in the tests generated by *default*. We inspected the execution counts of individual branches stemming from the operational use of the system. From this inspection, we found a single branch in the code of method `getExtensions` that has been executed as part of the usage scenarios from all participants.

This method is consistently involved in the test cases that kill the mutants that the tests generated by *default* fail to kill most of the time. This supports our initial assumption that using *commonality* can drive the search process to cover the code in a different way, possibly finding different kinds of faults.

6.3 Usefulness for Debugging

The end-goal of any test suite is to identify faults in source code and help the developer during her debugging activities. Our evaluation measured the fault-finding capabilities of the generated test suites, but did not investigate their usefulness for debugging. Previous research has confirmed that automatically generated tests can find faults in real software [2] and are useful for debugging [5].

However, there remain several challenges, like the understandability of the generated tests [2,14]. Since the *commonality* and *uncommonality* secondary objectives aim to influence how the lines in a method are covered, we expect that it will also have an impact on the understandability of the generated tests. Future research will include the assessment of the debugging capabilities of the generated tests (*e.g.*, understandability, performance, readability, *etc.*).

6.4 Threats to the Validity

Internal Validity. We repeated each execution of EvoSuite 30 times to take the randomness of the generation process into account in our data analysis. We have tested our implementation of the *commonality* and *uncommonality* secondary objectives to reduce bugs' chance of influencing our results.

[3] The results and complete manual analysis are available online [10–12].

External Validity. We gathered execution data needed from a small number of people in a relatively structured manner. We cannot guarantee that those executions are representative of all the usages of JabRef. However, we believe that the diversity of the tasks performed by our users is enough for this evaluation. Also, the evaluation was performed using only one case study. Future research includes the repetition of the assessment on other Java applications.

Construct Validity. We relied on the reports produced by EvoSuite for structural coverage and the reports produced by Pit for mutation analysis to compare our different secondary objectives. The usage of those standard metrics allows the comparison of our results with other search-based unit test generation approaches.

Conclusion Validity. Our conclusions were only drawn based on statistically significant results with $\alpha = 0.05$. We used the standard non-parametric Wilcoxon Rank Sum test for significance and the Vargha-Delaney statistic for the effect size.

7 Conclusion and Future Work

In this paper, we introduced the *commonality score*, denoting how close an execution path is from common or uncommon executions of the software in production, and the *commonality* and *uncommonality* secondary objectives for search-based unit test generation. We implemented our approach in EvoSuite and evaluated it on JabRef using execution data from real usages of the application. Our results are mixed. The *commonality* secondary objective leads to an increase of the commonality score, and the *uncommonality* secondary objective leads to a decrease of the score, compared to the *default* secondary objective (**RQ.1**). However, results also show that if the commonality score can have a positive impact on the structural coverage (**RQ.2**) and mutation score (**RQ.3**) of the generated test suites, it can also be detrimental in some cases. Future research includes a replication of our evaluation on different applications and using different algorithms (*e.g., DynaMOSA*) to gain a deeper understanding of when to apply *commonality* and *uncommonality* secondary objectives; the exploration and assessment of different definitions of commonality; and an assessment of the generated tests regarding their usefulness for debugging.

Acknowledgements. This research was partially funded by the EU Horizon 2020 ICT-10-2016-RIA "STAMP" project (No. 731529).

References

1. Allen, F.E.: Control flow analysis. ACM SIGPLAN Not. **5**(7), 1–19 (1970). https://doi.org/10.1145/390013.808479

2. Almasi, M.M., Hemmati, H., Fraser, G., Arcuri, A., Benefelds, J.: An industrial evaluation of unit test generation: finding real faults in a financial application. In: ICSE-SEIP 2017, pp. 263–272. IEEE, May (2017). https://doi.org/10.1109/ICSE-SEIP.2017.27
3. Campos, J., Ge, Y., Albunian, N., Fraser, G., Eler, M., Arcuri, A.: An empirical evaluation of evolutionary algorithms for unit test suite generation. IST **104**, 207–235 (2018). https://doi.org/10.1016/j.infsof.2018.08.010
4. Candido, J., Aniche, M., van Deursen, A.: Contemporary software monitoring: a systematic literature review (2019). http://arxiv.org/abs/1912.05878
5. Ceccato, M., Marchetto, A., Mariani, L., Nguyen, C.D., Tonella, P.: Do automatically generated test cases make debugging easier? An experimental assessment of debugging effectiveness and efficiency. TOSEM **25**(1), 1–38 (2015). https://doi.org/10.1145/2768829
6. Chen, B., Song, J., Xu, P., Hu, X., Jiang, Z.M.J.: An automated approach to estimating code coverage measures via execution logs. In: ASE 2018, no. 3, pp. 305–316. ACM Press (2018). https://doi.org/10.1145/3238147.3238214
7. Coles, H., Laurent, T., Henard, C., Papadakis, M., Ventresque, A.: PIT: a practical mutation testing tool for Java. In: ISSTA 2016, pp. 449–452. ACM (2016). https://doi.org/10.1145/2931037.2948707
8. Derakhshanfar, P., Devroey, X., Perrouin, G., Zaidman, A., Deursen, A.: Search-based crash reproduction using behavioural model seeding. STVR **30**(3), e1733 (2020). https://doi.org/10.1002/stvr.1733
9. Devroey, X., Perrouin, G., Cordy, M., Samih, H., Legay, A., Schobbens, P.Y., Heymans, P.: Statistical prioritization for software product line testing: an experience report. SoSyM **16**(1), 153–171 (2017). https://doi.org/10.1007/s10270-015-0479-8
10. Evers, B.: Unit test generation for common and uncommon behaviors. Master thesis, Delft University of Technology (2020). http://resolver.tudelft.nl/uuid:6d8a1835-9054-4e4a-a85f-99ac592978da
11. Evers, B., Derakhshanfar, P., Devroey, X., Zaidman, A.: Unit test generation for common and uncommon behaviors: dataset, June 2020. https://doi.org/10.5281/zenodo.3894711
12. Evers, B., Derakhshanfar, P., Devroey, X., Zaidman, A.: Unit test generation for common and uncommon behaviors: replication package, June 2020. https://doi.org/10.5281/zenodo.3897513
13. Fraser, G., Arcuri, A.: EvoSuite: automatic test suite generation for object-oriented software. In: ESEC/FSE 2011, p. 416. ACM (2011). https://doi.org/10.1145/2025113.2025179
14. Fraser, G., Arcuri, A.: EvoSuite: on the challenges of test case generation in the real world. In: ICST 2013, pp. 362–369. IEEE, March 2013. https://doi.org/10.1109/ICST.2013.51
15. Fraser, G., Arcuri, A.: A large-scale evaluation of automated unit test generation using EvoSuite. TOSEM **24**(2), 1–42 (2014). https://doi.org/10.1145/2685612
16. Fraser, G., Arcuri, A.: Achieving scalable mutation-based generation of whole test suites. EMSE **20**(3), 783–812 (2015). https://doi.org/10.1007/s10664-013-9299-z
17. Fraser, G., Zeller, A.: Exploiting common object usage in test case generation. In: ICST 2011, pp. 80–89. IEEE, March 2011. https://doi.org/10.1109/ICST.2011.53
18. Galeotti, J.P., Fraser, G., Arcuri, A.: Improving search-based test suite generation with dynamic symbolic execution. In: ISSRE 2013, pp. 360–369. IEEE, November 2013. https://doi.org/10.1109/ISSRE.2013.6698889

19. Gouraud, S.D., Denise, A., Gaudel, M.C., Marre, B.: A new way of automating statistical testing methods. In: ASE 2001, pp. 5–12. IEEE, November 2001. https://doi.org/10.1109/ASE.2001.989785
20. Grano, G., Ciurumelea, A., Panichella, S., Palomba, F., Gall, H.C.: Exploring the integration of user feedback in automated testing of Android applications. In: SANER 2018, pp. 72–83. IEEE, March 2018. https://doi.org/10.1109/SANER.2018.8330198
21. Inkumsah, K., Xie, T.: Improving structural testing of object-oriented programs via integrating evolutionary testing and symbolic execution. In: ASE 2008, pp. 297–306. IEEE, September 2008. https://doi.org/10.1109/ASE.2008.40
22. Kallepalli, C., Tian, J.: Measuring and modeling usage and reliability for statistical Web testing. TSE **27**(11), 1023–1036 (2001). https://doi.org/10.1109/32.965342
23. Kracht, J.S., Petrovic, J.Z., Walcott-Justice, K.R.: Empirically evaluating the quality of automatically generated and manually written test suites. In: QSIC 2014, pp. 256–265. IEEE, October 2014. https://doi.org/10.1109/QSIC.2014.33
24. Lakhotia, K., Tillmann, N., Harman, M., de Halleux, J.: FloPSy - search-based floating point constraint solving for symbolic execution. In: Petrenko, A., Simão, A., Maldonado, J.C. (eds.) ICTSS 2010. LNCS, vol. 6435, pp. 142–157. Springer, Heidelberg (2010). https://doi.org/10.1007/978-3-642-16573-3_11
25. Menzies, T., Zimmermann, T.: Software analytics: so what? IEEE Softw. **30**(4), 31–37 (2013)
26. Molina, U.R., Kifetew, F., Panichella, A.: Java unit testing tool competition - sixth round Urko. In: SBST 2018, pp. 22–29. ACM (2018). https://doi.org/10.1145/3194718.3194728
27. Panichella, A., Kifetew, F.M., Tonella, P.: Reformulating branch coverage as a many-objective optimization problem. In: ICST 2015, pp. 1–10. IEEE, April 2015. https://doi.org/10.1109/ICST.2015.7102604
28. Panichella, A., Kifetew, F.M., Tonella, P.: A large scale empirical comparison of state-of-the-art search-based test case generators. IST **104**, 236–256 (2018). https://doi.org/10.1016/j.infsof.2018.08.009
29. Panichella, A., Kifetew, F.M., Tonella, P.: Automated test case generation as a many-objective optimisation problem with dynamic selection of the targets. TSE **44**(2), 122–158 (2018). https://doi.org/10.1109/TSE.2017.2663435
30. Panichella, S., Panichella, A., Beller, M., Zaidman, A., Gall, H.C.: The impact of test case summaries on bug fixing performance: an empirical investigation, pp. 547–558. ACM (2016). https://doi.org/10.1145/2884781.2884847
31. Rojas, J.M., Fraser, G., Arcuri, A.: Seeding strategies in search-based unit test generation. STVR **26**(5), 366–401 (2016). https://doi.org/10.1002/stvr.1601
32. Salahirad, A., Almulla, H., Gay, G.: Choosing the fitness function for the job: automated generation of test suites that detect real faults. STVR **29**(4–5), e1701 (2019). https://doi.org/10.1002/stvr.1701
33. Silva, S., Dignum, S., Vanneschi, L.: Operator equalisation for bloat free genetic programming and a survey of bloat control methods. Genet. Program. Evolvable Mach. **13**(2), 197–238 (2012). https://doi.org/10.1007/s10710-011-9150-5
34. Sprenkle, S.E., Pollock, L.L., Simko, L.M.: Configuring effective navigation models and abstract test cases for web applications by analysing user behaviour. STVR **23**(6), 439–464 (2013). https://doi.org/10.1002/stvr.1496
35. Tonella, P., Ricca, F.: Statistical testing of Web applications. SMR **16**(1–2), 103–127 (2004). https://doi.org/10.1002/smr.284

36. Vargha, A., Delaney, H.D.: A critique and improvement of the CL common language effect size statistics of McGraw and Wong. J. Educ. Behav. Stat. **25**(2), 101–132 (2000). https://doi.org/10.3102/10769986025002101
37. Wang, Q., Brun, Y., Orso, A.: Behavioral execution comparison: are tests representative of field behavior? In: ICST 2017. IEEE, March 2017. https://doi.org/10.1109/icst.2017.36
38. Wang, Q., Orso, A.: Mimicking user behavior to improve in-house test suites. In: ICSE 2019. IEEE, May 2019. https://doi.org/10.1109/icse-companion.2019.00133
39. Winter, J., Aniche, M., Cito, J., van Deursen, A.: Monitoring-aware IDEs. In: ESEC/FSE 2019, pp. 420–431. ACM (2019). https://doi.org/10.1145/3338906.3338926

Using a Genetic Algorithm to Optimize Configurations in a Data-Driven Application

Urjoshi Sinha[1], Mikaela Cashman[1,2], and Myra B. Cohen[1](✉)

[1] Iowa State University, Ames, IA 50011, USA
{urjoshi,mcohen}@iastate.edu, mcashman.isu@gmail.com
[2] Oak Ridge National Laboratory, Oak Ridge, TN 37830, USA

Abstract. Users of highly-configurable software systems often want to optimize a particular objective such as improving a functional outcome or increasing system performance. One approach is to use an evolutionary algorithm. However, many applications today are data-driven, meaning they depend on inputs or data which can be complex and varied. Hence, a search needs to be run (and re-run) for all inputs, making optimization a heavy-weight and potentially impractical process. In this paper, we explore this issue on a data-driven highly-configurable scientific application. We build an exhaustive database containing 3,000 configurations and 10,000 inputs, leading to almost 100 million records as our oracle, and then run a genetic algorithm individually on each of the 10,000 inputs. We ask if (1) a genetic algorithm can find configurations to improve functional objectives; (2) whether patterns of best configurations over all input data emerge; and (3) if we can we use sampling to approximate the results. We find that the original (default) configuration is best only 34% of the time, while clear patterns emerge of other best configurations. Out of 3,000 possible configurations, only 112 distinct configurations achieve the optimal result at least once across all 10,000 inputs, suggesting the potential for lighter weight optimization approaches. We show that sampling of the input data finds similar patterns at a lower cost.

Keywords: Genetic algorithm · Data-driven · SSBSE

1 Introduction

Many scientific applications are heavily data-driven, meaning their function (or behavior) is dependent on the specific data used to run the application and the data is often complex and varied. At the same time, these systems are often highly-configurable; the end user can modify a myriad of configuration options that control how the system behaves. The options may induce simple changes such as controlling how output formatting is handled, or they can change underlying algorithms and the algorithm's parameters, returning entirely different results [2]. Some options may also change system performance, causing the

A. Aleti and A. Panichella (Eds.): SSBSE 2020, LNCS 12420, pp. 137–152, 2020.
https://doi.org/10.1007/978-3-030-59762-7_10

application to run faster, or to utilize fewer resources such as bandwidth and energy. There has been a lot of focus on highly-configurable software with the goal of optimizing or finding options to improve quality [7,8,20].

To assist with configuration selection, some state of the art techniques use prediction [20], building models of behavior of the configuration space. An alternative approach is to use optimization, or search-based techniques [5,17,21]. Instead of modeling the configuration space to ask how a particular configuration will behave, search-based methodologies work to find an optimal (or near-optimal) configuration for a particular input data set.

However, as we demonstrate in this paper, and something that has been eluded to by Nair et al. [16], optimization in data-driven systems carries additional challenges. In these systems, different configuration options may be optimal for different data in the input data set, making optimization a heavy-weight process – it needs to be run each time new data is utilized and this may be impractical in practice. In our case study, optimization on a single input takes minutes. Although an individual run over a single configuration takes milliseconds, this is magnified if optimization needs to be run for each input in the dataset. Instead, we want to re-optimize as little as possible. Furthermore, as configuration spaces grow, the time for optimization will continue to be amplified. Research on *transfer learning* for system performance in highly-configurable software has suggested that some models provide information which can be re-used as workloads change [7,8,10]. However, workloads define the load on a system; they are not system inputs, and they impact performance, not functionality. As far as we know, there has been little focus on the problem of optimization for a data-driven application, when the optimization goal is to improve system functionality.

In this paper we explore the problem of functional optimization for data-driven applications. We ask how different inputs from a data set change the functional results of configuration optimization in a scientific application. In order to establish ground truth to ask our questions, we design an experimental testbed that contains exhaustive data of 3,000 configurations (all possible configurations from our model) and 10,000 different input sets. While this is a small configuration space, it provides us with a rich set of data – an exhaustive set of data of almost 100 million records.[1] We use this database to first evaluate the quality of our genetic algorithm. We then ask if there is a pattern of *best configurations*. We see a reduction from 3,000 to 112 configurations (or only 3.7%) appearing as the best configurations, therefore we ask if we can sample the input data and achieve similar results. Our study finds that samples as small as 10 inputs can produce similar results, suggesting a light-weight optimization approach is possible.

[1] Each run of the application can return multiple answers leading to many more records than 3,000 $\times$ 10,000.

The contributions of this work are:

1. A study on using a genetic algorithm to optimize configurations to improve the functional behavior of a popular data-driven scientific software application;
2. An exhaustive database containing almost 100 million records for a set of 10,000 different inputs;
3. Results on sampling input data, suggesting that the observed patterns hold over even small random samples.

In the next section we present some motivation. In Sect. 3 we present an overview of our experimental database framework. We follow that with a case study in Sects. 4 and 5. We then present some related work (Sect. 6) and end in Sect. 7 with conclusions and future work.

2 Motivating Example

We present a small motivating example based on our case study application in this section. BLAST stands for the Basic Local Alignment Search Tool [1], and is a widely used data-driven application. Bioinformatics users utilize BLAST to ask questions about sequences of deoxyribonucleic acid (DNA) fragments that they observe in nature or in the lab. When using BLAST, the user selects a database of known sequences and then inputs their unknown sequence. The application returns *hits* which are a set of matches found in the database. Note, that most of these sequences are not exact matches, but are partial matches of the sequence. There are some key quality objectives that many users rely on. The *e-value* (expected value) is a number that describes how many times you would expect a match by chance in a database of that size. The lower the value is, the more significant the match. In many cases users filter for only e-values of zero (optimal). Another key metric called *percent identity* is a number that describes how similar the query sequence is to the target sequence (how many characters in each sequence are identical). The higher the percent identity is, the more significant the match. Users sometimes filter by a value of 100 or 99% (100% is optimal).

One common use case of a BLAST search is to obtain as many *quality hits* as possible so that the user can explore those and find out what is known about their input sequence. Underneath the covers BLAST uses a dynamic programming algorithm for matching, and this algorithm has been highly tuned over the years. The current version of BLAST [1] has over 50 configuration options that a user can modify. We only explore a part of this in our case study.

Suppose we want to optimize a BLAST query with the goal of obtaining the best e-value and percentage identity, while at the same time increasing the number of hits. An obvious first choice is to just run the default configuration. If we choose for instance, the second input sequence from this study, the default BLAST configuration returns 1 hit and has the max percentage identity and smallest e-value; it is the best we can do. If we optimize with a genetic algorithm, we will be unable to improve further; the default configuration is the best.

However, if we instead optimize using input sequence number 84, we see different behavior. The default configuration returns 3 hits with a median e-value of 0 and percentage identity of 100. But if we change 4 of the configuration options (away from the default) we can improve the number of hits to 8 while maintaining the minimal median e-value and maximum median percent identity possible. We have increased our quality hits by 5, therefore a user might want to consider using the new parameters. In our case study, only one-third of 10,000 input sequences returns the default configuration as the optimal result (using our objective value), meaning the other two-thirds have an opportunity to be improved.

After running these two experiments, we have a conundrum. If we use the results from the genetic algorithm of input sequence number 2, we won't optimize correctly for input sequence number 84, and vice versa. We may also want to consider other user preferences such as the distance from the default configuration. As we can see, tuning this configuration space is complex and may be data dependent.

As we started building a genetic algorithm for optimization of this program, we were left with many open questions. For instance: (1) How can we confirm a genetic algorithm is working, given the range of behavior for different input data? (2) Can we find patterns of genetic algorithm behavior across sequences? (3) Do we need to run our genetic algorithm for all different inputs? As we iterated to tune the fitness and landscape that can guide a genetic algorithm for this program, the data dependency made this time consuming and ad-hoc. We learned that data-driven applications are difficult to tune and reason about when building a search algorithm, and asked if instead, we can find a better approach. We hypothesized the need for an exhaustive data set that can be easily queried. We present and use such a framework in the following section.

3 A Framework for Data-Driven Exploration

We present our design of a Framework for a Data-driven Exploration in a search environment (FRDDE) that will help us understand the quality of our genetic algorithm, and the variability of optimizations within our dataset. Figure 1 shows an overview of FRDDE. We note that we only implemented this for a single data set in our case study, but the approach is general and we plan to build additional data sets for other data-driven systems as future work.

FRDDE begins (#1) with a model of the configurations and a set of inputs. The Cartesian product of these are generated as pairs (configurations, inputs) and then run against the application (#2). It is not necessary to define an exact fitness function yet, but to capture key measurements that can be used to build different fitness functions for this application. The relational data model is shown in Fig. 2. We have a table of the configuration models where the primary key (Pk) is the configuration number and the other fields are the individual configuration options and their possible choices. Table 2 (outputs) contains the results of running the configurations and inputs. In this table we have a three-way

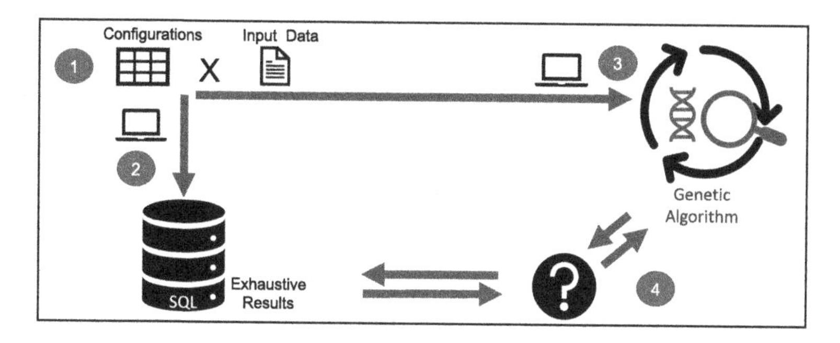

Fig. 1. Overview of the FRDDE framework. It starts with a set of configurations and input data (#1) and populates the database (#2). A genetic algorithm can then be run #3 and the database can be used to answer questions #4.

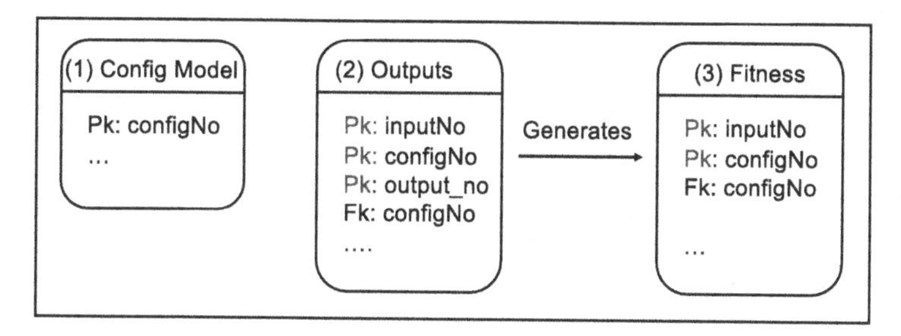

Fig. 2. FRDDE schema showing only the primary keys (Pk) and foreign keys (Fk). Table 1 contains the configuration model. It has one row for each of the configurations in the model. Table 2 is raw data from the application. Each returned hit has a record. It has a composite primary key, consisting of the configuration, input sequence and the number of the returned output (generated). The last table (Table 3) is a generated fitness table containing a fitness value along with the individual components of the fitness for each configuration and each input.

primary key which we see as inputNo, configNo, and outputNo. This three-way key allows for applications which can return more than one output per run (many data-driven applications such as the one we use, do not return a single result). This table has a foreign key (Fk) which can be joined with Table 1. The other fields contain detailed information relevant to the use case and fitness which can be used during exploration. The last table is an aggregate table built from Table 2. It has one record per inputNo and configNo and uses the configNo as a foreign key. We can use this to build different fitness values and to explore different parts of the search space.

The next phase (#3) of FRDDE is to run a genetic algorithm (or other evolutionary algorithm) using the sample application and configuration model. This can be run on all inputs or a sample of inputs. In the last phase (#4), we can interactively ask questions and compare results from the database against the runs of the genetic algorithm.

Once FRDDE is complete, the goal is to use what we have learned to expand the configuration space and or data set, where an exhaustive analysis is no longer possible, with information learned from the exploration. We present an example exploration with FRDDE in the next section. We leave the expansion of our configuration space for future work.

4 Case Study

We present a case study leveraging FRDDE.[2] While the long-term goal is to understand how useful FRDDE is in a data-driven search based environment, we perform this case study as an exemplar of its use. The questions we ask are:

RQ1: How effective and efficient is a genetic algorithm across all input data?
RQ2: What patterns, if any, emerge across all input data?
RQ3: How well does sampling capture patterns seen in the exhaustive data?

Table 1. Configuration Model for BLAST from [2]. The names of options are followed by the number of choices and their values. Default values are shown in bold.

Option name	Abbr.	No. of choices	Configuration values
dust	dust	3	yes, no, **'20 64 1'**
lcase_masking	lcase	2	true, **false**
soft_masking	soft	2	**true**, false
ungapped	ungap	2	true **false**
xdrop_gap	xdgap	5	0, 0.1, 0.5, **30**, 100
xdrop_gap_final	xdgfin	5	0, 0.1, 0.5, 10, **100**
xdrop_ungap	xdugap	5	0, 0.1, 0.5, **20**, 100

4.1 Object of Study

As our software subject we use the popular bioinformatics tool, BLAST. We use the nucleotide sequencer (BLASTn), version 2.6.0. This is the same subject studied by Cashman et al. [2] and was sensitive to changing configurations. We use their configuration model (shown in Table 1). The BLAST model has 7 configuration options. Three are binary, one has 3 choices, and the other three have 5 choices. This leads to an exhaustive configuration space of 3,000 configurations. This is not the complete configuration space for BLAST, but one that is tractable to study exhaustively. Their input data set contains 10,000 input sequences from the yeast genome. We use these as our input data set.

[2] Supplementary data website: https://github.com/LavaOps/ssbse-2020-FrDdE.

4.2 Genetic Algorithm Implementation

Our chromosome has seven genes (one for each configuration option). The alleles are configuration choices.

Fitness. The fitness for this case study is based on the scientific use case presented earlier. However, we can theoretically use alternative fitness functions within FRDDE. When using BLAST, a biologist may want to increase the number of *quality hits* they match from the database. Often, users restrict two quality values, e-value and the percentage identity. Instead of filtering these values, we use them for optimization, since a value that is close to the optimal may be of interest. We also include another component in our fitness, the distance from the default configuration. It has been documented that users prefer not to change too many settings [4], hence our goal is to stay close to the original configuration.

Since we are attempting to maximize the number of hits (and expect more than one), the e-value and percentage identity are aggregate values. We chose the median for this (we have also experimented with the mean and observe similar results). Putting these together our fitness is shown next. We are maximizing. The various weights for the components were determined heuristically.

$$w_{\text{e-val}} = \begin{cases} 10, & \text{if median e-value} = 0 \\ \dfrac{1}{\text{median e-value}}, & \text{otherwise} \end{cases} \tag{1}$$

$$w_{\text{distance}} = \begin{cases} 10, & \text{if distance} = 0 \\ \dfrac{7}{\text{distance}}, & \text{otherwise} \end{cases} \tag{2}$$

$$\textit{fitness} = \text{hits} \times 10 + w_{\text{e-val}} + \text{median(percent identity)}/2 + w_{\text{distance}} \tag{3}$$

Equation 1 shows the weighted e-value, (or $w_{\text{e-val}}$) which takes a value of 10 when the median e-value is 0. Otherwise it is the inverse. Equation 2 is the calculation of the weighted distance (w_{distance}). Under the default configuration this value is set to 10, otherwise we normalize by the number of configuration options (7). Last, Eq. 3 weights these together. Hits are weighted by 10 to make them as important as $w_{\text{e-val}}$ and distance. The percent identity is reduced by one half. Although not part of this study, we made use of the database to help tune the weights.

Genetic Algorithm Parameters. We heuristically tuned our algorithm. We ended with a population size of 16. When we seed the initial random population we generate all but one chromosome randomly. The last chromosome is made up of default values for the program to ensure its existence in each population. We use one point crossover selecting a random point (within the middle two thirds of the chromosome for each pair). We pair even and odd numbered chromosomes in rank order (e.g. the best with the third best and second best with the fourth best) to maintain some diversity.

We use a rank selection with elitism. This has an important effect for our use case. If the default configuration is the best fitness, this will propagate to be the best overall at the end. We tuned the mutation rate and settled on a value of 0.5 for this program. We randomly select 50% of the genes in the population and change the value randomly (with replacement). We do not mutate genes of the two fittest (elite) chromosomes. We experimented with different numbers of generations. We usually see convergence relatively quickly (within 20–30 generations), therefore we chose 50 which will balance the need to solver harder cases, with being tractable with respect to runtime.

4.3 Experimental Setup

All experiments are run on a parallel computing RedHat Enterprise Linux 7 Server with Intel(R) Xeon(R) Gold 6244 CPU 2.60 GHz nodes. We collected the exhaustive data by running BLASTn on 10,000 input sequences against the yeast database. The experiments were run such that each input sequence was tested under all 3,000 possible configuration options. These led to a total of 30 million BLASTn calls. 100 of these jobs were executed in parallel to make this scalable. Each job was allocated 1 node with 5 GB memory. The following outputs were collected post every BLASTn call: the sequence number, its corresponding configuration number, the blast sequence id, total hits generated, the percentage identity and e-value observed, and the start and end positions of the subject sequence against which the match was observed.

Next, we used a MariaDB fork of MYSQL, version 5.5.56 in a singularity container version 2.6.0. We constructed a database having three primary tables, which are concrete instantiations of the FRDDE tables. Additional temporary and derived tables were used for intermediate computations and to fetch results from the database.

In total we have 96,875,728 records in FRDDE. This is because we have a record for each hit. We also recorded the combination of configuration numbers and the sequences against which no hits were reported by BLASTn. All such sequence numbers along with their configurations were stored in a separate table. Next, we constructed the fitness table which is derived from the output table. We store the count of hits returned per configuration for a specific input sequence. We compute the median percentage identity and e-value over all hits recorded for a sequence under a given configuration number, the distance from the default configuration, and the computed fitness value. We annotated each possible distinct configuration set across its 3,000 possibilities with an integer number for ease of reference and access across tables. These annotations along with the combination of 7 individual configuration options such as dust = "20 64 1", specific to a configuration number, was also stored in the configuration table.

4.4 Threats to Validity

With respect to generalization we used only a single software subject, a single type of input file, and a single fitness for this study. However, we use 10,000 inputs and this is a common scientific application that is widely used and has been studied in other papers on configurable software. We used the same configuration space and input files as another paper to avoid bias due to our selection. With respect to threats to internal validity, we wrote programs to perform many of the parts of this experiment and acknowledge there could be errors. We did extensive checking of the results and compared the database against the genetic algorithm. We also make the data and artifacts from these experiments publicly available on our supplementary website so the community can validate and reuse our results.

5 Results

In this section we present results for each of our research questions in order.

5.1 RQ1: Effectiveness

To evaluate the effectiveness of the genetic algorithm we compare the configurations reported in the final population of the genetic algorithm to the best configurations found in the database. Of the 10,000 input sequences, 19 of them never find any hits in any configuration. We remove these from our analysis, leaving 9,981 input sequences. We then analyze for 9,981 input sequences if the fitness of the best configuration returned by the genetic algorithm is the same as the best fitness from the database. Results over two runs of the genetic algorithm can be seen in Table 2. The first result column provides the number of cases where we match the best fitness (9,721 and 9,713). This is also shown as a percentage (97.4% and 97.3%). For 3,382 of the inputs (34%), the default is the best fitness. We discuss this implication in RQ2. For the remaining sequences where a best configuration was not found by the genetic algorithm, we ask if the returned configuration has a higher fitness than the default. In all cases (260 and 268) the answer is yes. While we were not able to find the optimal, we still improved over the default. Hence, we conclude that our genetic algorithm is working well on this data set; finding an optimal solution for more than 97% of the inputs.

To confirm the genetic algorithm is useful even in this small configuration space, we randomly selected 500 inputs and randomly generated five sets of 1,000 chromosomes for each input. Our rationale is that each run of our genetic algorithm would evaluate a maximum of 800 chromosomes after 50 generations with our population size of 16 (we usually converge by generation 20–30). One of the chromosomes is seeded as the default configuration. 67.5% of the time, we find the best configuration option across all runs for all 500 inputs. There is also little variation (standard deviation of 8.7 for the number of times the maximum is found). However, for 185 inputs (or 37%), the default already is the best fitness. If we remove those inputs from our data, random finds the best fitness 48.38% of the time. However, in almost all cases, when the best fitness is not found, we can improve on the default (99.3%). We conclude that while random does relatively well and almost always finds the best or a better fitness than random, the genetic algorithm performs better overall. As the configuration size grows, we expect the gap in performance to increase.

To evaluate the efficiency we observe the runtime (in minutes) of the genetic algorithm. On average it takes approximately 1 min for the genetic algorithm to run on a single input sequence, with a maximum of 2.8 min. If we sum the runtimes across all of the genetic algorithms (these were run in parallel) it took 8.11 days for the first run and 7.84 days for the second run. Note, these are sequential times. The actual clock time was faster due to parallelization.

Summary RQ1: We conclude that the genetic algorithm is effective. While it is relatively efficient for a single input, it takes days to run the entire data set.

Table 2. Comparison of matches of maximum fitness of the genetic algorithm on 10,000 sequences compared against database (known maximum). 19 returned no hits in any configuration and are removed. Remaining # > default are runs which did not match the best fitness, but still improve on default. Times is shown for the genetic algorithm in minutes.

	No. of best (of 9981)	Percent	Remaining # > default	Avg. runtime	Std deviation	Min runtime	Max runtime
Run 1	9721	97.40	260	1.17	0.76	0.26	2.78
Run 2	9713	97.31	268	1.13	0.66	0.23	2.71

5.2 RQ2: Patterns

We utilize the database information to ask if there are patterns in the data. Of the possible 3,000 configurations, only 112 unique configurations achieve the best fitness for at least one of the 9,981 input sequences. This indicates that only 3.7% of the configurations are considered optimal for this fitness and set of inputs. We sorted these configurations by the times they occur. Since a maximal fitness for a particular input sequence may have more than one possible configuration, we normalize the counts so that each input sequence is accounted for once. For example, if the sequence has two configurations with the optimal fitness, we weight each of them by 1/2. Table 3 shows the top configurations, sorted in descending order by count of occurrences. These make up 92.4% of the best configurations. The first configuration (#1724) is the default. This occurs 3,382 or 34% of the time. 66% of the time, another configuration is optimal. We highlight the changes from the default in each of the other commonly seen configurations.

We can identify certain configuration option values that are important, such as ungap. Combining ungap (TRUE) with xdugap (0.1 or 0.5) account for the two next best configurations overall. Interestingly, neither of these options are set to these values in the default. We also see options like lcase which do not appear to have a strong effect. To analyze this further we look at the distribution

Table 3. Most frequent configurations (top 9 out of 112) appearing as the best configuration. The left most column is the configuration number, followed by the normalized times the configuration is seen. The rest of the columns indicate the configuration options selected. Bold indicates a change from the default configuration (#1724)

Default	Config no.	Count	dust	lcase	soft	ungap	xdgap	xdgfin	xdugap
YES	1724	3382	20 64 1	FALSE	TRUE	FALSE	30	100	20
–	1597	998	20 64 1	FALSE	TRUE	**TRUE**	30	100	**0.1**
–	1598	998	20 64 1	FALSE	TRUE	**TRUE**	30	100	**0.5**
–	1847	828	20 64 1	FALSE	**FALSE**	**TRUE**	30	100	**0.1**
–	1848	828	20 64 1	FALSE	**FALSE**	**TRUE**	30	100	**0.5**
–	1974	820	20 64 1	FALSE	**FALSE**	FALSE	30	100	20
–	1599	784	20 64 1	FALSE	TRUE	**TRUE**	30	100	20
–	2724	300	**No**	FALSE	TRUE	FALSE	30	100	20
–	1849	227	20 64 1	FALSE	**FALSE**	**TRUE**	30	100	20

of each configuration option's values in Fig. 3. We can see that in most cases the default values will work well for any given input (the black bars). This matches what we see in the best configurations as the default is the best overall. We see that options such as ungap, soft, and xdugap seem to have influence, which others (like lcase) have no influence.

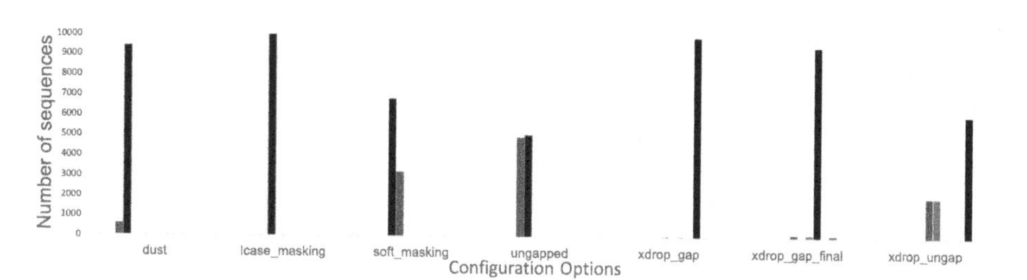

Fig. 3. Number of times each value for the seven configuration options appear (normalized) in the configuration with the highest fitness. Black solid bars indicate the default value for that configuration option. Other colors are alternative choices. Spaces indicate that the value does not appear.

Summary RQ2: We observe a set of patterns for the input data, demonstrating that it may be possible to sample the inputs. No single configuration works best across all input sequences, however we have reduced our configuration space from 3,000 to 112 (by 96.3%).

5.3 RQ3: Sampling

We now examine the effectiveness of sampling, i.e. can identify the same pattern of best configurations as we did in RQ2? We use random sample sizes of 10, 100

Table 4. Counts of the occurrences of the top configurations in the database. For each of the best configurations from Table 3 we show the number of times found in the five runs (R1–R5) for different sample sizes (10,100,1000). The percentage of the best configurations found by the GA sample relative to the ground truth best configurations found are displayed in the last row.

	DB	Sample Size 10					Sample Size 100					Sample Size 1,000				
ConfigID	**Occur.**	**R1**	**R2**	**R3**	**R4**	**R5**	**R1**	**R2**	**R3**	**R4**	**R5**	**R1**	**R2**	**R3**	**R4**	**R5**
1724	3382	3	4	5	4	4	34	24	35	30	46	336	328	348	352	340
1597	998	2	0	1	0	2	8	10	13	7	10	94	101	88	117	111
1598	998	1	1	1	0	2	12	15	7	8	6	117	92	100	86	92
1847	828	0	1	1	1	1	2	11	7	7	8	77	88	85	89	65
1848	828	0	0	1	1	1	11	9	7	9	5	71	85	83	75	90
1974	820	1	0	1	1	0	11	7	8	9	7	78	75	68	76	74
1599	784	1	1	0	0	0	5	14	9	12	5	78	58	69	43	74
2724	300	0	0	0	0	0	4	2	3	4	0	28	35	29	20	26
1849	227	0	0	0	2	0	2	0	1	3	1	10	21	15	22	17
% of Best		**80**	**70**	**100**	**90**	**100**	**89**	**92**	**90**	**89**	**88**	**88.9**	**88.3**	**88.5**	**88**	**88.9**

and 1,000. While the genetic algorithm's search space is not reduced, we can potentially avoid running it against all 10,000 inputs, which may provide a significant time savings. We repeat each sample 5 times. Table 4 reports how many times a sample run was able to identify any of the best (or top) 9 configurations seen in Table 3. Ideally, the majority of configurations reported by the sampled GAs would be from that set. In a sample size of 10 we see between 70–100% of reported configurations appearing from the top 9 configurations. As we move to a sample size of 100, 88–92% are from the top 9. We see slightly less (88.8%) in a sample size of 1,000. This may be due to more opportunities for noise; a sequence leading to a less common configuration is more likely to be selected.

Table 5. Average and standard deviation across five runs of each sample size. We report the percentage of occurrences of the default configuration. Uniqueness is measured as the percentage of unique configurations. We report the total runtime in hours.

	# Default		Uniqueness		Runtime (hrs)	
	AVG	STD	AVG	STD	AVG	STD
Sample size 10	40.00	7.07	60.00	7.07	0.06	0.00
Sample size 100	33.80	8.07	16.60	1.67	3.39	0.10
Sample size 1,000	34.08	0.95	4.38	0.27	15.94	2.91

Table 5 shows another view of this data. The first two columns show the average and standard deviation of the percent of times the default configuration is selected. Remember, the default appeared in 34% of the exhaustive data in the database. The next two columns, represents a metric we call Uniqueness. This shows convergence of a configuration pattern (lower is better). It is the percent of unique configurations observed as the best fitness. For instance, if we have a sample of 10 inputs, the maximum number of best configurations returned is 10 (all are unique). If we see 6 unique configurations, then uniqueness is 60%. As we move from samples of 10 to 1,000 uniqueness drops from 60% to 4.38%. The uniqueness from the database (RQ2) is 112/3,000 or 3.73%. The last two columns show the total sequential runtime to complete the sample, when we re-ran the genetic algorithm on each input in the sample. A sample of 10 takes on average about 3 min, 100 about 3.4 h, and a sample of 1,000 takes just under 16 h. To compute the complete data set on 10,000 inputs as in RQ1 takes approximately 8 days. Sampling can reduce the runtime to find the best configuration patterns by over 90% while still identifying over 80% of the best configurations reducing uniqueness.

Summary RQ3: Sampling input data provides similar patterns to the exhaustive data set while compressing the percentage of unique results and runtime significantly.

6 Related Work

In the closest work to ours, Nair et al. define the notion of data-driven search-based software engineering (DSE) [16]. They call for combining data mining and search-based software engineering and present a set of challenges and research in this domain. While they discuss techniques such as parameter tuning and transfer learning, they do not explicitly examine implications for using search under specific data inputs while tuning configurations as we do in this work.

The largest body of work using search algorithms for configurable software falls into two main categories. The first is in testing of software configurations where search is used to find samples [3,6,9] that are representative of configurations which are likely to find faults. The general idea of configurability in software which can lead to different faults is also well studied (e.g. [14,15,19,22]). There has also been extensive research on optimizing configurations that satisfy sets of constraints [5,17,18,21]. While all of this research uses search-based algorithms, none that we know of focus on the problem of changing input data.

Another line of research is for system performance which involves building prediction or influence models for configurable systems [20]. The goal is to build a model that describes the configuration space. This work differs in that it focuses primarily on quality attributes (not functional as in our work) and describes the configuration space for prediction, rather than optimization. Some research in this domain examines the notion of transfer learning [7,8], which attempts to use knowledge from one workflow for a different workflow. This has a similar flavor to our work, but the focus is on performance and we do not use learning techniques. Nair et al. proposed the idea of bellwether configurations; those which are indicative of good or bad performance across all workflows [10,11,17]. Interestingly, we see similar results where there are specific configurations that seem to matter across all input data, although our objective goals differ. With respect to functionality, Garvin et al. [4] documented the locality of failures within a configuration space, however they do not explore faults in a data-driven environment. Cashman et al. [2] presented problems related to configurability in bioinformatics tools. The work suggests the need for optimization, but does not go as far as to provide an optimization technique. We use their software subject (BLAST) and model in this work.

Finally, Landgon et al. have proposed the use of program data for evaluation as a fitness function in genetic improvement [12,13]. The difference with that work is that the data is the fitness function, where as we use a traditional fitness function, but are looking at results across different input data.

7 Conclusions and Future Work

In this paper we presented FrDdE, an approach for optimizing functional objectives in a highly-configurable data-driven application. We built a database of almost 100 million records using an exhaustive configuration space which is a part of the BLAST bioinformatics tool, and studied its behavior on 10,000 different input sequences. We find that our genetic algorithm is effective, and see patterns emerge across input sequences. While we don't find a single configuration that is best, the configuration space is reduced from 3,000 to 112 and only 9 configurations account for over 92% of the results. We demonstrate that sampling inputs can find similar patterns to the full data set which leads us to believe we can develop a light-weight technique for data-driven search based configuration optimization. Our ultimate goal is to use FrDdE to tune optimization techniques and to explore data sets before moving to larger configuration spaces where exhaustive enumeration is not possible.

In future work we plan to expand FrDdE further by (1) increasing the experimental configuration space to see if results learned still hold, (2) applying FrDdE to additional input data sets and (3) using FrDdE for different data-driven applications. We also plan to incorporate learning into FrDdE to see if we can learn more about the patterns of optimal configurations.

Acknowledgements. This work is supported in part by NSF Grant CCF-1901543 and by The Center for Bioenergy Innovation (CBI) which is supported by the Office of Biological and Environmental Research in the DOE Office of Science.

References

1. Altschul, S.F., Gish, W., Miller, W., Myers, E.W., Lipman, D.J.: Basic local alignment search tool. J. Mol. Biol. **215** (2018). https://blast.ncbi.nlm.nih.gov/
2. Cashman, M., Cohen, M.B., Ranjan, P., Cottingham, R.W.: Navigating the maze: the impact of configurability in bioinformatics software. In: International Conference on Automated Software Engineering, pp. 757–767. ASE, September 2018
3. Garvin, B.J., Cohen, M.B., Dwyer, M.B.: Evaluating improvements to a metaheuristic search for constrained interaction testing. Empir. Softw. Eng. (EMSE) **16**, 61–102 (2010)

4. Garvin, B.J., Cohen, M.B., Dwyer, M.B.: Failure avoidance in configurable systems through feature locality. In: Cámara, J., de Lemos, R., Ghezzi, C., Lopes, A. (eds.) Assurances for Self-Adaptive Systems. LNCS, vol. 7740, pp. 266–296. Springer, Heidelberg (2013). https://doi.org/10.1007/978-3-642-36249-1_10
5. Henard, C., Papadakis, M., Harman, M., Le Traon, Y.: Combining multi-objective search and constraint solving for configuring large software product lines. In: IEEE/ACM 37th IEEE International Conference on Software Engineering, vol. 1, pp. 517–528 (2015)
6. Henard, C., Papadakis, M., Perrouin, G., Klein, J., Heymans, P., Le Traon, Y.: Bypassing the combinatorial explosion: using similarity to generate and prioritize T-wise test configurations for software product lines. IEEE Trans. Softw. Eng. **40**(7), 650–670 (2014)
7. Jamshidi, P., Siegmund, N., Velez, M., Kästner, C., Patel, A., Agarwal, Y.: Transfer learning for performance modeling of configurable systems: an exploratory analysis. In: International Conference on Automated Software Engineering (ASE), pp. 497–508, November 2017
8. Jamshidi, P., Velez, M., Kästner, C., Siegmund, N.: Learning to sample: exploiting similarities across environments to learn performance models for configurable systems. In: Joint Meeting on European Software Engineering Conference and Symposium on the Foundations of Software Engineering, pp. 71–82. ESEC/FSE (2018)
9. Jia, Y., Cohen, M.B., Harman, M., Petke, J.: Learning combinatorial interaction test generation strategies using hyperheuristic search. In: IEEE/ACM 37th IEEE International Conference on Software Engineering, vol. 1, pp. 540–550 (2015)
10. Krishna, R., Menzies, T., Fu, W.: Too much automation? The bellwether effect and its implications for transfer learning. In: 31st IEEE/ACM International Conference on Automated Software Engineering (ASE). pp. 122–131 (2016)
11. Krishna, R., Menzies, T.: Bellwethers: a baseline method for transfer learning. IEEE Trans. Softw. Eng. **45**(11), 1081–1105 (2019)
12. Langdon, W.B.: Big data driven genetic improvement for maintenance of legacy software systems. SIGEVOlution Newsl. ACM Spec. Interes. Group Genet. Evol. Comput. **12**(3), 6–9 (2019)
13. Langdon, W.B., Krauss, O.: Evolving sqrt into 1/x via software data maintenance. In: Coello, C.A.C. (ed.) GECCO 2020: Genetic and Evolutionary Computation Conference, Companion Volume, pp. 1928–1936. ACM, July 2020
14. Medeiros, F., Kästner, C., Ribeiro, M., Gheyi, R., Apel, S.: A comparison of 10 sampling algorithms for configurable systems. In: International Conference on Software Engineering (ICSE), pp. 643–654. ACM, May 2016
15. Meinicke, J., Wong, C.P., Kästner, C., Thüm, T., Saake, G.: On essential configuration complexity: measuring interactions in highly-configurable systems. In: International Conference on Automated Software Engineering (ASE), pp. 483–494. ACM, September 2016
16. Nair, V., et al.: Data-driven search-based software engineering. In: IEEE/ACM 15th International Conference on Mining Software Repositories (MSR), pp. 341–352 (2018)
17. Nair, V., Menzies, T., Siegmund, N., Apel, S.: Using bad learners to find good configurations. In: Joint Meeting on Foundations of Software Engineering, pp. 257–267. ESEC/FSE (2017)
18. Oh, J., Batory, D., Myers, M., Siegmund, N.: Finding near-optimal configurations in product lines by random sampling. In: Joint Meeting on Foundations of Software Engineering, p. 61–71. ESEC/FSE (2017)

19. Qu, X., Cohen, M.B., Rothermel, G.: Configuration-aware regression testing: an empirical study of sampling and prioritization. In: International Symposium on Software Testing and Analysis, pp. 75–86. ISSTA, ACM (2008)
20. Siegmund, N., Grebhahn, A., Kästner, C., Apel, S.: Performance-influence models for highly configurable systems. In: European Software Engineering Conference and ACM SIGSOFT Symposium on the Foundations of Software Engineering (ESEC/FSE), pp. 284–294. ACM Press, August 2015
21. Xiang, Y., Zhou, Y., Zheng, Z., Li, M.: Configuring software product lines by combining many-objective optimization and sat solvers. ACM Trans. Softw. Eng. Methodol. **26**(4), 1–46 (2018)
22. Yilmaz, C., Dumlu, E., Cohen, M.B., Porter, A.: Reducing masking effects in combinatorial interaction testing: a feedback driven adaptive approach. IEEE Trans. Softw. Eng. **40**(1), 43–66 (2014)

Measuring and Maintaining Population Diversity in Search-Based Unit Test Generation

Nasser Albunian[1(✉)], Gordon Fraser[2(✉)], and Dirk Sudholt[1(✉)]

[1] The University of Sheffield, Sheffield, UK
{nmalbunian1,d.sudholt}@sheffield.ac.uk
[2] University of Passau, Passau, Germany
gordon.fraser@uni-passau.de

Abstract. Genetic algorithms (GAs) have been demonstrated to be effective at generating unit tests. However, GAs often suffer from a loss of population diversity, which causes the search to prematurely converge, thus negatively affecting the resulting code coverage. One way to prevent premature convergence is to maintain and increase population diversity. Although the impact of population diversity on the performance of GAs is well-studied in the literature, little attention has been given to population diversity in unit test generation. We study how maintaining population diversity influences the Many-Objective Sorting Algorithm (MOSA), a state-of-the-art evolutionary search algorithm for generating unit tests. We define three diversity measures based on fitness entropy, test executions (phenotypic diversity), and Java statements (genotypic diversity). To improve diversity, we apply common methods that fall into two groups: niching (such as fitness sharing and clearing) and non-niching (such as diverse initial populations). Our results suggest that increasing diversity does not have a beneficial effect on coverage in general, but it may improve coverage once the search stagnates.

Keywords: Search-based test generation · Population diversity

1 Introduction

As software testing is a laborious and error-prone task, automation is desirable. Genetic Algorithms (GAs) are frequently employed to generate tests, especially in the context of unit testing object oriented software. However, a common general issue when applying GAs is premature convergence: If the individuals of the search population all become very similar and lack diversity [6,25,26], then the search may converge on a local optimum of the objective function. This reduces the effectiveness of the GA, and in the case of search-based test generation, premature convergence would imply a reduced code coverage.

To avoid such premature convergence, it is important to maintain diversity in the population. Different techniques have been proposed to achieve this at the

A. Aleti and A. Panichella (Eds.): SSBSE 2020, LNCS 12420, pp. 153–168, 2020.
https://doi.org/10.1007/978-3-030-59762-7_11

genotype and the phenotype levels [6,26]. For example, diversity can be achieved by scaling an individual's fitness based on the density of its niche or by eliminating duplicate individuals from the population. While diversity maintenance has been extensively investigated within different domains of evolutionary algorithms (e.g., [6]), much less is known about diversity in search-based unit test generation.

We empirically investigate the impact of population diversity on the generation of unit tests for Java programs. More specifically, we aim to see whether increasing population diversity leads to a better GA performance, i.e., generating unit tests that achieve higher code coverage. We first adapt common diversity measurements based on phenotypic and genotypic representation to the search space of unit test cases (Sect. 3). We then study the effects of different diversity maintenance techniques (Sect. 4) on population diversity and code coverage (Sect. 5).

2 Search-Based Unit Test Generation

In the context of generating tests for object-oriented programs, a solution is represented as a test case τ which consists of a sequence of calls $\tau = \langle s_1, s_2, \dots, s_n \rangle$ on the class under test (CUT) [9]. That is, each s_j is an invocation of a constructor of the CUT, a method call on an instance of the CUT, a call on a dependency class in order to generate or modify dependency objects, or it defines a primitive value (e.g., number, string, etc.) As the ideal test case size is not known a priori, the number of statements in a test case is variable and can be changed by the search operators. Consequently, crossover and mutation not only modify the individual statements of a sequence, but can also remove or insert statements.

The fitness functions commonly used are based on code coverage [21]. We focus on branch coverage as it is one of the most common coverage criteria in practice [9], and fitness functions for other criteria are typically based on branch coverage fitness calculations. In the many-objective representation of the unit test generation problem [17], each branch in the CUT forms a single objective. The fitness function of a test τ on branch b_i is $f(\tau, b_i) = al(b_i, \tau) + \alpha(bd(b_i, \tau))$ [13], where bd is the branch distance, α is a normalization function mapping the branch distance to the range $[0, 1]$, e.g. $x/(x+1)$, and al is the approach level, the distance between the closest control dependency of the target node executed by a test and the target node in the control dependency graph.

A common approach lies in evolving sets of unit tests [9] using single-objective optimization, or individual test cases using many-objective optimization [17]. The Many-Objective Sorting Algorithm (MOSA) has been shown to generally perform best [3]. MOSA (Algorithm 1) starts with an initial population of randomly generated test cases (line 4), and applies standard genetic operators (line 7). To generate the next generation, parents and offspring are combined (line 8) and sorted using the preference criterion and non-dominance relation (line 9). The preference criterion identifies a subset of test cases that have the lowest fitness values for uncovered branches, which are assigned rank 0, while the traditional non-dominated sorting used by NSGA-II is applied to rank the remaining test cases into further

Algorithm 1: Many-Objective Sorting Algorithm (MOSA)

```
1  Input: Population size n, Stopping criterion C
2  Output: An archive of best test cases T
3  t ← 0 ;                                         ▷ current iteration
4  P_t ← GenerateRandomPopulation(n)
5  T ← UpdateArchive(P_t)
6  while ¬C do
7      P_o ← GenerateOffspring(P_t)
8      P_u ← P_t ∪ P_o
9      F ← PreferenceSorting(P_u)
10     r ← 0
11     P_{t+1} ← {}
12     while |P_{t+1}| + |F_r| ≤ n do
13         AssignCrowdingDistance(F_r)
14         P_{t+1} ← P_{t+1} ∪ F_r
15         r ← r + 1
16     end
17     CrowdingDistanceSort(F_r)
18     P_{t+1} ← P_{t+1} ∪ F_r ;                   ▷ size n − P_{t+1}
19     T ← UpdateArchive(T, P_t)
20     t ← t + 1
21 end
22 return T
```

fronts. Selection is then applied based on the assigned ranks starting at the first front, until reaching the population size n (lines 12–16). When the number of selected test cases exceeds the population size n, the individuals of the current front F_r are sorted based on the crowding distance (line 17) and only those individuals with higher distance are selected. At the end of each generation, MOSA updates an archive with test cases that cover uncovered branches with the lowest possible length (line 19).

3 Measuring Population Diversity

Population diversity refers to the variety in a population based on the differences at the genotype (i.e., structural) or phenotype (i.e., behavioural) levels. The genotypic diversity measures the structural (i.e., syntactic) differences among the individuals of a population. In contrast, the phenotypic diversity is based on the behavioural (i.e., semantic) differences in the population's individuals.

However, these levels differ among different domains [6], e.g., the structure of an individual in the case of genetic programming (GP) is not similar to the one with other evolutionary algorithms (EAs). For example, McPhee and Hopper [14] considered the number of different nodes in a GP as structural difference among the individuals. In the case of phenotypic diversity, the fitness of the population's individuals can be mainly used to measure the behavioural differences among the

Table 1. Two examples (cases) of two test cases (TC_1 and TC_2) that vary in how often they execute the predicates (p_j) of the CUT.

Case 1	p_1	p_2	p_3	p_4
TC_1	2	1	0	3
TC_2	3	4	2	1

Case 2	p_1	p_2	p_3	p_4
TC_1	1	1	0	2
TC_2	1	2	0	1

individuals where the diversity rate is based on the spread of fitness values [10]. A well-known fitness-based measure is the entropy measure, first proposed by Rosca [23]. The entropy represents the amount of disorder of a population, where an increase in entropy leads to an increase in diversity in the population. Rosca defines diversity based on the entropy as $E(P) = -\sum_k p_k \cdot \log p_k$, where the population P is partitioned according to the fitness value which will result in a proportion of the population p_k that is occupied by the partition k.

In order to determine the influence of the diversity of populations of test cases, a prerequisite is to measure diversity. For this, we adapted three techniques based on the phenotypic and genotypic levels: We measure the phenotypic diversity based on the fitness entropy and test execution traces, and we define a genotypic measurement based on the syntactic representation of test cases.

3.1 Fitness Entropy

The entropy measure adapts the aforementioned principle of fitness entropy. It constructs *buckets* that correspond to the proportions of population that are partitioned based on the fitness values of test cases τ_s in the population μ as:

$$\text{Bucket}(f) \leftarrow |\{\tau_i \mid \text{fitness}(\tau_i) = f\}| \tag{1}$$

where f is the fitness value that partitioning is based on and τ_i is each individual in the population whose fitness value equals to f. In this case, each bucket of fitness holds the number of individuals that are in the same fitness interval (e.g., the interval of fitness values that are similar in the first five decimal points). The entropy is then calculated based on each bucket of fitness as:

$$\text{Entropy} = \sum_{i=1}^{B} \frac{\text{Bucket}_i}{\mu} \cdot \log\left(\frac{\text{Bucket}_i}{\mu}\right) \tag{2}$$

where B is the number of buckets. However, in a multiobjective context, the fitness entropy is applied on a set of buckets that is constructed for each objective, and then all entropies are added up to calculate the overall entropy.

3.2 Predicate Diversity

As the fitness value in test case generation is mainly based on the branch distance (and other similar measurements), there is the potential issue that fitness entropy

```
1  @Test public void test1() {
2    String string0 = "foo";
3    String string1 = "bar";
4    Foo foo0 = new Foo(string0, string1);
5    String string2 = foo0.bar();
6  }
```

```
1  @Test public void test2() {
2    Bar bar0 = new Bar();
3    String string0 = "bar";
4    Foo foo0 = new Foo(bar0, string0);
5    String string1 = foo0.bar();
6    Foo foo1 = new Foo(string1, string0);
7    String string2 = foo1.bar();
8  }
```

Fig. 1. Two automatically generated example test cases to illustrate statement difference.

is dominated by a few statements that achieve the best coverage. For example, the fitness value considers only the minimal branch distance for each branch, but ignores all other executions of the same branch. Therefore, we define an alternative phenotypic diversity measurement that takes more execution details into account. The idea behind this measure is to quantify the diversity of the individuals based on an execution profile of the conditional statements in the class under test. To illustrate this, assume two individuals (TC_1 and TC_2) and a class under test (CUT) that has four conditional statements (p_j). Each individual test case covers each predicate in the CUT as shown in Table 1.

The diversity in this case is measured based on how often each predicate is executed by each individual, e.g., p_3 is covered 2 times by TC_2, while it is not covered by TC_1 as shown in Case 1. Predicate diversity is calculated by counting the number of times each predicate p_j in CUT is covered by each individual TC_i, resulting in vectors V_1 and V_2. The distance between TC_1 and TC_2 is calculated using the Euclidean distance between V_1 and V_2, which is also calculated for each pair of individuals in the population. The use of Euclidean distance as a population diversity measure is shown to be effective in measuring the behavioural diversity, and controlling the evolution process [1,20]. The diversity of the two cases shown in the example are 4.243 for Case 1 and 1.414 for Case 2, which means that the first two test cases are more different than the second two. Therefore, the overall population diversity is the average of all pairwise distances between all individuals, and is calculated as follows:

$$\text{diversity}(P) = \frac{\sum_{i=0}^{|P|} \sum_{j=0, j\neq i}^{|P|} \text{dist}(T_i, T_j)}{|P|\,(|P|-1)} \tag{3}$$

where P is the population of individual test cases and dist is the Euclidean distance between a pair of test cases.

3.3 Statement Diversity

Genotypic diversity aims to measure the structural differences among the individuals of a population. In our case, the genotypes are the sequences of statements. We measure syntactic difference based on the profile of statements, with normalised variable names. This is important since identical statements at different positions of tests will have different variable names. For example, consider the two test cases in Fig. 1: Line 5 in `test1` and Lines 5 and 7 in `test2` are the same except for variable names. To normalise a statement, all variable names are replaced with a placeholder.

To calculate the distance between two test cases `test1` and `test2`, we determine the set of normalised statements contained in both test cases, and then create two vectors representing the number of occurrences of each statement (i.e., for the tests in the example: $V_1 = (2, 2, 1, 1, 0, 0)$ and $V_2 = (1, 1, 1, 2, 1, 2)$, where the 2's in V_2 result from Lines 5 and 7 being counted as the same statement). The distance between two test cases is calculated as Euclidean distance between these two vectors (i.e. 2.82), and the overall diversity is the average distance between all pairs of test cases in the population that is calculated similar to Eq. 3.

4 Maintaining Population Diversity

Techniques for maintaining population diversity are typically classified as niching and non-niching techniques [6,25]. Niching techniques try to divide the population into subpopulations to locate multiple optimal solutions. Non-niching techniques maintain diversity in other ways, for example, by increasing the population size, changing the selection pressure, or applying replacement restrictions.

Fitness Sharing (FS) is the most popular niching technique [24] and was proven to be effective in pseudo-Boolean optimisation [15]. It aims to find multiple peaks in the solution space by defining niches around peaks where individuals share the same resource (i.e., fitness value). The idea is to decrease the value of the resource that is shared by the individuals of a niche when the number of individuals is high, and increase it when there are few individuals in a niche, which gives these individuals higher probability to be selected for next generations. The shared fitness of each individual is

$$f_i^{'} = \frac{f_i}{m_i} \quad \text{with} \quad m_i = \sum_{j=1}^{\mu} sh(d_{ij}) \tag{4}$$

where m_i is the niche count, which is defined by measuring the distance among the individuals: here μ is the population size and d_{ij} is the distance between individual i and individual j (e.g., Euclidean distance [24]). The function sh measures the distance between each two individuals in the population as follows:

$$sh(d_{ij}) = \begin{cases} 1 - (d_{ij}/\sigma_s)^{\alpha}, & \text{if } d < \sigma_s \\ 0, & \text{otherwise} \end{cases} \tag{5}$$

where σ_s is the peak radius (i.e., sharing radius) and α is the parameter that regulates the form of the sharing function, commonly equal to 1.

Since the fitness function is minimised in our case, the division in Eq. (4) is replaced with a multiplication, i.e., $f_i^{'} = f_i \cdot m_i$. This will maximise the shared fitness of the individuals that are dominant in the population and make them less attractive for selection, encouraging other, dissimilar individuals to be selected.

The niche count can be based on any type of distance measurement. In the basic version, the distance is defined as difference between fitness values; for predicate diversity the distance between the predicate execution vectors is used to determine niches, and for statement diversity the distance in statement counts.

Clearing (CL): Clearing is similar to fitness sharing, except that it shares the available resources among the best individuals of each subpopulation rather than all individuals of a subpopulation. That is, it keeps the fitness of the best individuals (i.e., *dominants*) in each subpopulation as they are, and the fitness of the other individuals is cleared (e.g., set to zero). This technique is found to be promising for solving challenging multimodal functions [5] and, in addition, outperforms fitness sharing in dealing with genetic drift [19]. Since we are minimising and the optimal fitness is zero, we set the fitness of cleared individual to a higher fitness value other than zero (e.g., `Integer.MAX_INT`). In order to define a niche, we use the distance between the predicate execution vectors for predicate diversity, and the distance in statement counts for statement diversity. Both fitness sharing and clearing are calculated for each objective, and when the niche count is based on fitness (i.e., fitness-based sharing), the distance is computed based on fitness values of each objective. However, the two techniques are applied in two steps of Algorithm 1: They are applied on the initial population (after line 4), and on the union of parents and offspring population (after line 8).

Diverse Initial Population (DIP): The initial population is known to have an impact on convergence [11,27], and its diversity can potentially enhance the performance of the GA [7]. The initial population is diversified by generating a population of random individuals with a size m larger than the intended population size n, and then selecting the most distant n individuals from the population of m individuals based on a diversity measure [28]. We modify Line 4 in Algorithm 1 to generate random individuals of size m and then select only the most n distant individuals to form P_t.

Adaptive Crossover and Mutation Rates (AR): Adaptively changing mutation and crossover rates [6] based on the diversity level is thought to help avoiding convergence to a local optimum. The crossover probability is increased when diversity is high to allow for more exploitation, whereas the mutation probability is increased when diversity is low to allow for more exploration [12]. The crossover probability is adapted as:

$$P_c = \left[\left(\frac{PD}{PD_{\max}} \cdot (K_2 - K_1)\right) + K_1\right] \tag{6}$$

where K_2 and K_1 define the range of P_c, PD is the current diversity level, and $PD_{\max}$ is the possible maximum diversity level. The mutation probability is adjusted using the following equation:

$$P_m = \frac{PD_{\max} - PD}{PD_{\max}} \cdot K, \tag{7}$$

where K is an upper bound on P_m. Since a variable size representation tends to have multiple different mutation types, i.e., adding, changing, removing statements, the probability of these three operations is adapted by increasing by a random value when the mutation probability increases, and vice versa.

Duplicate Elimination (DE): The purpose of this technique is to remove the similarity between individuals of the population, which has been shown to enhance population diversity and GA performance [4,22]. Two individuals are considered similar when their distance to each other is zero. To ensure enough diversity in the population, the eliminated individual is replaced with a new generated individual. However, the two similar individuals are evaluated, and the one with the best fitness is kept. This technique is applied after generating the offspring, and more specifically on the union of parents and offspring (after line 8 in Algorithm 1).

Diversity-Based Ranking (DR): In the ranking assignment, test cases are selected to form the first non-dominated front based on their objective values such that a test case x is preferred over a test case y if $f_i(x) < f_i(y)$ where $f_i(x)$ denotes the objective score of test case x_i for branch b_i. When two test cases result in a similar lowest fitness value for a given branch b_i, one of them is chosen randomly to be included in the first non-dominated front. Instead of the random selection, we modify the selection to be based on the diversity such that the test case with high distance from other individuals in the population is preferred to be selected. The distance can be based on the predicate execution vectors (predicate-based) or the statement counts (statement-based).

Diversity-Based Selection (DS): Once a rank is assigned to all candidate test cases, the crowding distance is used to make a decision about which test case to select. The basic idea behind the crowing distance is to compute the Euclidean distance between each pair of individuals in a front based on their objective value. In this case, the test cases having a higher distance from the rest of the population are given higher probability of being selected. To investigate the influence of distance-based measures on the selection, we replace the crowding distance with our two measures (i.e., statement-based and predicate-based measures) to calculate the distance between any pair of individuals in each front.

All the previously mentioned techniques can be generalised to any form of Evolutionary Algorithms (EAs), except the last two techniques (i.e., DR and DS) that are designed specifically for MOSA.

5 Empirical Study

The goal of our study is to investigate the evolution of unit tests, and whether maintaining the population diversity during the search has an influence on the performance of GAs. We therefore aim to answer the following research questions:

RQ1: How does population diversity change throughout evolution in MOSA?
RQ2: How effective are diversity maintenance techniques in MOSA?
RQ3: What are the effects of increasing population diversity in MOSA?

For experiments we use EvoSuite [9], which generates JUnit test suites for a given Java CUT and target coverage criterion using different evolutionary algorithms, with MOSA being the most effective algorithm for JUnit test generation [3]. Since open source Java code contains many classes that are either trivially easy to cover, or impossible to cover by EvoSuite, we used the selection of 346 complex classes from the DynaMOSA study [18].

To better understand the influence of the population diversity on the generation of JUnit tests, we conducted an experiment that involves (i) applying each of the three diversity measures defined in Sect. 3 on each CUT to measure the diversity level throughout the evolution, (ii) applying each of the diversity maintaining techniques defined in Sect. 4 on each CUT to promote the diversity throughout the evolution. Each of diversity maintaining techniques is integrated into MOSA where its performance compared to the performance of the default MOSA, i.e., without using diversity techniques. Note that each of the diversity techniques is run separately, and therefore the total number of runs is 8 (i.e., a single MOSA run with each of the 7 techniques and one run of default MOSA).

We used the following values for the parameters of the diversity techniques based on preliminary experiments: The sharing radius $\sigma_s = 0.1$; the number of dominants for clearing is set to 1; the parameters of the diverse initial population are $m = 80$ and $n = 50$; the parameters of adaptive mutation rate are $K_1 = 0.6$ and $K_2 = 0.8$ and the adaptive crossover rate was set to $K = 0.8$.

5.1 RQ1—How Does Population Diversity Change Throughout Evolution in MOSA?

A first step towards understanding the influence of population diversity on the evolution is to measure how the diversity is changed throughout the evolution. For that, we measure the diversity in MOSA to get an idea of whether MOSA is able to maintain a high level of diversity during the search. However, as different CUTs result in different patterns of coverage during the evolution, it is necessary to look at the diversity based on the different coverage patterns.

Past experiments with EvoSuite considered search durations of 1–2 minutes; in order to study the effects on convergence we use a substantially larger search budget. We first performed runs of 30 min to observe the development of coverage (the best coverage achieved in the current population) over a long period of time and we used this data to classify CUTs into four disjoint groups as follows:

- **Evolving** contains CUTs where the coverage after 30 min is higher (by more than 0.01) than after 10 min (93 CUTs).
- **Flat** is the set of CUTs where the coverage never changes (60 CUTs).
- **Stagnating** contains CUTs for which the coverage stagnates after 10 min: it is higher than after 2 min, but increases by less than 0.01 from 10 min to 30 min (43 CUTs).
- **Plateauing** contains the remaining CUTs for which the coverage after two minutes is constant (115 CUTs).

We then started 30 runs for each CUT with a search budget of 10 min and applied our proposed measures defined in Sect. 3 to measure the diversity level for each of the four groups, as shown in Fig. 2. It is obvious that the diversity behaviour is different among the four groups, and the difference can be seen with the three diversity measures. For the *evolving* group, coverage keeps growing throughout the entire 10 min, and this group also shows a continuous growth of entropy and phenotype diversity. In terms of genotype diversity, there is a reduction after an initial sharp growth phase, but less than in all other groups. For the *flat* group, the phenotype diversity is overall lowest; this is because only very few predicates are covered in the first place, as shown in the coverage plot, and the low entropy. For the *stagnating* group, once the coverage increase slows down, all three diversity measurements go down as well. For the *plateauing* group, once the search converges all diversity measurements drop sharply, and notably the genotypic diversity is lowest of all groups. Overall it seems that, as long as coverage grows, MOSA does well at maintaining diversity. Once coverage stagnates, the population loses diversity. To some extent, this can be explained by EvoSuite's ranking mechanism: If two individuals have the same fitness value, then the shorter of the two is preferred; this is also used in MOSA's rank-based preference sorting. Shorter individuals by construction will have less diversity.

> ***RQ1****: MOSA maintains high diversity while coverage increases, but diversity drops once a maximum coverage has been reached.*

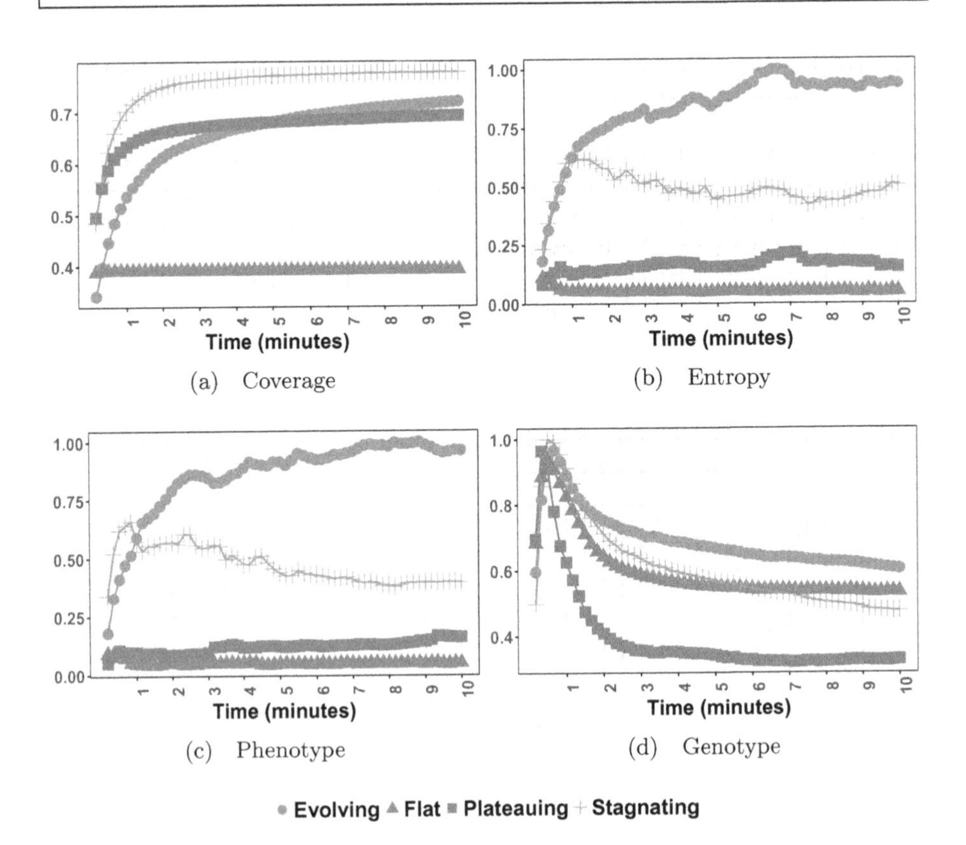

(a) Coverage (b) Entropy

(c) Phenotype (d) Genotype

Fig. 2. Average values for coverage and population diversity throughout evolution in MOSA for the four groups of CUTs.

5.2 RQ2—How Effective Are Diversity Maintenance Techniques in MOSA?

In RQ1 we saw a general tendency that diversity drops once the coverage stops increasing. We therefore would like to see which diversity maintenance techniques succeed at increasing the population diversity during evolution. For that, we apply the techniques mentioned in Sect. 4 on all classes.

Table 2. Average diversity over time when applying diversity maintenance techniques based on different distance measures

Technique	Phenotype	Genotype	Entropy
MOSA	0.7206	0.6825	0.6761
AR (predicate)	0.5719	0.7854	0.9126
AR (statement)	0.6708	0.7702	0.7005
CL (predicate)	0.8740	0.9470	0.8769
CL (statement)	0.8388	0.9075	0.8503
DE (predicate)	0.7696	0.9350	0.8675
DE (statement)	0.7020	0.9051	0.8095
DIP (predicate)	0.7675	0.6593	0.6976
DIP (statement)	0.7179	0.6851	0.7084
DR (predicate)	0.7996	0.7749	0.7387
DR (statement)	0.6142	0.7775	0.6798
DS (predicate)	0.8737	0.9331	0.7422
DS (statement)	0.8368	0.9053	0.8543
FS (fitness)	0.8893	0.7402	0.9534
FS (predicate)	0.8877	0.9022	0.9353
FS (statement)	0.8881	0.9364	0.8836

Table 2 summarises the results of the average diversity among the four groups with each technique based on different distance measures (i.e., predicate, statement, and fitness). We can clearly see that the three diversity measures indicate that many of the diversity maintenance techniques are able to promote diversity higher than MOSA. The entropy measure indicates that diversity is the lowest with MOSA compared to all other techniques although the difference is negligible when compared to the statement-based Diversity-based Selection (DS), and fitness-based fitness sharing increases entropy the most. A similar trend can be observed for phenotypic diversity, where fitness-based fitness sharing results in the overall highest diversity. In contrast to entropy, not all techniques succeed in increasing phenotypic diversity; for example, predicate-based Adaptive Rates (AR) results in the lowest diversity. The genotype measure indicates a wide range of effectiveness, with Clearing (CL) based on predicate distance and the statement-based fitness sharing as the most successful in promoting genotypic diversity. However, several other techniques lead to a reduction on genotypic diversity, surprisingly in particular Diverse Initial Population (DIP).

> ***RQ2:*** *Most diversity maintenance techniques succeed at increasing diversity, but there are exceptions. Fitness sharing achieves the most consistent increase.*

5.3 RQ3—What Are the Effects of Increasing Population Diversity in MOSA?

To investigate the impact of diversity on the performance of MOSA, we look at the achieved coverage and the average size of the individuals in the population throughout the evolution. We focus on fitness sharing (using all three diversity metrics), since RQ2 suggested that this is the most effective technique to increase diversity. In addition to the naive application of fitness sharing at all times, we also consider an *adaptive* version, where we apply fitness sharing only when diversity drops below a certain threshold, and once the diversity level exceeds the threshold, the diversity technique is not applied. We empirically determined a threshold of 60%. Figure 3 shows the results of the best coverage in the population and the average length of all test cases in the population for MOSA with and without fitness sharing (FS (fitness/predicate/statement)) during the evolution.

For the *flat* group, there is a small difference in coverage as MOSA and fitness-based fitness sharing result in slightly higher coverage than the other techniques. The length plot shows how MOSA removes all redundancy from the population, while adding diversity leads to larger individuals. In particular, genotype-based fitness sharing has quite dramatic effects on size. Fitness-based sharing (adaptive and non-adaptive) has the smallest effects on size. For the *evolving* group, MOSA achieves the highest coverage, while maintaining a somewhat constant population size. Generally, fitness sharing slightly increases size, quite dramatically so for genotype-based fitness sharing. Adaptive fitness-based sharing even leads to smaller individuals than MOSA. Non-adaptive fitness sharing using genotype and phenotype diversity leads to a notably lower coverage. For the *stagnating* group, the non-adaptive fitness sharing using genotype and phenotype diversity again lead to a notably lower coverage and larger size. This time, however, adaptive fitness-based sharing consistently leads to a higher average coverage than MOSA, and even smaller individuals than MOSA. The *plateauing* group shows similar results to the *stagnating* group, with larger coverage improvement of adaptive phenotype-based sharing. While genotype and phenotype based non-adaptive sharing again lead to lower coverage initially, in this group the size remains large but constant, and the coverage catches up and even overtakes MOSA in the end.

Adaptive fitness sharing leads to higher coverage when the search in MOSA stagnates (e.g., when coverage does not increase). To see whether adaptive fitness sharing is always beneficial, Table 3 shows the number of classes where it

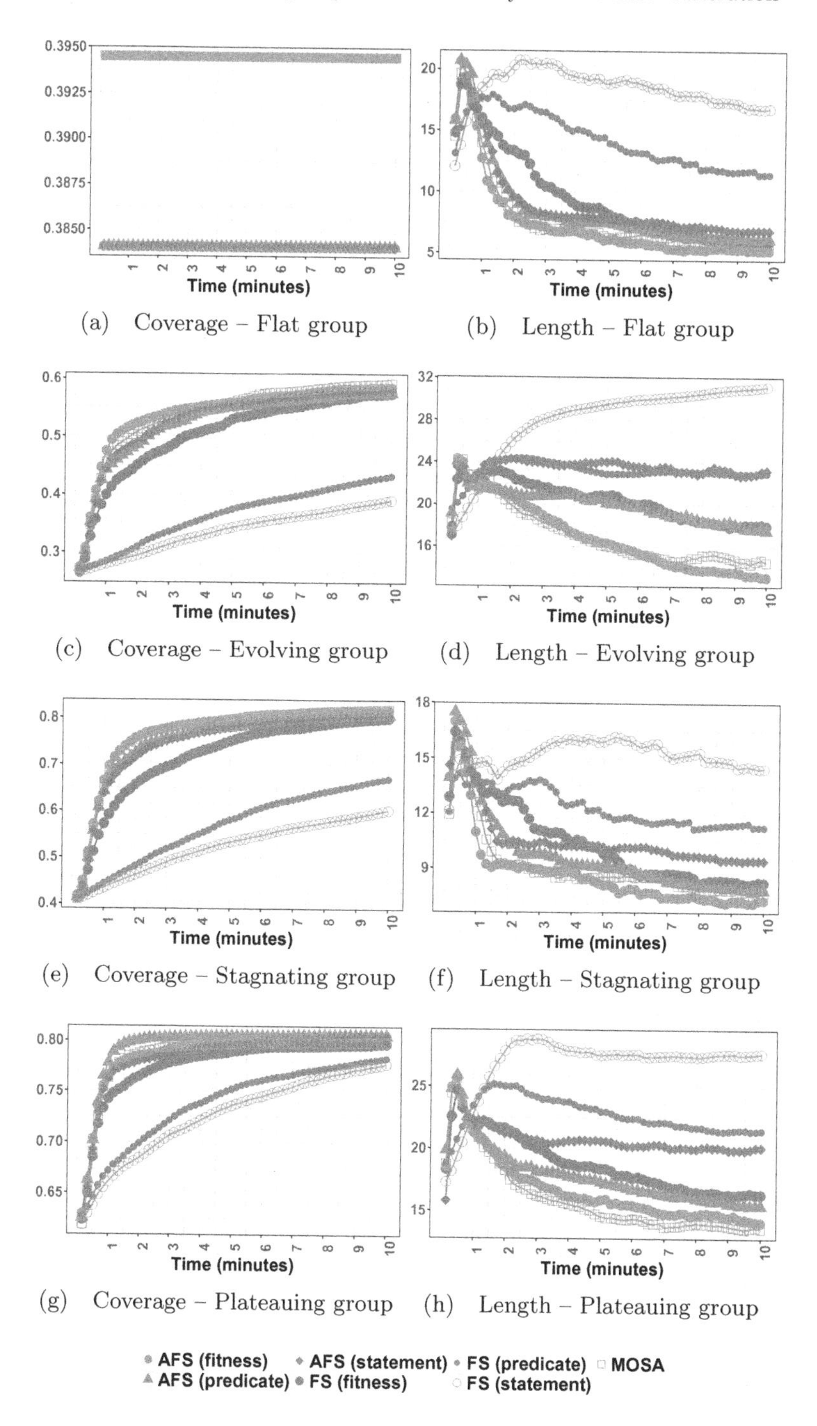

Fig. 3. Coverage and length over time with MOSA, fitness sharing (FS), and adaptive fitness sharing (AFS) per four groups of CUTs

Table 3. Number of classes where adaptive fitness sharing has an increased/decreased/equal coverage compared to MOSA, the average effect size $\hat{A}_{12}$ and the number of classes for which this comparison is statistically significant ($\alpha = 0.05$).

Technique	Increased coverage			Decreased coverage			Equal
	#classes	#sig.	$\hat{A}_{12}$	#classes	#sig.	$\hat{A}_{12}$	#classes
Fitness	99	6	0.52	87	27	0.48	125
Predicate	52	2	0.51	141	39	0.48	118
Statement	77	4	0.51	116	33	0.49	118

increases, decreases, or results in equal coverage with MOSA. Adaptive fitness-based sharing (AFS-fitness) increases coverage on 99 classes and decreases it on 87 classes, albeit having only 6 significant increases as opposed to 27 significant decreases. For adaptive sharing based on predicate (AFS-predicate) and statement (AFS-statement) differences we found more decreases than increases.

The negative effect of diversity on coverage can be explained by the increase in length: Execution of longer tests takes more time, thus slowing down the evolution. Within the time limit of 10 min, MOSA executed 629 generations on average. Always applying fitness sharing decreases the average number of generations (457, 344, and 339 generations with FS-fitness, FS-predicate, and FS-statement respectively), but these figures improve when using the adaptive approach (631, 538, and 519 generations with AFS-fitness, AFS-predicate and AFS-statement, respectively). Consequently, applying adaptive fitness sharing *can* be beneficial, but the question of when it does so is more subtle. Identifying the context in which it can help will be the focus of future research.

RQ3: *Promoting diversity generally leads to larger tests and reduces coverage. However, when coverage stops growing, adding diversity may improve MOSA.*

6 Related Work

There have been several studies that considered diversity when generating test cases [8,16]; however, the aim is to increase the diversity of the tests within the final test suite, rather than the individuals in the search population. More recently, Vogel et al. [28] studied the population diversity when evolving tests for mobile applications where they used a distance-based genotypic measure to quantify the diversity of the population. Applying four diversity maintenance techniques on a multi-objective algorithm shows that diversity does not have an effect on coverage, but it finds more faults. However, the negative impact of increasing diversity on the length is also noticed in previous studies [2,28].

7 Conclusions

Measuring the diversity of generated unit test cases based on entropy, genotypic, and phenotypic levels suggest that diversity maintenance techniques are very effective at promoting diversity throughout the evolution. Looking at their effect on the performance of MOSA, we see that increasing diversity leads to reduced coverage, and a possible increase in the length. However, preliminary results of the adaptive approach suggest that adaptive fitness sharing with a fitness distance metric has a positive impact on coverage on CUTs on which standard MOSA tends to stagnate or plateau during early stages of the evolution. Further work is required to better understand the reasons behind this effect.

References

1. Adra, S.F., Fleming, P.J.: Diversity management in evolutionary many-objective optimization. IEEE Trans. Evol. Comput. **15**(2), 183–195 (2010)
2. Albunian, N.M.: Diversity in search-based unit test suite generation. In: Menzies, T., Petke, J. (eds.) SSBSE 2017. LNCS, vol. 10452, pp. 183–189. Springer, Cham (2017). https://doi.org/10.1007/978-3-319-66299-2_17
3. Campos, J., Ge, Y., Albunian, N., Fraser, G., Eler, M., Arcuri, A.: An empirical evaluation of evolutionary algorithms for unit test suite generation. Inf. Softw. Technol. **104**, 207–235 (2018)
4. Chaiyaratana, N., Piroonratana, T., Sangkawelert, N.: Effects of diversity control in single-objective and multi-objective genetic algorithms. J. Heuristics **13**(1), 1–34 (2007)
5. Covantes Osuna, E., Sudholt, D.: On the runtime analysis of the clearing diversity-preserving mechanism. Evol. Comput. **27**, 403–433 (2019)
6. Črepinšek, M., Liu, S.H., Mernik, M.: Exploration and exploitation in evolutionary algorithms: a survey. ACM Comput. Surv. (CSUR) **45**(3), 1–33 (2013)
7. Diaz-Gomez, P.A., Hougen, D.F.: Empirical study: initial population diversity and genetic algorithm performance. In: Proceedings of AIPR 2007, pp. 334–341 (2007)
8. Feldt, R., Poulding, S., Clark, D., Yoo, S.: Test set diameter: quantifying the diversity of sets of test cases. In: Proceedings of ICST (2016), pp. 223–233. IEEE (2016)
9. Fraser, G., Arcuri, A.: Whole test suite generation. IEEE Trans. Softw. Eng. **39**(2), 276–291 (2013)
10. Jackson, D.: Promoting phenotypic diversity in genetic programming. In: Schaefer, R., Cotta, C., Kołodziej, J., Rudolph, G. (eds.) PPSN 2010. LNCS, vol. 6239, pp. 472–481. Springer, Heidelberg (2010). https://doi.org/10.1007/978-3-642-15871-1_48
11. Maaranen, H., Miettinen, K., Penttinen, A.: On initial populations of a genetic algorithm for continuous optimization problems. J. Glob. Optim. **37**(3), 405 (2007)
12. Mc Ginley, B., Maher, J., O'Riordan, C., Morgan, F.: Maintaining healthy population diversity using adaptive crossover, mutation, and selection. IEEE Trans. Evol. Comput. **15**(5), 692–714 (2011)
13. McMinn, P.: Search-based software test data generation: a survey. Softw. Test. Verif. Reliab. **14**(2), 105–156 (2004)
14. McPhee, N.F., Hopper, N.J.: AppGP: an alternative structural representation for GP. In: Proceedings of CEC 1999, vol. 2, pp. 1377–1383. IEEE (1999)
15. Oliveto, P.S., Sudholt, D., Zarges, C.: On the benefits and risks of using fitness sharing for multimodal optimisation. Theor. Comput. Sci. **773**, 53–70 (2019)

16. Palomba, F., Panichella, A., Zaidman, A., Oliveto, R., De Lucia, A.: Automatic test case generation: what if test code quality matters? In: Proceedings of ISSTA (2016), pp. 130–141 (2016)
17. Panichella, A., Kifetew, F.M., Tonella, P.: Reformulating branch coverage as a many-objective optimization problem. In: Proceedings of ICST 2015, pp. 1–10. IEEE (2015)
18. Panichella, A., Kifetew, F.M., Tonella, P.: Automated test case generation as a many-objective optimisation problem with dynamic selection of the targets. IEEE Trans. Softw. Eng. **44**(2), 122–158 (2017)
19. Pétrowski, A.: A clearing procedure as a niching method for genetic algorithms. In: Proceedings of ICEC 1996, pp. 798–803. IEEE (1996)
20. Robič, T., Filipič, B.: DEMO: differential evolution for multiobjective optimization. In: Coello Coello, C.A., Hernández Aguirre, A., Zitzler, E. (eds.) EMO 2005. LNCS, vol. 3410, pp. 520–533. Springer, Heidelberg (2005). https://doi.org/10.1007/978-3-540-31880-4_36
21. Rojas, J.M., Campos, J., Vivanti, M., Fraser, G., Arcuri, A.: Combining multiple coverage criteria in search-based unit test generation. In: Barros, M., Labiche, Y. (eds.) SSBSE 2015. LNCS, vol. 9275, pp. 93–108. Springer, Cham (2015). https://doi.org/10.1007/978-3-319-22183-0_7
22. Ronald, S.: Duplicate genotypes in a genetic algorithm. In: Proceedings of CEC/WCCI 1998, pp. 793–798. IEEE (1998)
23. Rosca, J.P.: Entropy-driven adaptive representation. In: Proceedings of Workshop on Genetic Programming: From Theory to Real-world Applications, vol. 9, pp. 23–32 (1995)
24. Sareni, B., Krahenbuhl, L.: Fitness sharing and niching methods revisited. IEEE Trans. Evol. Comput. **2**(3), 97–106 (1998)
25. Shir, O.M.: Niching in evolutionary algorithms. In: Rozenberg, G., Bäck, T., Kok, J.N. (eds.) Handbook of Natural Computing, pp. 1035–1069. Springer, Heidelberg (2012). https://doi.org/10.1007/978-3-540-92910-9_32
26. Squillero, G., Tonda, A.: Divergence of character and premature convergence: a survey of methodologies for promoting diversity in evolutionary optimization. Inf. Sci. **329**, 782–799 (2016)
27. Toğan, V., Daloğlu, A.T.: An improved genetic algorithm with initial population strategy and self-adaptive member grouping. Comput. Struct. **86**(11–12), 1204–1218 (2008)
28. Vogel, T., Tran, C., Grunske, L.: Does diversity improve the test suite generation for mobile applications? In: Nejati, S., Gay, G. (eds.) SSBSE 2019. LNCS, vol. 11664, pp. 58–74. Springer, Cham (2019). https://doi.org/10.1007/978-3-030-27455-9_5

New Ideas and Emerging Results

Search@Home: A Commercial Off-the-Shelf Environment for Investigating Optimization Problems

Erik M. Fredericks[1(✉)] and Jared M. Moore[2]

[1] Department of Computer Science and Engineering, Oakland University, Rochester, MI 48309, USA
fredericks@oakland.edu

[2] School of Computing and Information Systems, Grand Valley State University, Allendale, MI 49401, USA
moorejar@gvsu.edu

Abstract. Search heuristics, particularly those that are evaluation-driven (e.g., evolutionary computation), are often performed in simulation, enabling exploration of large solution spaces. Yet simulation may not truly replicate real-world conditions. However, search heuristics have been proven to be successful when executed in real-world constrained environments that limit searching ability even with broad solution spaces. Moreover, searching *in situ* provides the added benefit of exposing the search heuristic to the exact conditions and uncertainties that the deployed application will face. Software engineering problems can benefit from *in situ* search via instantiation and analysis in real-world environments. This paper introduces *Search@Home*, an environment comprising heterogeneous commercial off-the-shelf devices enabling rapid prototyping of optimization strategies for real-world problems.

Keywords: Real-world systems · Evolutionary search · In-situ search · Search-based software engineering

1 Introduction

Commercial off-the-shelf (COTS) microcomputers (e.g., Raspberry Pi, Arduino, BeagleBone, etc.[1]) are democratizing access to many-core distributed computing environments. Such devices are generally constrained in terms of available processing power and hard drive space, limiting their ability to individually carry out complicated tasks quickly. Moreover, such devices will generally consume far less power than a typical device used for simulation (e.g., server blade). We posit that a microcomputing environment is therefore beneficial in terms of rapidly prototyping search heuristics, in real-world conditions, yet may require

[1] See https://www.raspberrypi.org/, https://www.arduino.cc/, and https://beagleboard.org, respectively.

A. Aleti and A. Panichella (Eds.): SSBSE 2020, LNCS 12420, pp. 171–177, 2020.
https://doi.org/10.1007/978-3-030-59762-7_12

additional time to complete complex computing tasks. Specifically, we highlight search-based software engineering (SBSE) as an attractive domain for real-world search.

Optimization performed in real-world situations allows the system to analyze relevant information from data specific to its operating context (i.e., combination of system and environmental parameters) rather than simulating such parameters [2]. Self-adaptation has been applied to cyber-physical systems enabling reconfiguration at run time in response to uncertainty, including those systems considered to be safety critical [8]. Online evolutionary optimization has been previously performed in fields such as robotics, where a (1 + 1) evolutionary strategy searches for optimal neural network configurations [3]. Given the difficulties in performing speculative, evaluation-driven optimizations at run time, continuous optimization methods such as Markov chains and Bayesian methods have also been applied [4,10]. Regardless of the method, an online optimization strategy must consider the implications of updating an executing system within its production environment.

This paper introduces and demonstrates *Search@Home*, a framework for quickly prototyping in-place search-based software engineering (SBSE) techniques using COTS hardware. *Search@Home* is intended to provide a low-cost testbed that can be deployed in a target environment to rapidly prove out online search heuristics (e.g., run-time requirements monitoring/optimization). While a longer evaluation time is to be expected with low-power hardware, the benefits of performing search in a real-world environment far outweigh the speed gains of using a simulation that may be misconfigured or inaccurately describe the environment-to-be. We next describe background and related work, demonstrate *Search@Home* on a proof-of-concept optimization problem, and outline future experiments.

2 Real-World Systems

This section presents relevant information on microcomputers and how SBSE techniques can be effective within constrained environments

Microcomputers: Consumer-grade microcomputers have been widely propagated at inexpensive price points with the recent explosion of interest in Maker-related topics (e.g., hacking home electronics, 3D printing, etc.), many of which are targeted at STEM education and non-production projects. Moving from simulation to reality in constrained environments requires addressing concerns in:

- ***Power***: Beyond power optimization, microcomputers often supplied by sub-standard power cables, resulting in *undervoltage* (i.e., slow/erratic behavior).
- ***Temperature***: Heat can negatively impact devices without active (i.e., fans) or passive (i.e., heat sinks) cooling, leading to CPU throttling.
- ***Memory***: Microcomputers are constrained with the amount of available memory for handling computing tasks. While the newest Raspberry Pi (4B)

has up to 8 GB of available RAM, older models have significantly less memory. Edge devices exist specifically for heavy-duty computing (e.g., Google Coral[2]), however they are generally used for only specialized purposes.
- ***Disk space***: Depending on the device, permanent storage (e.g., EEPROM, ROM, flash memory, etc.) is often at a premium in terms of availability. Devices such as the Arduino and BeagleBone rely on programmable memory space for long-term storage, whereas the Raspberry Pi uses a microSD card for its storage.
- ***Timing constraints***: While the devices used in this paper do not use real-time operating systems, timing constraints must be explicitly handled by the engineer, else faults can occur when software deadlines are violated.

Online/Hardware-Based Optimization: One common theme across optimization algorithms is that there exists no "free lunch" as there are always limiting factors in the application or environment [13]. Limitations for online optimization will be readily-noticeable in run time, memory overhead, etc. In constrained systems these impacts are exceedingly noticeable given their lower operating capabilities. Care must be taken when performing optimization in constrained systems.

In situ optimization has been applied to wireless sensor network applications, where run-time reconfiguration and programming models enable optimization [11]. Li *et al.* minimized power consumption of test suites in low power systems using an integrated linear programming approach [6]. Continuous optimization is a common feature in other domains as well. Wang and Boyd applied an online optimization technique (a primal barrier method) to model-predictive control applications [12]. Mars and Hundt combined static and dynamic optimization strategies to direct online reconfiguration based on scenarios [7]. Optimization has also been applied online in a data-driven capacity, where uncertainty models can inform decision making [2].

Search heuristics have also been deployed *in silica*. Genetic algorithms (GA) have been deployed to field-programmable gate arrays (FPGA) for hardware-based optimization [9], enabling rapid prototyping of fast, low-power search. Energy management is another concern for metaheuristics, specifically those involving smart power grids [5]. With additional hardware modules, *Search@Home* can be extended to use *in silica* search and act instead as a controller, and moreover, monitor energy consumption with appropriate sensors.

3 *Search@Home* Overview

For the purposes of this paper (and its initial implementation), *Search@Home* was implemented as an IoT environment comprising two Raspberry Pi 3B devices, a Raspberry Pi 4B (4 GB model), a spare 2010-era netbook (acting as a point of entry to the network), and a wireless router (flashed with Tomato

[2] See https://www.coral.ai/.

firmware) to provide a sandboxed network. Devices can be interchangeable – as long as the device can connect to the network it can participate in the environment. For example, an Arduino Duo could be included as long as an appropriate communication channel to the rest of the network is established (i.e., a WiFi shield is installed).

We applied a string optimization problem demonstrating the feasibility of deploying a search algorithm to a heterogeneous collection of devices. The algorithm was developed with Python 3.7 and each Raspberry Pi used Raspbian as its operating system. Given that Raspbian is a full Linux-based operating system, libraries such as `DEAP` and `DisPy` can facilitate development of multiple search heuristics or distributed cluster computing, respectively.[3] To allow for replication, we have made our parts list, source code, and results publicly available on GitHub.[4]

String Search Configuration: The string search application leverages a standard GA to search for a given string, where this particular GA was released as a GitHub.[5] To increase the difficulty of this task, we specified a complex string (`OPT`) comprising `ASCII` characters sampled between indices 32 and 64 and a total length of 84.[6] `OPT` is defined in Eq. 1:

$$\begin{aligned} OPT = {}&'12th\ Symposium\ for\ Search - Based\ Software \\ &Engineering \mid http://ssbse2020.di.uniba.it/' \end{aligned} \tag{1}$$

The fitness function defined for this study is defined as follows in Eq. 2, where `ord` represents a character's Unicode integer value:

$$ff_{string} = \sum_{i=0}^{len(OPT)} |ord(TARGET[i]) - ord(OPT[i])| \tag{2}$$

This particular GA uses single-point crossover, single-point mutation, and a weighted-fitness selection function (i.e., fitness directly correlated to the probability that an individual is selected). Mutation is automatically applied to the children generated by the crossover operation, where a random gene is modified. The GA was configured to run for a maximum of 500,000 generations, with a population size of 500, a crossover rate of 0.5, and a mutation rate of 1.0. The GA can converge early if the correct string is discovered. We performed 25 experimental replicates to ensure statistical significance, using the Wilcoxon-Mann-Whitney U-test with a significance level of $p < 0.001$. We compared the GA to random search as specified by Arcuri *et al.* [1].

For this feasibility study, we are interested in the amount of time required before convergence to the expected solution and the number of generations necessary to reach that value, as the number of generations is set very high to ensure

[3] See https://deap.readthedocs.io/ and http://dispy.sourceforge.net/, respectively.

[4] See https://github.com/efredericks/SearchAtHome.

[5] See https://gist.github.com/josephmisiti/940cee03c97f031188ba7eac74d03a4f.

[6] There exist $2.7 * 10^{126}$ possible combinations based on string length and characters.

convergence. To focus on execution time, we intentionally did not introduce parallelism or distributed processing, however such a procedure can be used on a microcomputer (e.g., Python's `multiprocessing` package). Figure 1(a) compares the number of generations required for the algorithm to converge and Fig. 1(b) compares the amount of time (seconds) required to reach convergence between a current-generation laptop[7] and the Raspberry Pis. As can be seen from these figures, there exists no difference in the number of generations required to converge to the optimal solution. However, a significant difference exists between the execution times required for convergence between each device ($p < 0.001$). Moreover, each experimental replicate resulted in convergence. These results are expected given the disparity in processing capability. However, the Raspberry Pi was able to successfully execute optimization in all cases within a reasonable amount of time, proving feasibility of optimization on constrained devices.

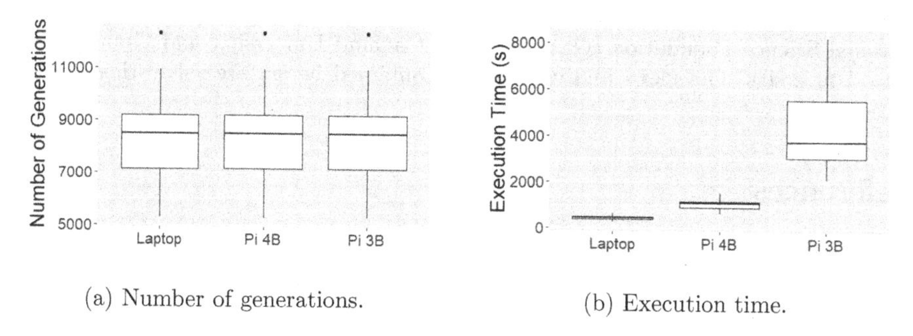

(a) Number of generations.

(b) Execution time.

Fig. 1. Comparison of string search results between laptop and Raspberry Pis.

SBSE Implications: We now highlight three future experiments for *in situ* optimization research.

1. Distributed processing of search algorithms
2. Power concerns resulting from search
3. Implications of SBSE in production environments

Item (1) can be an interesting study of offloading SBSE tasks (e.g., requirements monitoring, fitness evaluation, etc.) to distributed nodes. Item (2) demonstrates the implications of modeling power consumption as a first-class citizen in a software model (e.g., non-functional requirements). Item (3) uses optimization *in situ* to investigate the interaction of search, software artifacts, and expressed behaviors.

4 Discussion

This paper presents *Search@Home*, an open-source framework for enabling *in situ* SBSE research within constrained environments. *Search@Home* uses

[7] Intel Core i7 quad-core 2.8 GHz 64-bit processor, 16 GB RAM, 1 TB hard drive space.

inexpensive COTS hardware providing an environment in which students and researchers can quickly and effectively prototype and deploy applications that would benefit from using real-world data to support an online search procedure. For this paper, we targeted evolutionary computation, however extension to other optimization domains (e.g., continuous optimization) is feasible as well.

We demonstrate the effectiveness of *Search@Home* on a string search exemplar to demonstrate basic search feasibility, where optimal solutions are discovered in a reasonable amount of time. Future research directions for this project include incorporation of low-cost robotics environments (e.g., iRobot Roomba, Turtlebot, Lego Mindstorms, etc.), usage of a compute cluster (e.g., Beowulf cluster, high-performance compute cluster, etc.), and incorporation of cloud technologies offsetting the cost of heavy evaluations (e.g., Google Cloud Functions, Amazon Web Services Lambda functions, etc.).

Acknowledgements. This work has been supported in part by grants from the National Science Foundation (CNS-1657061), Oakland University, and Grand Valley State University. The views and conclusions contained herein are solely those of the authors.

References

1. Arcuri, A., Briand, L.: A practical guide for using statistical tests to assess randomized algorithms in software engineering. In: Proceedings of the 33rd International Conference on Software Engineering, ICSE 2011, pp. 1–10. ACM (2011)
2. Bertsimas, D., Thiele, A.: Robust and data-driven optimization: modern decision making under uncertainty. In: Models, Methods, and Applications for Innovative Decision Making, pp. 95–122. INFORMS (2006)
3. Bredeche, N., Haasdijk, E., Eiben, A.E.: On-line, on-board evolution of robot controllers. In: Collet, P., Monmarché, N., Legrand, P., Schoenauer, M., Lutton, E. (eds.) EA 2009. LNCS, vol. 5975, pp. 110–121. Springer, Heidelberg (2010). https://doi.org/10.1007/978-3-642-14156-0_10
4. Calinescu, R., Grunske, L., Kwiatkowska, M., Mirandola, R., Tamburrelli, G.: Dynamic QoS management and optimization in service-based systems. IEEE Trans. Softw. Eng. **37**(3), 387–409 (2010)
5. Lezama, F., Soares, J., Vale, Z.: A platform for testing the performance of metaheuristics solving the energy resource management problem in smart grids. Energy Inform. **1**(1), 35 (2018)
6. Li, D., Jin, Y., Sahin, C., Clause, J., Halfond, W.G.: Integrated energy-directed test suite optimization. In: Proceedings of the 2014 International Symposium on Software Testing and Analysis, pp. 339–350 (2014)
7. Mars, J., Hundt, R.: Scenario based optimization: a framework for statically enabling online optimizations. In: 2009 International Symposium on Code Generation and Optimization, pp. 169–179. IEEE (2009)
8. Muccini, H., Sharaf, M., Weyns, D.: Self-adaptation for cyber-physical systems: a systematic literature review. In: Proceedings of the 11th International Symposium on Software Engineering for Adaptive and Self-Managing Systems, pp. 75–81 (2016)
9. Peker, M.: A fully customizable hardware implementation for general purpose genetic algorithms. Appl. Soft Comput. **62**, 1066–1076 (2018)

10. Shahriari, B., Swersky, K., Wang, Z., Adams, R.P., De Freitas, N.: Taking the human out of the loop: a review of Bayesian optimization. Proc. IEEE **104**(1), 148–175 (2015)
11. Taherkordi, A., Loiret, F., Rouvoy, R., Eliassen, F.: Optimizing sensor network reprogramming via in situ reconfigurable components. ACM Trans. Sens. Netw. (TOSN) **9**(2), 1–33 (2013)
12. Wang, Y., Boyd, S.: Fast model predictive control using online optimization. IEEE Trans. Control. Syst. Technol. **18**(2), 267–278 (2009)
13. Wolpert, D.H., Macready, W.G.: No free lunch theorems for optimization. IEEE Trans. Evol. Comput. **1**(1), 67–82 (1997)

Replications and Negative Results

Exploring the Use of Genetic Algorithm Clustering for Mobile App Categorisation

Afnan A. Al-Subaihin[1(✉)] and Federica Sarro[2]

[1] CCIS, King Saud University, Riyadh, Saudi Arabia
aalsubaihin@ksu.edu.sa
[2] University College London, London, UK
f.sarro@ucl.ac.uk

Abstract. Search-based approaches have been successfully used as clustering algorithms in several domains. However, little research has looked into their effectiveness for clustering tasks commonly faced in Software Engineering (SE). This short replication paper presents a preliminary investigation on the use of Genetic Algorithm (GA) to the problem of mobile application categorisation. Our results show the feasibility of GA-based clustering for this task, which we hope will foster new avenues for Search-Based Software Engineering (SBSE) research in this area.

Keywords: Software clustering · Mobile applications · Replication study

1 Introduction

Automatic software categorisation is an ongoing research problem which aims to find similarity among software artefacts [4]. Such similarity can aid, for example, in detecting malicious software [8], requirements discovery [3,19] and mining similar software behaviour and analytics [2,14,18]. Mobile applications (apps) present a particularly interesting domain since the app store categorisation has been deemed unsuitable and alternative automated categorisations are advocated [2,8]. Furthermore, app stores boast various metadata that can be leveraged for this task [9,12].

Al-Subaihin et al. [1,2] have shown that categorising mobile apps according to their functionalities can provide a better categorisation than the current app store ones. However, their work focuses on comparing the effectiveness of various feature extraction techniques from textual corpora, and thus only uses one clustering algorithm: hierarchical clustering.

In this paper, we carry out a partial replication of the original study [1] to investigate whether the results can be improved using a Genetic Algorithm-based clustering algorithm, as evolutionary approaches were shown to be successful as clustering techniques in other application domains [10]. Specifically, we investigate four of the five research questions posed in the original study, but we shift the focus on the clustering approach rather than the feature extraction method.

A. Aleti and A. Panichella (Eds.): SSBSE 2020, LNCS 12420, pp. 181–187, 2020.
https://doi.org/10.1007/978-3-030-59762-7_13

Firstly, as a sanity check, we measure the degree of difference between the two clustering solutions. Then we report the degree of improvement on the original app store clustering. We also report partial results of investigating the best value of K for the GA-clustering algorithm (GAC) and, finally, we report the efficiency of using the GA-clustering technique compared to hierarchical clustering. Our results reveal that using a GAC produces significantly better clustering of mobile applications than those observed in the app store categorisation, and those produced by hierarchical clustering. However, GAC fails to surpass the quality of the hierarchical clustering solution at higher K values. To the best of our knowledge, our study is the first to explore the viability of GA-based clustering solutions for mobile app categorisation, and, more in general, few studies have investigated search-based clustering in SE research (e.g., [5,6,11,15]). This study shows the viability of search-based clustering solutions for SE tasks, and we hope can open further avenues for SBSE research.

2 Replication Study Design

The original study extracted features from mobile apps descriptions using four different techniques and compared their effectiveness for clustering apps by using these features and only one clustering approach, i.e. hierarchical clustering [1]. The study found that extracting features using Latent Dirichlet Allocation (LDA) consistently performs well among the investigated feature extraction techniques. Therefore, this replication uses LDA as a feature extraction technique, investigates the effectiveness of GA as a clustering technique, and compares it to the LDA-based hierarchical clustering results as reported in the original study. In the following, we report further details on the empirical study design.

2.1 Research Questions

We investigate four of the five research questions from the original study.

RQ 0. How similar is the GA-based clustering solution to its hierarchical counterpart? Investigating how similar GAC results are to the hierarchical clustering solution is a sanity check before proceeding further in this study. If the results are identical or very similar, there is no value in investigating GAC further. As in the original study, the similarity is measured by using the Jaccard index, which is a commonly used measure for the agreement of two partitions. Jaccard index ranges from 0 (complete dissimilarity) to 1 (identical).

RQ 1. Can GAC improve on current app store categorisation and hierarchical clustering? This research question compares the quality of the GAC results to the original app store, which has 24 categories (K = 24). The answer will confirm whether the use of GAC can actually improve the status quo (app store categorisation) and the state-of-the-art (original study). As done in the original study, we measure the quality of the clustering solutions using the Silhouette score [17]. A Silhouette score is assigned to each data point (i.e. app) in the dataset based on its similarity to the apps in the same cluster and its dissimilarity to apps in other clusters. The Silhouette score of an entire clustering solution is the mean scores of all data points in the dataset, and it ranges from −1 (complete mis-assignments) to 1 (perfect assignments).

RQ 2. How does the choice of K affect the clustering quality of GAC? Selecting a suitable K (i.e. number of clusters) is an ongoing problem in cluster analysis. This RQ explores how much the choice of K affect the quality of the resulting clusters. Due to the large cost of running GAC, we initially test the quality of the randomly generated populations at the possible values of K. This is then further explored by running GAC over three different values of K (23, 98 and 397), and by comparing it to the Hierarchical clustering solution. As in RQ1, we use the Silhouette score to measure the clustering solution quality.

RQ 3. How efficient is GAC compared to Hierarchical clustering? It is well known that a GA can be costly to fine-tune and evolve. Therefore, it is important to report its efficiency in terms of run-time in order to properly weigh benefits over its costs.

2.2 Dataset

In order to answer these RQs, we used the same dataset as the original study [1][1]. This dataset contains 12,664 Android mobile applications belonging to 24 categories, which have been randomly sampled from the Google Play app store. A detailed description of how this data was collected can be found elsewhere [1].

2.3 GA Approach

In this study, we opted to use the GA clustering approach proposed by Maulik and Bandyopadhyay [13]. This algorithm was shown to be able to find a global optimum, and its variations are widely available as code libraries. In our experiment we used the GAMA R package (v. 1.0.3) [16], and modified it to enhance the initial population generation and the penalty function (see Sect. 2.3 for the problem at hand[2]).

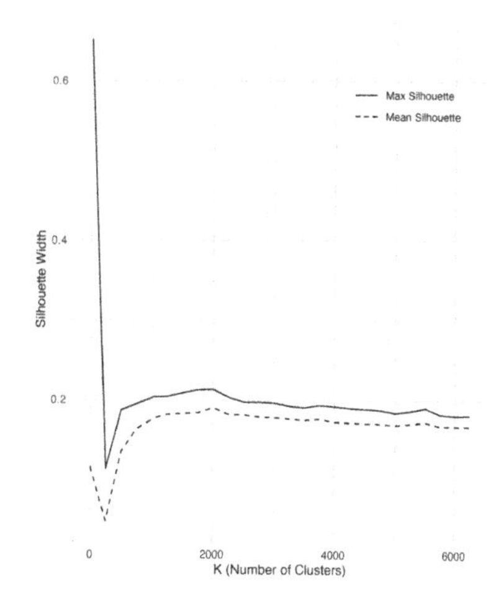

Fig. 1. RQ 2: Max (solid line) and mean (dashed line) silhouette scores (y-axis) of 500 random solutions at different values of K (x-axis) starting from k = 2 to k = dataset size/2 and a step of 250.

Solution Representation and Evaluation. The dataset of mobile apps is represented such that each mobile application is coded in terms of its LDA topics. The original study used 273 topics,

[1] http://clapp.afnan.ws/data/.

[2] Modified GAMA code can be found here: https://github.com/afnan-s/gama.

thus, the dataset is a 12,664 by 273 matrix with each cell containing the relatedness of the app to the topic. Each individual is a clustering solution, which is encoded as a vector of real values, each representing a cluster centre[3] (with K known a priori). As the original study used the Silhouette score to measure the quality of a clustering solution, we have used this measure as fitness function for GAC. Upon generating the initial population of cluster centres randomly, they are evolved using linear rank selection, blend crossover and non-uniform mutation [13,21].

Empty Clusters Problem. Upon generating the initial random population (i.e. random cluster centres), GAC looks at each gene and generates a uniformly random number between the upper and lower bounds found in the dataset. However, as our dataset represents topic relatedness (i.e. each gene in the individual is the relatedness of that individual to one specific topic), the dataset is a very sparse matrix, and the initial population of random centres can be very far from the actual data points in the dataset. As a result, many of the initial random solutions have mainly empty clusters as finding viable random cluster centres that are sufficiently close to the data points is unlikely. In order to address this limitation, we have modified the random population generation such that it uniformly samples from the set of pre-observed values for each gene. In addition, we have adopted a penalty function that deducts the fitness of an individual proportionally to the number of empty clusters it contains.

Table 1. RQ 1: Silhouette width scores of existing app store categorisation, hierarchical clustering and GAC ($k = 24$, which is the number of categories in the app store).

	Min.	Max.	Mean	Median
Existing categorisation	−0.54	0.59	0.003	−0.01
Hierarchical clustering	−0.64	0.99	0.02	−0.01
GAC	−0.49	1.00	0.52	0.55

Parameter Tuning and Setting. We have explored running the GA with population sizes as low as 5, 10, 25, and 100. However, the cost significantly increases as the population size increases, since each individual consists of 273 (number of topics/genes) values multiplied by K (number of cluster centres). We have found that a good compromise is using population = 500 and generations = 1000, crossover rate = 0.9, mutation rate = 0.1. We also investigated decreasing crossover rate and increasing mutation rate, however, this did not produce better results.

3 Results and Discussion

This section presents the results for each of the investigated RQs, in addition to a discussion of the comparison to the original study and possible implications.

RQ 0. Similarity of Clustering Solutions. When comparing the GAC clustering solution to the hierarchical clustering solution at K = 24 (with GAC having 1 empty cluster), the Jaccard similarity score is 0.37. This suggests that the

[3] The cluster centre is the arithmetic mean of all the points belonging to the cluster.

solutions bear some similarity, however, as will be reported in following RQs, the GAC solutions are of significantly higher quality. This shows that GAC's solution does not stray much from the hierarchical clustering one, but indeed improves upon it.

Table 2. **RQ 2**: Mean Silhouette scores at three different levels of K using random cluster centres, hierarchical clustering, and GAC.

Categorisation solution	K = 23	K = 98	K = 397
Random - Mean	−0.12	−0.05	0.13
Random - Best	0.13	0.05	0.19
Hierarchical cluster	0.03	0.13	0.3
GAC cluster	0.46	0.13	0.17

RQ 1. Evaluation of Clustering Quality at Low K. Table 1 shows summary statistics of the silhouette scores of each of the three clustering solutions. We observe that the GAC algorithm is able to produce significantly higher quality segmentation of the dataset at the same granularity of the app store. While in the original study, the hierarchical clustering solution improved upon the existing app store classification by 1.7%, the GAC solution improved it by 51.5% (note that 100% improvement means reaching a Silhouette score of 1). Indeed, the quality of the GAC-based solution exceeds that of the hierarchical-based one at its best selected K. We conclude that, upon requiring a coarser granularity clustering technique, GAC is a more suitable solution than a hierarchical technique, for this dataset.

RQ 2. Best Overall K. Upon studying the resulting clustering solutions of GAC for higher cluster numbers (i.e., higher K), the algorithm fails to produce solutions with high mean silhouette scores at the given parameters, as shown in Table 2. This could provide evidence that, for the studied dataset, GAC may not be suitable for finer clustering granularity, though very competent at coarser ones. In order to gain further insight regarding the silhouette trend as K increases, we have investigated the mean silhouette score for random initial populations at different levels of K. The results (see Fig. 1) suggest that silhouette scores are higher at lower K values and remain stagnant as K increases.

Table 3. **RQ 3**: Running time of the two clustering algorithms (measured on a standard laptop with an Intel Core i7 3.1 GHz and 16 GB RAM; d = days, h = hours, m = minutes, s = seconds).

No.	Steps	Hierarchical	GAC
1	Data preprocessing	5.4 d	5.4 d
2	DTM construction	3.0 h	3.0 h
3	Distance matrix	21.0 s	1.3 m
4	Clustering	6.0 s	1.7 d

RQ 3. Efficiency. Using LDA to represent the dataset of mobile app descriptions requires an upfront cost to search for the best LDA parameters that represent the data (further details can be found in [1]). Therefore, both GAC and Hierarchical clustering cost exactly the same for the first two steps (see Table 3). As GAC uses squared euclidean distance, as opposed to cosine distance used by the Hierarchical algorithm, it requires slightly longer to calculate the distance matrix. The major difference, however, can be observed in the time taken to produce the clustering solution. While Hierarchical clustering can produce a dendrogram in 6 s, from which a solution can be produced at any desired K (by cutting the denrogram), GAC

can require several days on the same machine to produce a solution at any given K[4]. This shows that GAC might be too costly an option especially for larger K values, especially when considering the time taken to tune the parameters. However, due to the large improvement of the cluster quality over low K (from 0.02 produced by Hierarchical to 0.52 produced by GAC), this trade-off might be worthwhile. Moreover, the use of parallelisation when running GA can mitigate these costs [7,20].

4 Conclusion and Future Work

This paper reports the initial results of our replication in which we investigate the efficacy of adopting a GA-based clustering approach to find a latent segmentation of mobile apps in the app store. We have found that GAC can be costly to run over a dataset comprising of a sparse matrix as typical of text analysis datasets. However, given a low enough K, GAC can exceed the results of hierarchical clustering with low enough cost given the improvement. On the other hand GAC did not produce clustering solutions over larger values of K that have higher quality than random search, possibly rendering it an unsuitable choice for finer granularity clustering. We plan to continue the line of investigation to fully replicate the original study and extend it. This includes investigating other similarity measurement techniques: vector space model, collocation- and dependency-based feature extraction methods as they might produce different results when combined with GAC. Also, we aim to further tune the GAC parameters to increase the confidence of the findings. Additionally, applying other search-based approaches may yield interesting results, including solving the empty cluster problem when generating random cluster centres by using a multi-objective GA that aims to maximise both the cluster quality and achieving different desired granularities (K). Our study sheds light on the feasibility of GA-based clustering for the SE task of mobile app categorisation. We hope this will foster further avenues for SBSE research in this area, as well as for many other clustering and classification tasks in SE [4].

References

1. Al-Subaihin, A., Sarro, F., Black, S., Capra, L.: Empirical comparison of text-based mobile apps similarity measurement techniques. EMSE **24**(6), 3290–3315 (2019)
2. Al-Subaihin, A.A., et al.: Clustering mobile apps based on mined textual features. In: ESEM 2016 (2016)
3. AlSubaihin, A., Sarro, F., Black, S., Capra, L., Harman, M.: App store effects on software engineering practices. IEEE TSE (2019). https://doi.org/10.1109/TSE.2019.2891715
4. Auch, M., Weber, M., Mandl, P., Wolff, C.: Similarity-based analyses on software applications: a systematic literature review. JSS **168** (2020). https://doi.org/10.1016/j.jss.2020.110669. Article No. 110669

[4] Running time for GAC with k = 24, population = 500, generations = 1000.

5. Ceccato, M., Falcarin, P., Cabutto, A., Frezghi, Y.W., Staicu, C.A.: Search based clustering for protecting software with diversified updates. In: SSBSE 2016 (2016)
6. Doval, D., Mancoridis, S., Mitchell, B.: Automatic clustering of software systems using a genetic algorithm. In: Proceedings of STEP 1999. IEEE Computer Society (1999)
7. Ferrucci, F., Salza, P., Sarro, F.: Using hadoop mapreduce for parallel genetic algorithms: a comparison of the global, grid and island models. ECJ **26**(4), 535–567 (2018)
8. Gorla, A., Tavecchia, I., Gross, F., Zeller, A.: Checking app behavior against app descriptions. In: ICSE 2014 (2014)
9. Harman, M., Al-Subaihin, A., Jia, Y., Martin, W., Sarro, F., Zhang, Y.: Mobile app and app store analysis, testing and optimisation. In: MOBILESoft 2016 (2016)
10. Hruschka, E., Campello, R., Freitas, A., de Carvalho, A.: A survey of evolutionary algorithms for clustering. IEEE TCMCC **39**(2), 133–155 (2009)
11. Huang, J., Liu, J., Yao, X.: A multi-agent evolutionary algorithm for software module clustering problems. Soft Comput. **21**(12), 3415–3428 (2016). https://doi.org/10.1007/s00500-015-2018-5
12. Martin, W., Sarro, F., Jia, Y., Zhang, Y., Harman, M.: A survey of app store analysis for software engineering. IEEE TSE **43**(9), 817–847 (2017)
13. Maulik, U., Bandyopadhyay, S.: Genetic algorithm-based clustering technique. Pattern Recogn. **33**(9), 1455–1465 (2000)
14. Nayebi, M., Farrahi, H., Lee, A., Cho, H., Ruhe, G.: More insight from being more focused: analysis of clustered market apps. In: WAMA 2016 (2016)
15. Praditwong, K., Harman, M., Yao, X.: Software module clustering as a multi-objective search problem. IEEE TSE **37**(2), 264–282 (2011)
16. Rodrigues, J., Vasconcelos, G., Tin'os, R.: GAMA: Genetic Approach to Maximize Clustering Criterion (2019). https://github.com/jairsonrodrigues/gama
17. Rousseeuw, P.J.: Silhouettes: a graphical aid to the interpretation and validation of cluster analysis. JCAM **20**, 53–65 (1987)
18. Sarro, F., Harman, M., Jia, Y., Zhang, Y.: Customer rating reactions can be predicted purely using app features. In: RE 2018 (2018)
19. Sarro, F., Al-Subaihin, A.A., Harman, M., Jia, Y., Martin, W., Zhang, Y.: Feature lifecycles as they spread, migrate, remain, and die in app stores. In: RE 2015 (2015)
20. Sarro, F., Petrozziello, A., He, D.Q., Yoo, S.: A new approach to distribute MOEA pareto front computation. In: GECCO 2020 (2020)
21. Scrucca, L.: GA: a package for genetic algorithms in R. J. Stat. Softw. **53**(4), 1–37 (2013)

Impact of Test Suite Coverage on Overfitting in Genetic Improvement of Software

Mingyi Lim, Giovani Guizzo, and Justyna Petke(✉)

Department of Computer Science, University College London, London, UK
{mingyi.lim.17,g.guizzo,j.petke}@ucl.ac.uk

Abstract. Genetic Improvement (GI) uses automated search to improve existing software. It can be used to improve runtime, energy consumption, fix bugs, and any other software property, provided that such property can be encoded into a fitness function. GI usually relies on testing to check whether the changes disrupt the intended functionality of the software, which makes test suites important artefacts for the overall success of GI. The objective of this work is to establish which characteristics of the test suites correlate with the effectiveness of GI. We hypothesise that different test suite properties may have different levels of correlation to the ratio between overfitting and non-overfitting patches generated by the GI algorithm. In order to test our hypothesis, we perform a set of experiments with automatically generated test suites using EvoSuite and 4 popular coverage criteria. We used these test suites as input to a GI process and collected the patches generated throughout such a process. We find that while test suite coverage has an impact on the ability of GI to produce correct patches, with branch coverage leading to least overfitting, the overfitting rate was still significant. We also compared automatically generated tests with manual, developer-written ones and found that while manual tests had lower coverage, the GI runs with manual tests led to less overfitting than in the case of automatically generated tests. Finally, we did not observe enough statistically significant correlations between the coverage metrics and overfitting ratios of patches, i.e., the coverage of test suites cannot be used as a linear predictor for the level of overfitting of the generated patches.

Keywords: Genetic Improvement · Search-based software engineering · Overfitting

1 Introduction

Genetic Improvement (GI) uses automated search to improve existing software [18]. GI navigates the search space of mutated program variants in order to find one that improves the desired property. This technique has been successfully used to fix bugs [1,14], add an additional feature [3,19], improve runtime [12], energy [7], and reduce memory consumption [5,23].

A. Aleti and A. Panichella (Eds.): SSBSE 2020, LNCS 12420, pp. 188–203, 2020.
https://doi.org/10.1007/978-3-030-59762-7_14

In the vast majority of GI work, each software variant is evaluated using a test suite, which is treated as a proxy for correctness. Although this assumption cannot prove absence of bugs in the evolved software, it has been good enough to evolve useful patches that have been adopted into development [13]. On the other hand, the generated patches have been criticised for overfitting, i.e., passing the tests used during the GI search process, but not producing actual fixes that generalise to unseen scenarios [20]. This triggers the question about:
What feature should a given test suite have to aid the GI process in producing useful, correct patches?

The current state-of-the-art uses existing test suites, together with the given program, as input to the GI process. In the case of functional improvement, such as program repair, such a test suite would contain some failing tests that reveal a given property (such as a bug, or a feature not yet present in the software) that needs amending. The objective is to find a semantic change to the program so that the evolved software would pass all the provided test cases. In both functional and non-functional improvement branches of GI the test suite serves as an *oracle* for whether the evolved software has the desired semantics. Note, however, that in order to achieve improvement with respect to a functional property, its semantics needs to change, by definition. In contrast, in non-functional improvement we want to preserve the semantics of the original software, whilst improving a property of choice, such as running time. This has consequences in how test suites can be used in GI.

In order to improve a program using GI, we need a test suite that will faithfully capture the desired software behaviour. However, finding such test suites is a non-trivial task. Frequently in GI work the test suite needs to be usually manually improved before the GI process can begin [3]. If the test suite is too weak, GI will keep on deleting code (it can access) that contains uncovered functionality. Therefore, traditional software metrics, such as branch coverage, have been used to estimate how good a given test suite is before inputting them into the GI process [3]. This can be very costly, especially in the functional improvement case, where one has to devise test cases manually, as an automated approach treats the current implementation as the test case oracle [4]. However, in non-functional improvement, automated test case generation tools can be utilised.

Regardless of whether a given test suite has been generated manually or automatically, the question still remains: which features should it have that would lead to least overfitting when used within a GI process? Smith et al. [20] took the first step by investigating the amount of overfitting by comparing manual vs. automated tests for the purpose of test-based automated program repair. Assiri and Bieman [2] sampled from existing test suites to show that statement-covering and random test suites lead to introduction of new faults in the automatically 'repaired' software. More recently, Yi et al. [24] tried to correlate various test metrics in existing test suites with their ability to lead to a non-overfitting patch.

In this work we aim to measure the correlation between traditional test suite metrics and the given test suite's impact on overfitting in the GI process. In contrast to previous work, we focus on non-functional improvement (runtime, in particular), and the Java program space. Hence, we use EvoSuite [11] to automatically generate test suites that achieve a given test suite coverage. We thus consider one test suite metric at a time, in order to provide a more systematic view of the metrics' impact. In this study we additionally measure the level of overfitting during the GI search process. We also re-run GI 20 times for each test suite – program pair (due to the non-deterministic nature of the GI search process we use). This way we can investigate a larger space of plausible software variants in order to establish a correlation between a given test suite coverage measure and the amount of overfitting. We also provide a replication package, available at: https://github.com/justynapt/ssbse2020RENE.

2 Background

Genetic Improvement (GI) uses automated search to improve existing software [18]. In a typical scenario, the input to GI is a program and a test suite. GI then uses a set of mutation operators and a meta-heuristic, such as genetic programming, to evolve thousands of software variants, to be evaluated using a given fitness measure. In functional improvement fitness is based on the test suite alone, while in non-functional improvement an additional evaluation needs to be made against a property of choice, such as running time. The process runs until a given criterion is met. For the purpose of program repair, for instance, the search can be stopped when a program variant passes all the given tests, or only after a specified number of generations of the search algorithm of choice.

There have been several metrics presented in the literature to evaluate the strength of a test suite [8]. We focus on those that are implemented in the arguably most successful automated test case generation tool for Java, i.e., EvoSuite [11]. This tool implements a total of 8 coverage metrics. In this work we focus on 4, as the other ones are either not applicable to the benchmarks we use (for instance, our programs have a single method with no exceptions thrown, so there's no need to consider these) or is not fully supported for our purpose (output diversity measure cannot currently be automatically calculated for an existing test suite using EvoSuite). Therefore, we consider line, branch, conditional branch, and weak mutation coverage.

Line and branch test suite coverage metrics are self-explanatory, i.e. the test suite aims to cover the largest number of lines or branches of the program, respectively. Conditional branch coverage aims to cover all branches with the right conditions, e.g., for an IF statement with an OR condition on 2 Boolean variables, 4 tests would have to be generated to cover all conditions, while branch coverage would only need 2 tests (for the *false* and any *true* evaluation of the *if* condition). In Mutation Analysis [17], a mutant program is said to be killed (i.e. covered) if its output differs from the output of the original program. The more mutants killed by a test suite, the better it is in revealing the faults. Unlike conventional mutation, Weak Mutation does not compare the final output of the mutated program, but rather compares intermediate states to decide whether a mutant is dead or not [15]. Offutt and Lee showed that weak mutation can produce stronger test suites [15,16] than strong mutation though.

3 Methodology

Our aim is to investigate the impact of various test suite coverage metrics on overfitting in the GI process. We also want to know if automatically generated tests, using such metrics, could yield to low overfitting rate and thus be used for the purpose of non-functional software improvement in GI.

In contrast to previous work, we do not analyse only the resulting patch of each GI run, but also all the valid (i.e., test-suite adequate) patches generated during the GI search process. We analyse all of them because, even though they have been discarded during the search process (as not leading to better improvement than the final patch found), they are still valid and could still be used as feasible solutions. This gives us better statistical power during analysis, and thus more conclusive evidence from thousands of patches as opposed to a few hundreds.

Furthermore, we focus on Java programs, as they have not been investigated in this context before. We also use a non-functional property, namely, program's execution time as the goal for improvement.

3.1 Research Questions

In order to answer the question about which criteria should a given test suite satisfy in order to lead to least overfitting in the genetic improvement process, we generate test suites that achieve maximum coverage with respect to a given metric and compare them with respect to the amount of overfitting when input into the GI process. In particular, we pose the following research questions:

RQ1 (Validity) Given a particular test suite, can GI find a valid non-overfitting patch?

We want to know whether GI is able to find a non-empty, potentially runtime-improving patch[1] in the first place, given a particular input test suite. As done in related work [20], this question focuses on the final patch output by the GI process. This step checks how well GI can find a non-overfitting patch, regardless of how much overfitting might have occurred during the search process, and forms a baseline comparison with previous work (albeit in Java rather than C program space).

RQ2 (Overfitting) How does the overfitting rate vary with the input test suite during search?

We want to know how often produced patches overfit to the training test suite. Given that GI usually uses a heuristic approach, we conduct repeated runs and additionally report on the overfitting rates during the different runs. As mentioned at the beginning of this section, for this RQ (and RQs 3–4 too), we look at all valid patches generated during the whole GI search process.

[1] We use the word 'potentially' here, as although the patch might improve upon our training and test set, it does not mean the runtime improvement will generalise to all possible usages of software. Manual check is thus necessary.

RQ3 (Metric vs. Overfitting) How does the non-overfitting rate correlate with the changes in coverage?

This question is designed to answer how fragile is the GI process to the change in the coverage for a given test suite. Although we do not test for causation, we are interested to discover if any of the coverage measures can be used as a reliable predictor for the overall ratio of non-overfitting patches. If so, one can aim at improving their test suite with regards to that specific measure, in order to reduce the amount of overfitting.

RQ4 (Automated vs. Manual) How often do the automatically generated test suites overfit with respect to the manually generated ones?

The same question was asked by Smith et al. [20], though in the C domain and in the automated program repair context. We want to check if the same conclusions hold in our scenario. In order to answer this question, we perform a cross-validation using the automatically generated test suites using a set of coverage criteria, against manually curated test suites.

With the above research questions in mind we set up our experiments. The next subsections describe the datatset, tools, and experimental procedure in more detail.

3.2 Dataset and Tools

We used the genetic improvement toolbox Gin v2.0 [6] in our experiments. Gin fulfills all our requirements: it is open-source, targets Java programs, uses runtime improvement as fitness by default, and its second release provides integration with EvoSuite [11].

Unlike in the automated program repair field, there is no standard benchmark for runtime improvement using GI. We also require the programs to be relatively small, so we could run thousands of experiments in reasonable time and avoid the, often very costly, profiling stage of the GI process [6], targeting the whole software instead. With those restrictions in mind, we chose to use the set of 9 sort algorithms and the triangle example provided with the first release of Gin [22] in our study, for which improvements have previously been found using GI [9].

All experiments were run on a Mac Mini with a 3.2 GHz 6-core Intel Core i7.

3.3 Experimental Procedure

We will now outline the details of the empirical study aimed at answering our research questions.

For each program and for each test suite coverage criterion (i.e., branch, line, conditional branch, and weak mutation) we generate a test suite using EvoSuite. Additionally, we generate a test suite that aims to cover all 4 coverage criteria at once. Next, we input the program and the given test suite to the GI process.

Gin uses a simple hill climber by default. It first generates a random mutation (which could be a delete, copy or replace operation), applies it to the code and evaluates it. If the change is beneficial in terms of runtime, it is retained, otherwise it is retained with 50% probability. The process continues for 100 iterations. Since this is a heuristic approach we repeat the GI cycle 20 times. We extended the algorithm to make changes at the statement rather than the default line-level (to allow for known improvements from previous work to be found). Moreover, we used Gin's PatchAnalyser to evaluate generalisability and runtime improvement of each generated non-empty patch on a held-out test suite.

In order to get variation in the coverage percentage for the various metrics, from each automatically generated test suite we sample uniformly at random 25%, 50%, and 75% of its tests, creating new test suites of varying coverage. We repeat the GI process with these as well. In order to check for overfitting we use the manual test suite, provided with the programs, as an oracle.

Altogether, we ran 20 rounds of GI on each of the 10 programs and 20 generated test suites (5 coverage criteria [4 single + 1 combined] $\times$ 4 samples), for a total of 400 GI runs per program. Finally, we also generated a test suite with the 4 coverage criteria as goals to treat as an oracle for GI runs on the manual test suite, to compare the impact of manual vs automated test suites on overfitting in GI. Therefore, a total of 4 200 GI runs was conducted, with 420 000 patches generated (4 200 $\times$ 100 steps of each local search run).

Pseudo-code for our experimental procedure is found in Algorithm 1. Note that whenever a patch passed the training suite during a GI run, it was evaluated against a test suite.

Algorithm 1: Pseudo-code for the experimental procedure. Each GI run consists of 100 steps of local search.

```
for each Program P do
    for Coverage metric C from [line, branch, conditional branch, weak
    mutation, all4] do
        Generate test suite T using C as coverage goal
        for perc in [100%, 75%, 50%, 25%] do
            T' = Select perc of tests from T
            for i = 1; i <= 20; i++ do
                Run GI with P and T' as input
            end
        end
        for i = 1; i <= 20; i++ do
            Run GI with P and T_manual as input
        end
    end
end
```

4 Results

In this section we present the results of our experiments and provide answers to research questions posed in Sect. 3.1. We deem a resultant patch as overfitting if it fails on the held-out test suite. For the runs where the manual test set was used as input, we generated tests using EvoSuite with the four coverage goals previously considered, i.e., branch, line, conditional branch, and weak mutation. For all the other test suites we used the manual test suite as the test set, to check for overfitting.

The experiments took a total of 16 h to complete. All the data (and the modified Gin code to facilitate the experiments) is available as a replication package on GitHub: https://github.com/justynapt/ssbse2020RENE.

4.1 RQ1 – Validity

To recap, for each of the 10 subject programs we generated 21 test suites: 4 satisfying 100% coverage of the test suite criterion; one that aimed to satisfy all 4 goals at once; and the manual one; the 5 automatically generated test suites were sampled at 100% 75%, 50% and 25%. This yielded 210 experimental scenarios. In answer to **RQ1**, a patch was found in all scenarios. That is, for each (test suite, program) pair GI found a non-empty patch in at least one run. Moreover, in 203 scenarios a patch was found in at least one of the GI runs that generalised to the held-out test suite (Table 1).

Table 1. Number of all and non-overfitting patches found in at least one of the 20 repeated GI runs for each program.

Program	LoC	Test sizes	Test suites	Patch found	Non-overfitting
SortMerge	52	1–8	21	21	19
Triangle	40	1–10	21	21	21
SortQuick	32	1–9	21	21	21
SortBubbleDouble	24	1–7	21	21	21
SortRadix	24	1–7	21	21	21
SortSelection	19	1–7	21	21	21
SortBubbleLoops	17	1–7	21	21	21
SortSelection2	17	1–7	21	21	19
SortBubble	15	1–7	21	21	19
SortInsertion	14	1–7	21	21	20
Total	254	1–10	210	210	203

With these results, we can positively answer **RQ1** and state that, within our experimental setup, GI can indeed find valid and non-overfitting patches.

4.2 RQ2 – Overfitting

We provide more detailed results on the rate of overfitting throughout the search process (i.e., including intermediate solutions) in Tables 2 and 3, and Fig. 1. Here we report on all intermediate patches found during all 4 200 GI runs.

Table 2. Number of all intermediate and non-overfitting patches found during improvement for all 4 200 GI runs, aggregated by each test suite type. Summative results shown for the 10 programs investigated.

Criterion	Sample%	Patch found	Non-overfitting	Ratio
Branch	100	4 407	1 481	0.34
	75	3 907	1 443	0.37
	50	4 160	1 338	0.32
	25	4 651	1 486	0.32
Line	100	4 989	1 366	0.27
	75	5 009	1 412	0.28
	50	4 907	1 141	0.23
	25	5 122	1 374	0.27
W. Mutation	100	4 983	1 196	0.24
	75	4 455	1 356	0.30
	50	4 498	1 447	0.32
	25	4 902	1 138	0.23
C-Branch	100	4 719	1 253	0.27
	75	4 666	1 526	0.33
	50	4 826	1 167	0.24
	25	4 995	1 257	0.25
All 4 criteria	100	5 360	1 413	0.26
	75	4 620	1 625	0.35
	50	4 670	1 234	0.26
	25	5 098	1 312	0.26
Manual	100	2 491	2 410	0.97

The ratios yielded by the generation criteria average from approximately 0.23 to 0.37, with the branch criterion yielding a better ratio overall. Moreover, there is no apparent trend in the changes in sample percentages to the ratio, i.e., in some cases even sampling as few as 25% of the test cases yielded a non-overfitting ratio similar to that of using the whole test suite. However, when we group the data by program (Table 3), a higher variation in ratios becomes apparent. This may be an indication that the overfitting ratio is more dependent on the program being improved, rather than on the criteria used to create the test suites.

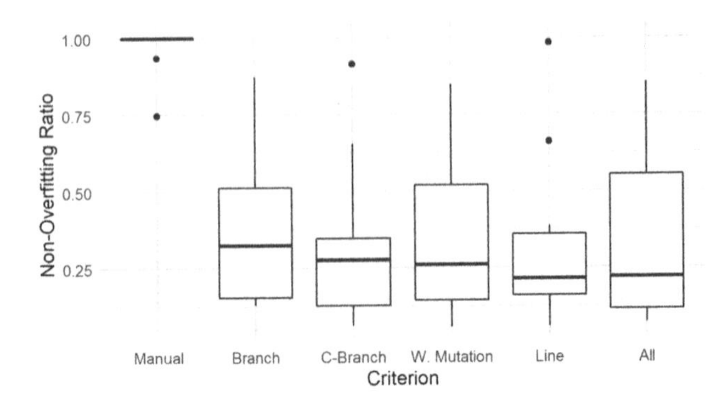

Fig. 1. Boxplot with the ratio between non-overfitting patches/total patches found (y-axis) per generation criterion (x-axis). Each box presents the ratios for the 10 programs.

Table 3. Number of all intermediate and non-overfitting patches found during improvement for all GI runs for each program. Data presented for all test suites investigated.

Program	Patch found	Non-overfitting	Ratio
SortBubble	12 131	3 213	0.26
SortBubbleDouble	12 632	8 537	0.68
SortBubbleLoops	12 466	3 366	0.27
SortInsertion	10 129	2 048	0.20
SortMerge	5 228	2 588	0.50
SortQuick	3 005	2 690	0.90
SortRadix	9 622	1 090	0.11
SortSelection	11 548	1 472	0.13
SortSelection2	12 081	1 187	0.10
Triangle	8 593	3 184	0.37

We applied the Fisher's exact test [10] on the data in order to determine if the differences in proportion of overfitting and non-overfitting patches obtained by the test generation criteria are statistically significant. Figure 2 presents the ranks of each generation criterion over the 10 programs. The results of the Fisher's exact test were used to perform the rank computation, such that two criteria are considered "statistically tied" $p \geq 0.05$ for their pair comparison (i.e, the difference in their proportions of non- and overfitting patches is not statistically significant). In such a case, the rank of a criterion is given by the average of the ranks of all criteria to which it ties (including its own). We adopted this analysis because it would be infeasible to report all the 150 p-values (10 programs $\times$ 15 pairwise combinations), and because it can easily depict rank superiority with statistical significance. We refer to this method herein as "statistical rank".

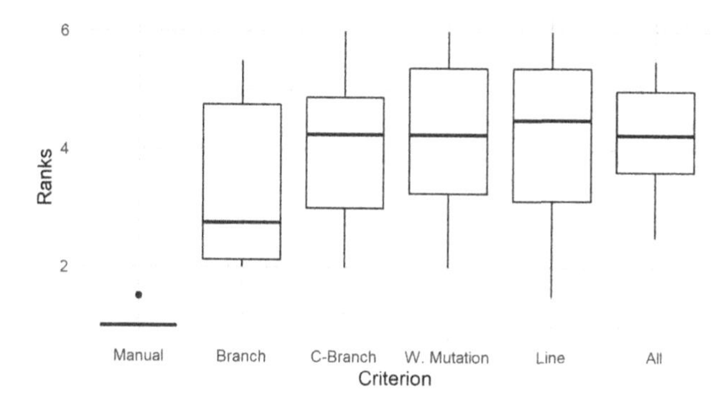

Fig. 2. Boxplot with the statistical ranks of non-overfitting ratio (y-axis) per generation criterion (x-axis). Each box presents the ranks for the 10 programs. Statistical ties are averaged.

The branch criterion presented the best results among all criteria, with a mean rank of 3.35 (median 2.75). Line coverage presented the worst results, with an average rank of 4.20 (median 4.50). Surprisingly, using all available criteria to guide the creation of test suites yields the second worst ratio and rank (mean and median 4.20 and 4.25), even though it provides the best coverage.

Considering the analysis of this section, the answer to **RQ2** is somewhat mixed. First, the differences in non-overfitting ratio are not that striking as shown in Fig. 1 and Table 2, but the results by program in Table 3 vary quite widely. However, when considering the statistical ranks of the criteria, test suites generated with branch coverage are slightly (but significantly) better than the others in the proportions. Hence, using only branch coverage as opposed to the other options can give the engineer a slight advantage on the result of GI in regards to overfitting, although it depends on the program being improved.

4.3 RQ3 – Metrics vs. Overfitting

To answer this question, we collected all the results from all GI executions and applied the Spearman's rank correlation coefficient test (Spearman's ρ) [21]. We use this non-parametric correlation test because we cannot assume the normal distribution of data. In fact, we checked for normality and could not reject the hypothesis that the data does not follow a normal distribution.

We aim at assessing if there is any correlation between the non-overfitting ratio and the coverage metrics of the test suites. In order to cater for the different confounding variables, we have also applied the correlation test on the data by sampling them based on their program, criteria, and sampling percentages. By isolating these variables and making them constant in each sample, we can unveil whether any of them can have some influence on the outcome of the correlation. Table 4 presents the correlation results.

Only 24 out of 100 tested correlations showed statistical significance (p-value < 0.05), and only 4 of those showed large correlation (Spearman's $\rho > 0.5$). For

Table 4. Spearman's ρ correlation coefficient results. Each sub-table shows the results for a different grouping. Each row represents a group of test suites and each column represents a measure of coverage. A given cell $[R, C]$ shows the correlation result for all test suites in the group of row R, between their ratio of non-overfitting and their measure of column C. For instance, the first cell [SortBubble, Size] indicates a Spearman's correlation $\rho = 0.29$ between the ratio of non-overfitting and the size of test suites applied to program SortBubble. '*' highlights significant correlations (p-value < 0.05).

Grouped by program					
Groups	Size	Branch Cov.	Line Cov.	Weak Mut. Cov.	C-Branch Cov.
SortBubble	0.29	0.40	*0.45	0.29	0.40
SortBubbleDouble	0.42	−0.11	−0.14	−0.31	−0.11
SortBubbleLoops	0.27	0.37	*0.53	*0.54	0.37
SortInsertion	0.07	0.27	0.17	0.44	0.27
SortMerge	−0.18	0.19	0.10	0.15	−0.15
SortQuick	−0.23	−0.24	−0.22	−0.27	−0.22
SortRadix	0.42	*0.62	*0.49	*0.46	*0.62
SortSelection	−0.08	−0.14	−0.31	−0.31	−0.14
SortSelection2	0.10	0.15	0.09	0.09	0.15
Triangle	0.05	0.11	0.04	0.14	0.11

Grouped by test suite generation criterion					
Groups	Size	Branch Cov.	Line Cov.	Weak Mut. Cov.	C-Branch Cov.
All 4 Criteria	0.22	0.15	0.07	0.07	−0.03
Branch	0.24	−0.04	−0.04	0.02	−0.05
C-Branch	*0.37	−0.01	−0.23	0.05	−0.02
Line	0.21	*0.35	0.22	*0.41	0.19
W. Mutation	0.21	*0.40	*0.46	0.21	−0.05

Grouped by test suite sample %					
Groups	Size	Branch Cov.	Line Cov.	Weak Mut. Cov.	C-Branch Cov.
25%	*0.28	0.21	0.17	0.25	−0.04
50%	*0.29	*0.35	*0.28	*0.36	0.12
75%	*0.41	*0.30	0.19	0.25	0.10
100%	*0.28	−0.17	*−0.30	−0.22	−0.21

No data grouping					
Groups	Size	Branch Cov.	Line Cov.	Weak Mut. Cov.	C-Branch Cov.
Whole Data	*0.24	*0.18	0.10	*0.17	0.03

instance, when GI is applied to SortRadix, the levels of Branch coverage and the C-Branch coverage of the test suites showed large correlation to the ratio of non-overfitting patches generated using the test suites. This also happened to Line and Weak Mutation coverage of the test suites when applied to SortBubbleLoops.

However, due to the low frequency of significant results and stronger correlations, our answer to **RQ3** is rather negative. We could not find any consistency from a specific coverage metric in linearly predicting the level of non-overfitting ratio yielded by the test suites during improvement. Furthermore, the correlation is significant only for a few programs, which may indicate that the properties of the programs are more important variables for the overfitting of GI.

4.4 RQ4 – Automated vs. Manual

Table 5 presents the non- and overfitting results for the manual created test suites and to the test suites created with all 4 criteria (test suites that yielded the best coverage overall). To recap, we perform a cross-validation, where the patches generated with the aid of one type of test suite (automatically generated/manual) are validated with the other type of test suites.

The first observation is that automatically generated test suites always have the best coverage, for all coverage metrics and in all programs, with only two ties. Moreover, automatically generated test suites always lead to more valid patches during GI optimisation. However, manually created test suites always produce more non-overfitting patches, despite generating fewer patches overall.

Table 5. Coverage measures and cross-validation overfitting results for Manually vs. Automatically (abbreviated as "Aut.") generated test suites. T. Suite – type of test suite. Branch, Line, Weak Mutation (abbreviated "W. Mut."), and C-Branch – coverage obtained by the test suites. Patches – number of valid patches found during improvement. Non-Overf. – number of non-overfitting patches found during improvement. Ratio – ratio of number of non-overfitting by number of patches. Best values are highlighted in bold.

Program	T. Suite	Branch	Line	W. Mut.	C-Branch	Patches	Non-Overf.	Ratio
SortBubble	Manual	86%	88%	97%	86%	276	**276**	**1.00**
	Aut.	**100%**	**100%**	**99%**	**100%**	**537**	212	0.39
SortBubbleDouble	Manual	85%	71%	81%	85%	618	**618**	**1.00**
	Aut.	**92%**	**79%**	**83%**	**92%**	**643**	497	0.77
SortBubbleLoops	Manual	89%	89%	97%	89%	345	**344**	**1.00**
	Aut.	**100%**	**100%**	**98%**	**100%**	**523**	192	0.37
SortInsertion	Manual	86%	86%	97%	86%	218	**218**	**1.00**
	Aut.	**100%**	**100%**	**99%**	**100%**	**660**	89	0.13
SortMerge	Manual	94%	97%	96%	6%	118	**118**	**1.00**
	Aut.	**100%**	**100%**	**97%**	**67%**	**439**	29	0.07
SortQuick	Manual	93%	95%	**97%**	7%	152	**152**	**1.00**
	Aut.	**100%**	**100%**	**97%**	**100%**	**153**	132	0.86
SortRadix	Manual	93%	93%	97%	93%	122	**114**	**0.93**
	Aut.	**100%**	**100%**	**98%**	**100%**	**428**	50	0.12
SortSelection	Manual	86%	92%	97%	86%	178	**178**	**1.00**
	Aut.	**100%**	**100%**	**98%**	**100%**	**788**	5	0.04
SortSelection2	Manual	86%	91%	96%	86%	178	**178**	**1.00**
	Aut.	**100%**	**100%**	**97%**	**100%**	**704**	49	0.07
Triangle	Manual	89%	92%	**98%**	89%	286	**214**	**0.75**
	Aut.	**100%**	**96%**	**98%**	**100%**	**485**	128	0.26

Consequently, the proportion of non-overfitting patches is considerably and significantly better for manual test suites (Fisher's exact test [10], p-value < 0.05).

These results indicate that in fact, better coverage does not translate to better patches regarding overfitting. This analysis serves as further evidence for the lack of correlation between coverage and non-overfitting.

Finally, answering **RQ4**, manually created test suites generate patches that overfit significantly less than patches generated with automatically generated test suites. The non-overfitting ratio of manual test suites is almost always 1.0 of the valid patches, whereas the ratio for automatically generated test suites varies from 0.04 to 0.86 but never greater than their counterpart.

5 Threats to Validity

In this section we discuss threats to validity of the presented work.

Firstly, the programs investigated are quite small thus the results might not generalise. There were several reasons for choosing this small set. There's no standard benchmark for GI for runtime, yet improvements for the programs we use have been found in previous work. We also investigated the rate of overfitting during the search process, yielding 420 000 patches generated, roughly a quarter of those being re-run to check for overfitting. With this small sample the experiments took a non-trivial but feasible 16 h to complete. Moreover, similar size programs with similar size test suites were used in previous work [20]. We believe that for a preliminary study these were good enough.

Next, we used EvoSuite and Gin, thus inherited any limitations of the tools. For instance, EvoSuites generation of test suites is non-deterministic, thus multiple runs might yield different results. However, Gin's test case generator[2] sets EvoSuite to produce the same results for each seed[3]. We have then re-done the experiments with a few different seeds and found the results consistent with the ones reported in this paper. We did not conduct enough to report on statistical significance of those though. Furthermore, different results might be obtained with other mutation strategies than the default in Gin's local search implementation. However, these mutation operators are currently standard ones in test-based GI.

6 Conclusions

In this paper we evaluate the levels of non- and overfitting patches obtained by different test suites during the GI optimisation process. The goal of our experiments is to show the differences between manually and automatically generated test suites, and the correlations between coverage and the ratios of non-overfitting patches.

---

[2] See gin.util.TestCaseGenerator at https://github.com/justynapt/ssbse2020RENE.

[3] Following advice given here: https://github.com/EvoSuite/evosuite/issues/48.

Our results unveiled that, regardless of the criterion used to guide the automatic generation of test suites, the ratios of overfitting differ slightly between each other, with branch coverage being significantly better than the other criteria, but only by a small margin. Moreover, we could only find 4 significant and large correlations amongst a set of 100 tested correlations, which is not enough scientific evidence to consider any of the tested coverage measures as accurate predictors for the ratio of non-overfitting patches. Finally, our results showed that even though automatically generated test suites cover significantly more the programs under test and generate more valid patches throughout the search process, manually curated test suites yield a significantly better proportion of non-overfitting patches with almost no overfitting at all. This shows that classical automated test suite measures seem to have no bearing on how good a test suite is for the purpose of applying genetic improvement. Thus, a question of which characteristics a test suite should have so that useful, non-overfitting variants are produced during the GI search process, remains unanswered.

As future work, we intend to extend the study with larger programs as well as evaluate whether the properties of the programs under improvement play a bigger role in the overfitting of patches than the properties of tests suites.

Acknowlegements. This work was funded by the EPSRC grant EP/P023991/1 and the ERC grant 741278 Evolving Program Improvement Collaborators (EPIC). The authors would also like to thank Prof. Gordon Fraser from University of Passau for consultation on the output diversity metric.

References

1. An, G., Kim, J., Yoo, S.: Comparing line and AST granularity level for program repair using pyggi. In: Petke, J., Stolee, K.T., Langdon, W.B., Weimer, W. (eds.) Proceedings of the 4th International Genetic Improvement Workshop, GI@ICSE 2018, pp. 19–26. ACM (2018). https://doi.org/10.1145/3194810.3194814
2. Assiri, F.Y., Bieman, J.M.: An assessment of the quality of automated program operator repair. In: Seventh IEEE International Conference on Software Testing, Verification and Validation, ICST 2014, pp. 273–282. IEEE Computer Society (2014). https://doi.org/10.1109/ICST.2014.40
3. Barr, E.T., Harman, M., Jia, Y., Marginean, A., Petke, J.: Automated software transplantation. In: Young, M., Xie, T. (eds.) Proceedings of the 2015 International Symposium on Software Testing and Analysis (ISSTA 2015), pp. 257–269. ACM (2015). https://doi.org/10.1145/2771783.2771796
4. Barr, E.T., Harman, M., McMinn, P., Shahbaz, M., Yoo, S.: The oracle problem in software testing: a survey. IEEE Trans. Software Eng. **41**(5), 507–525 (2015). https://doi.org/10.1109/TSE.2014.2372785
5. Basios, M., Li, L., Wu, F., Kanthan, L., Barr, E.T.: Darwinian data structure selection. In: Leavens, G.T., Garcia, A., Pasareanu, C.S. (eds.) Proceedings of the 2018 ACM Joint Meeting on European Software Engineering Conference and Symposium on the Foundations of Software Engineering, ESEC/SIGSOFT 2018, pp. 118–128. ACM (2018). https://doi.org/10.1145/3236024.3236043

6. Brownlee, A.E.I., Petke, J., Alexander, B., Barr, E.T., Wagner, M., White, D.R.: Gin: genetic improvement research made easy. In: Auger, A., Stützle, T. (eds.) Proceedings of the Genetic and Evolutionary Computation Conference, GECCO 2019, pp. 985–993. ACM (2019). https://doi.org/10.1145/3321707.3321841
7. Bruce, B.R., Petke, J., Harman, M., Barr, E.T.: Approximate oracles and synergy in software energy search spaces. IEEE Trans. Software Eng. **45**(11), 1150–1169 (2019). https://doi.org/10.1109/TSE.2018.2827066
8. Chekam, T.T., Papadakis, M., Le Traon, Y., Harman, M.: An empirical study on mutation, statement and branch coverage fault revelation that avoids the unreliable clean program assumption. In: 2017 IEEE/ACM 39th International Conference on Software Engineering (ICSE), pp. 597–608 (2017)
9. Cody-Kenny, B., Lopez, E.G., Barrett, S.: locoGP: improving performance by genetic programming Java source code. In: Langdon, W.B., Petke, J., White, D.R. (eds.) Genetic Improvement 2015 Workshop, pp. 811–818. ACM (2015). https://doi.org/10.1145/2739482.2768419
10. Fisher, R.A.: On the interpretation of chi-squared from contingency tables, and the calculation of P. J. R. Stat. Soc. **85**(1), 87–94 (1922). https://doi.org/10.2307/2340521
11. Fraser, G., Arcuri, A.: Evolutionary generation of whole test suites. In: Núñez, M., Hierons, R.M., Merayo, M.G. (eds.) Proceedings of the 11th International Conference on Quality Software, QSIC 2011, pp. 31–40. IEEE Computer Society (2011). https://doi.org/10.1109/QSIC.2011.19
12. Langdon, W.B., Harman, M.: Optimizing existing software with genetic programming. IEEE Trans. Evol. Comput. **19**(1), 118–135 (2015). https://doi.org/10.1109/TEVC.2013.2281544
13. Langdon, W.B., Lam, B.Y.H., Petke, J., Harman, M.: Improving CUDA DNA analysis software with genetic programming. In: Silva, S., Esparcia-Alcázar, A.I. (eds.) Proceedings of the Genetic and Evolutionary Computation Conference, GECCO 2015, pp. 1063–1070. ACM (2015). https://doi.org/10.1145/2739480.2754652
14. Le Goues, C., Nguyen, T., Forrest, S., Weimer, W.: GenProg: a generic method for automatic software repair. IEEE Trans. Software Eng. **38**(1), 54–72 (2012). https://doi.org/10.1109/TSE.2011.104
15. Offutt, A.J., Lee, S.D.: How strong is weak mutation? In: Howden, W.E. (ed.) Proceedings of the Symposium on Testing, Analysis, and Verification, TAV 1991, Victoria, British Columbia, Canada, 8–10 October 1991, pp. 200–213. ACM (1991). https://doi.org/10.1145/120807.120826
16. Offutt, A.J., Lee, S.D.: An empirical evaluation of weak mutation. IEEE Trans. Software Eng. **20**(5), 337–344 (1994). https://doi.org/10.1109/32.286422
17. Offutt, A.J., Untch, R.H.: Mutation 2000: uniting the orthogonal. In: Wong, W.E. (ed.) Mutation Testing for the New Century, pp. 34–44. Springer, Boston (2001). https://doi.org/10.1007/978-1-4757-5939-6-7
18. Petke, J., Haraldsson, S.O., Harman, M., White, D.R., Woodward, J.R.: Genetic improvement of software: a comprehensive survey. IEEE Trans. Evol. Comput. (2017). https://doi.org/10.1109/TEVC.2017.2693219
19. Petke, J., Harman, M., Langdon, W.B., Weimer, W.: Using genetic improvement and code transplants to specialise a C++ program to a problem class. In: Nicolau, M., et al. (eds.) EuroGP 2014. LNCS, vol. 8599, pp. 137–149. Springer, Heidelberg (2014). https://doi.org/10.1007/978-3-662-44303-3_12

20. Smith, E.K., Barr, E.T., Le Goues, C., Brun, Y.: Is the cure worse than the disease? Overfitting in automated program repair. In: Proceedings of the 2015 10th Joint Meeting on Foundations of Software Engineering, ESEC/FSE 2015, pp. 532–543 (2015). https://doi.org/10.1145/2786805.2786825
21. Spearman, C.: The proof and measurement of association between two things. Am. J. Psychol. **15**(1), 72–101 (1904). https://doi.org/10.2307/1422689
22. White, D.R.: GI in no time. In: Proceedings of the Genetic and Evolutionary Computation Conference (GECCO 2017), pp. 1549–1550. ACM (2017). https://doi.org/10.1145/3067695.3082515
23. Wu, F., Weimer, W., Harman, M., Jia, Y., Krinke, J.: Deep parameter optimisation. In: Silva, S., Esparcia-Alcázar, A.I. (eds.) Proceedings of the Genetic and Evolutionary Computation Conference, GECCO 2015, pp. 1375–1382. ACM (2015). https://doi.org/10.1145/2739480.2754648
24. Yi, J., Tan, S.H., Mechtaev, S., Böhme, M., Roychoudhury, A.: A correlation study between automated program repair and test-suite metrics. Empirical Softw. Eng. **23**(5), 2948–2979 (2017). https://doi.org/10.1007/s10664-017-9552-y

Bet and Run for Test Case Generation

Sebastian Müller(✉), Thomas Vogel, and Lars Grunske

Software Engineering Group, Humboldt-Universität zu Berlin, Berlin, Germany
{muelerse,thomas.vogel,grunske}@informatik.hu-berlin.de

Abstract. Anyone working in the technology sector is probably familiar with the question: "Have you tried turning it off and on again?", as this is usually the default question asked by tech support. Similarly, it is known in search-based testing that metaheuristics might get trapped in a plateau during a search. As a human, one can look at the gradient of the fitness curve and decide to restart the search, so as to hopefully improve the results of the optimization with the next run. Trying to automate such a restart, it has to be programmatically decided whether the metaheuristic has encountered a plateau yet, which is an inherently difficult problem. To mitigate this problem in the context of theoretical search problems, the *Bet and Run* strategy was developed, where multiple algorithm instances are started concurrently, and after some time all but the single most promising instance in terms of fitness values are killed. In this paper, we adopt and evaluate the *Bet and Run* strategy for the problem of test case generation. Our work indicates that use of this restart strategy does *not* generally lead to gains in the quality metrics, when instantiated with the best parameters found in the literature.

Keywords: Search based testing · Test case generation · Bet and Run

1 Introduction

Software testing plays an important role in providing evidence for the quality of software [17]. Due to the costs and complexity of writing tests manually, automating the generation of tests by metaheuristic search techniques is an active field of research [16,17,21]. A popular example for such a search-based approach to automated *test case generation (TCG)* is *EvoSuite* [5,6]. EvoSuite has shown its practical relevance by producing tests that achieve "good levels" of code coverage [7, p. 1] and detect real faults in open-source software [8]. The metaheuristic used by EvoSuite has evolved over time according to the state of the art, which is currently the *Dynamic Many-Objective Sorting Algorithm (DynaMOSA)* [18].

Due to the stochastic nature of metaheuristics, such algorithms are approximate and therefore "find in a reasonable computation time a solution that is as good as possible, but not necessarily optimal" [2, p. 242]. Exploring the search space stochastically, a metaheuristic typically yields different results with every run, and might get trapped in an inferior part of the search space (e.g., a plateau or local optimum) which leads to suboptimal results. As for any metaheuristic, this also applies to DynaMOSA in the context of TCG.

A. Aleti and A. Panichella (Eds.): SSBSE 2020, LNCS 12420, pp. 204–219, 2020.
https://doi.org/10.1007/978-3-030-59762-7_15

One way to improve the search results is to restart the metaheuristic whenever it is trapped in a plateau or local optimum to try out another run [12,13]. However, such a restarting approach requires an appropriate strategy to be integrated in the metaheuristic that determines the trap and restarts the search [10]. To avoid such a metaheuristic-specific strategy and explicit restarts, Friedrich et al. [10] proposed *generic Bet and Run*, an approach that starts multiple short sample runs of a metaheuristic, evaluates the intermediate results of these runs after a certain point in time, and bets on the most promising run to continue the search until the search budget is used. This approach is generic as it is independent of the metaheuristic being used. This is in contrast to the original *Bet and Run* that is intertwined with the metaheuristic [4]. Thus, by starting multiple instances of a metaheuristic and selecting the most promising one, *Bet and Run* turns the problem of repeatedly restarting an instance on its head. First, there is no need to determine when a metaheuristic is trapped in an inferior part of the search space to trigger a restart. Second, the effect of repeated restarts is still obtained by starting and trying out multiple runs of the metaheuristic. The *generic Bet and Run* has been successfully evaluated on two theoretical problems (Traveling Salesperson and Minimum Vertex Cover) that are structurally different, so that Friedrich et al. [10] expect that *Bet and Run* is generally helpful.

In this paper, we adopt the idea of the *generic Bet and Run* strategy, transfer it to the TCG problem using EvoSuite/DynaMOSA, and evaluate its general feasibility for TCG and its effectiveness in comparison to the state of the art, being EvoSuite/DynaMOSA. In our *Bet and Run* approach, we split the total time budget for generating test cases for a subject into two phases. In the first phase, we start multiple instances of EvoSuite/DynaMOSA concurrently, each with the same configuration but a different seed. Since *Bet and Run* is generic, we can treat EvoSuite/DynaMOSA as a black box without having to change the metaheuristic. At the end of the first phase, we sample the intermediate results (test cases) in terms of the fitness score provided by EvoSuite. After the first phase, we bet on the most promising instance that continues generating test cases during the second phase, whereas all of the other instances will be terminated. Accordingly, variants of *Bet and Run* in terms of how many instances will be started concurrently, and how the total time budget is split into the two phases, are possible and will be investigated in this paper. EvoSuite as the state-of-the-art approach serves as our baseline. It starts a single instance of DynaMOSA to generate test cases for a subject over the total time budget. We evaluate the effectiveness—in terms of achieved code coverage—of our *Bet and Run* approach in a head-to-head comparison with this baseline, giving both approaches the same time budget to generate tests for 107 Java classes.

To the best of our knowledge, no previous study has applied a generic *Bet and Run* approach to a search-based software engineering problem, let alone TCG. Several TCG approaches in the literature, however, have considered restarts of search. The Alternative Variable Method (AVM), a variant of hill climbing, restarts the search with a randomly selected solution candidate to overcome local optima, that is, when the fitness cannot be improved (cf. [1,9,11]). Chan et al. [3] use a restart (complete reset) as one way to forget test cases in adaptive random testing, which should reduce overheads of restricting the search space. Mathesen et al. [14] propose a metaheuristic whose global search proposes locations where the local search is restarted if a local minimum has been found. Finally, a genetic algorithm has been proposed that produces the offspring (test suites) randomly in a generation rather than by evolution if the current population lacks diversity, which has an effect of restarting the search [22]. All of these approaches provide metaheuristic-specific restart strategies that are intertwined with the metaheuristic. In contrast, *Bet and Run* lifts the restart strategy to a generic level that is independent of the metaheuristic being used.

The main goal of this paper is to evaluate the feasibility and effectiveness of a *generic Bet and Run* approach for the TCG problem. For this purpose, we investigate the following two research questions:

RQ1 Can *generic Bet and Run* be adapted to work on the TCG problem?
RQ2 Does *generic Bet and Run* show a significant improvement in the quality of the generated tests as measured by coverage metrics?

Accordingly, the primary contributions of this paper are:

(1) As the first study, we investigate the applicability of *generic Bet and Run* on the TCG problem using EvoSuite and DynaMOSA. This will answer RQ1.
(2) We conduct an empirical study to compare *Bet and Run* and default EvoSuite/DynaMOSA on real-world 107 Java classes from the *SF110* corpus [7].
(3) We provide a statistical analysis of the effectiveness of *Bet and Run* and EvoSuite/DynaMOSA. This will answer RQ2.

2 Test Case Generation with EvoSuite and DynaMOSA

The context of our work is the automated *test case generation (TCG)* problem that is about generating good quality tests for a given software. Particularly, we focus on generating unit tests, for which we use *EvoSuite*. EvoSuite is a popular search-based "tool that automatically generates test cases with assertions for classes written in Java code." [5, p. 416]. The metaheuristic used in EvoSuite has evolved over time from whole test suite generation [6] to *DynaMOSA* [18].

The *Dynamic Many-Objective Sorting Algorithm (DynaMOSA)* [18] is the state-of-the-art many-objective genetic algorithm to solve the test case generation problem by redefining it into a many-objective problem. Conceptually, DynaMOSA works on each statement (i.e., target to cover) of the class under test individually, instead of trying to generate a test suite for all statements simultaneously. Thus, it breaks the complex task of generating a test suite for an entire class into more manageable smaller pieces. DynaMOSA also only computes test cases for targets that can be reached immediately: All branches that are still nested below other uncovered control flow nodes are temporarily ignored. Thereby, DynaMOSA reduces the number of targets that are to be covered simultaneously and thus decreases computational complexity.

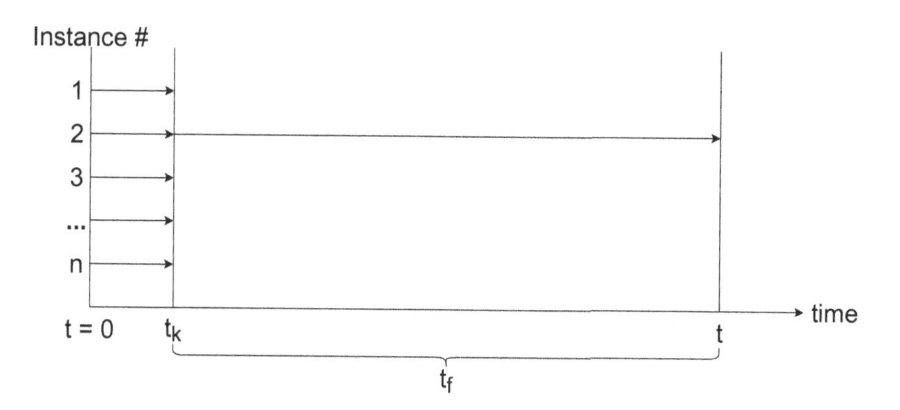

Fig. 1. Example: Behavior of *Bet and Run* with n instances and a given run time $t_{total} = t_f + n \cdot t_k$ with a maximum initial run time t_k for each instance. In this case instance #2 showed the most promise at t_k and was left running.

3 Bet and Run for Test Case Generation

Any metaheuristic can get stuck in local optima. Once trapped, such algorithms do not tend to break free of such a plateau for a while. Friedrich et al. [10] adopted and evaluated the approach *Bet and Run* for two general theoretical computer science problems. *Bet and Run* was initially presented by Fischetti and Monaci [4] for a sequential tree search method. Friedrich et al.'s approach is no longer metaheuristic-specific and makes use of the typical multi-processing architecture of modern hardware. The strategy is characterized by first starting a number of algorithm instances that are identical in terms of call parameters but contain different starting populations. After some time only the instance that showed the most promise in terms of fitness values is then kept running.

Currently in the field of TCG, no restart strategies are automatically applied. Unless the end-users intervene themselves, the optimization algorithm is simply run once, and once only. But since the typically used optimization algorithms in state-of-the-art tools such as EvoSuite are employing heuristics, the observations above hold true in this field as well. Therefore, *Bet and Run* strategies are generally applicable to the *Test Case Generation* problem as well.

In their 2018 paper [18], Panichella et al. state that DynaMOSA quickly increases test suite quality, once it starts to generate tests. Therefore, DynaMOSA is a good candidate for *Bet and Run*, as we can select the most promising instance early in the search process.

In our study, we adapt the general *Bet and Run* approach as follows:

Definition 1. Bet and Run adaptation to *TCG*: *Let t_{total} be the full time budget available to the optimization.*

(i) Starting phase. *First we start n instances of DynaMOSA simultaneously, for some fixed n. While all typical run parameters are held fixed, the seed for the random number generator used for the optimization is varied across all n instances.*
(ii) Evaluation phase. *After the initially selected but fixed time t_k has passed, we evaluate all n instances for their fitness in terms of the overall fitness score computed by EvoSuite for the tests generated so far.*
(iii) Elitism phase. *We then kill the least fit $n-1$ instances. If there are two or more candidates with the exact same fitness score, we continue the first of those candidates.*
(iv) Run phase. *Let the single remaining (and so far most promising) instance continue until t_{total} is fully used up. That is, the remaining instance runs for $t_f := t_{total} - n \cdot t_k$.*

This means n, t_k, and t_{total} are free parameters of our adaptation of the general *Bet and Run* approach. An intuitive visualization is shown in Fig. 1.

Notation: Let p be the percentage so that $t_k := p \cdot t_{total}$. We then say $RESTARTS_p^n$ when we use n instances of the algorithm during the *starting phase* and $t_k = p \cdot t_{total}$ as the evaluation time.

4 Implementation

In this section we provide an overview of the TCG project EvoSuite [5], as well as the adaptations made by us, so that we may answer RQ1. We used EvoSuite from a self-compiled *1.0.7* snapshot from the *git* repository[1]. This was done as the `-generateMO` API option to use the state-of-the-art DynaMOSA algorithm is not available yet in the official 1.0.6 release version of EvoSuite.

To answer RQ1, the first requirement is to make use of the EvoSuite framework so as to extend its functionality with the generic *Bet and Run* restart strategy. As this approach works on top of the tool itself, we do not need to change any code of EvoSuite directly. Instead, we may use it as a blackbox tool, that is, by using its command-line API we are able to get it to run tasks for us.

Therefore, it suffices to concentrate on providing a runner script implementing the *Bet and Run* strategy, along with a set of tools needed for saving and evaluating raw data. Apart from those scripts, all other program functionalities (i.e., genetic operators, metrics, problem encoding, etc.) are those implemented in the EvoSuite tool.

In most real-world tools, there is one limitation present: One cannot simply "pause" a given algorithm run at some arbitrary point during computation, evaluate the fitness at that point, and then decide whether to continue that run

[1] Download of EvoSuite after commit https://github.com/EvoSuite/evosuite/commit/e26a6b725539370aa22976f041f3304972e201a1 on January 10, 2020.

or not. Instead, many tools provide a timeout function, that allows the user to specify a runtime, after which the optimization is stopped and evaluated. For implementing the *Bet and Run* approach, one therefore needs to fully restart the most promising instance after the given evaluation. This, however, leads to two main strategies of coping with the tool limitation in terms of remaining runtime:

- It is possible to give the remaining instance after the full restart a maximum runtime of $t_f + 1 \cdot t_k$, in order to emulate the theoretical idea of the generic *Bet and Run* approach, where "pausing" the initial n instances would be possible. However, this implies that *Bet and Run* has t_k more computation time, when compared to the normal algorithm run without the restart strategy.
- A second possibility is to give the most promising instance after the full restart a maximum runtime of t_f. In this case, the time limit ensures, that both the normal algorithm *and Bet and Run* do not take more than t_{total}.

In this paper, we decided to use the latter approach, so as to not give *Bet and Run* any more runtime than the baseline is given. This helps to make the results between both approaches comparable.

To run the prototype implementation of *Bet and Run*, EvoSuite as a stand-alone, blackbox tool only requires a runner script to build the command-line API calls. This paper provides two such runner scripts: the *evosuiteRunner*, and the *evosuiteBARRunner*. Both of those runners are automatically called by the main *experiments.py*. While the first runner simply provides an interface for running the standard EvoSuite implementation on many classes in parallel (i.e., the $RESTARTS^1_{100\%}$), the second one includes all changes necessary for the *Bet and Run* variants. Both runners also collect data provided by EvoSuite and write the raw values to a checkpoint file per class. Additionally, both scripts are able to parse this raw data and thus compress the verbose output to the metrics.

All used scripts, parameter and settings files, as well as the used EvoSuite sources are made available publicly with the online resources of this paper[2].

5 Evaluation

In this section, we present our experimental set up, lay out the methodology with which we analysed the raw data gathered, present and discuss the results of our analysis, and finally we briefly discuss threats to validity. We provide both of the interesting raw data sets with the online resources of this paper.

5.1 General Experimental Set-Up

All experiments of this paper, and the algorithms they are based upon, are implemented in either *Java OpenJDK 1.8.0* or *Python 3.6.10*. Additionally, for building the EvoSuite files, a *Maven* installation is required. All other tool versions are listed in the online resources accompanying this paper.

[2] https://www.doi.org/10.5281/zenodo.3903206.

Computational Resources: The experiments were run concurrently on a *Dell R920* compute server with 4 *Intel Xeon E7-4880 v2* processors (60 cores) @ 2.5 GHz, 1024 GB RAM, and a *Suse Leap 15* OS (running *Linux* kernel 4.12.14). For both experimental approaches we used 40 parallel worker threads.

Baseline Selection: To investigate both the applicability (RQ1) as well as the effectiveness (RQ2) of the *Bet and Run* strategy, DynaMOSA is the natural baseline to compare with. Not only is it the current state-of-the-art algorithm in the field of *Test Case Generation*, it is also the algorithm extended in this study: We implemented the restart strategy using the EvoSuite tool (by use of its command-line API) and its implementation of the DynaMOSA algorithm.

Data Set Generation: Finding a set of classes that is both large enough to be representative of the state of the art in software development, as well as comparable to the studies in the field presents a certain challenge, as the adapted data sets are often not made easily available for the public.

The *SF100* corpus of classes is a collection of 100 Java projects from www.sourceforge.net. The data set was designed to be statistically representative. In 2014 Fraser et al. [7] accounted for the fact that *SourceForge* is home to a large number of older (and thus stale) projects, by revising the data set and adding the 10 most popular projects to it. This resulted in the data set now known as *SF110*[3]. This benchmark has been used in many studies in the field [19,20], and most relevant to this study, it is also the main source of subjects for the study by Panichella et al. [18], in which they introduced DynaMOSA. Thus, to be comparable to their results, we chose to adopt the *SF110* benchmark. As we could not find the exact list of classes used by Panichella et al. in their 2018 study, we followed their class selection algorithm:

1. Compute McCabe's cyclomatic complexity [15] for each method in the SF110 dataset using the *CKJM* library[4].
2. Discard any project that only contains trivial methods (i.e., classes that only contain methods with a cyclomatic complexity of 1).
3. Randomly sample from the remaining projects.

Applying the above algorithm to the *SF110* corpus of classes resulted in a dataset of 107 different, non-trivial classes from 68 projects as provided by the *SF110* corpus of classes. The projects vary widely in size and difficulty: from a mere 18 branches on average (with 122 statements) in project `lavalamp`, to an average of 2515 branches in a medial 5887 statements in project `jmca`. This dataset was used for all experiments of this study. A complete list of all selected classes are made public with the online resources accompanying this paper.

[3] http://www.evosuite.org/experimental-data/sf110/.
[4] http://gromit.iiar.pwr.wroc.pl/p_inf/ckjm/.

5.2 Experimental Set-Up for RQ1

$RESTARTS^{1}_{100\%}$ is the standard run of a given algorithm without any restarts – and is thus ideally suited as a baseline. Friedrich et al. [10] have shown that $RESTARTS^{40}_{1\%}$ is the parameter set that yields the best results for both theoretical problems in their study. Of the non-trivial settings, we thus initially try $RESTARTS^{40}_{1\%}$, $RESTARTS^{20}_{2\%}$, and $RESTARTS^{8}_{5\%}$.

Therefore, to answer RQ1 and understand whether *Bet and Run* actually can work in the field of TCG, we ran an experiment using the above-mentioned computational settings, baseline, and dataset to test the four *Bet and Run* strategies.

The parameters for the *Bet and Run* experiments were thus set as follows: $n = 1$, 40, 20, and 8 initial runs, as well as $t_k = 100\%$, 1%, 2%, and 5% of t_{total}. The total time budget (t_{total}) was set to 5 min. In any run of the *Bet and Run* strategies t_{total} is split into $n \cdot t_k$ and t_f for the two phases (cf. Definition 1). Seeds for the initial state of the random number generator were selected at random for both the baseline and the initial phase of all employed *Bet and Run* strategy executions. For the *Bet and Run* experiments, we set the timeout t_k for all initial instance runs via the `-Dglobal_timeout` API parameter. On reaching time t_k, we evaluate each instance using the overall fitness score as provided by EvoSuite. This fitness score is an approach level (i.e., a distance) and means "lower is better". After that evaluation, we only run the single most promising candidate by restarting that instance with the timeout now set to t_f, in an effort to not give *Bet and Run* more runtime than the baseline. All four approaches were tested 10 times each in order to account for the pseudo-random nature of the employed DynaMOSA algorithm. A single experimental run of any given class took between approximately 60 s to 5 min. The entire experiment for answering RQ1 took about 3 full days per tested approach, when we used 40 concurrent worker threads.

5.3 Answering RQ1

In our first experiment, neither $RESTARTS^{40}_{1\%}$ nor $RESTARTS^{20}_{2\%}$ managed to generate any usable data. This is true for *all* 107 classes in the dataset. This problem is due to the fact that the requested test case generation time of the initial phase (t_k) – as required by the respective restart strategy – was too short for DynaMOSA to have managed to produce any tests while within the actual generation phase of EvoSuite. The restart strategy thus could not select any seed for the continued computation past t_k and therefore failed. In the $RESTARTS^{8}_{5\%}$ experiment, this particular problem happened to two of the selected classes: *Evaluation* of project *weka*, and *MethodWriter* of project *jiprof*. For all other classes $RESTARTS^{8}_{5\%}$ did manage to generate at least one test suite over the ten experimental runs and thus qualified for further evaluation.

Answer to RQ1: The *Bet and Run* approach can indeed be adapted to yield usable results within the field of TCG, as shown by our experimental runs of $RESTARTS^{8}_{5\%}$.

5.4 Experimental Set-Up for RQ2

Using the result from RQ1, we select $RESTARTS^{8}_{5\%}$ for the statistical comparison to the baseline. To determine the effectiveness of the selected *Bet and Run* strategy, we keep the dataset, computational set-up, and the parameters of both strategies the same as they were set to in the experiment for RQ1. We do, however, in order to generate a robust statistical evaluation repeat the execution of each strategy 30 times. Furthermore, we also log for evaluation the following fitness metrics upon fully using the time budget for each class.

Fitness Metrics: EvoSuite provides a number of different fitness metrics for each run by default: An overall fitness score, as well as Line, Branch, Exception, Weak Mutation, Output, Method, Method No Exception, and C Branch coverages. Additionally, we also track per class and per experimental run if EvoSuite encountered any internal errors during computation.

The overall fitness score is an aggregated *approach level* of the test suite to covering the remaining targets. As this approach level is a distance, it is to be read as "lower is better". Any coverage metric is a percentage of reached coverage and all are thus to be read as "higher is better." The according definitions used by EvoSuite may be found in [19], as well as in the source code of the tool. The number of internal EvoSuite errors, which is also tracked, is an absolute value, thus "lower is better".

The full experiment for RQ2 took another 18 days. Note, that these times do not include script run times for the subsequent evaluation of the raw data gathered.

5.5 Evaluation Methodology for RQ2

The first step in answering RQ2, is to provide an overview over the actual behavior of the collected measures with regards to the two experiments. For *mean* values given in this paper the reader can safely assume "higher is better" for every metric, with the exception of the two measures *Time* and *Fitness Score*. The two excluded indicators need to be handled differently: The *Fitness Score* is an *approach level* of the test suite (cf. Sect. 5.4), where "lower is better". The measure *Time* has only informative character and shows how long each of the final instances spent in the actual generation phase.

In the subsequent statistical analysis (see Sect. 5.7), we employ the non-parametric *Wilcoxon Rank Sum* hypothesis test to determine statistical significance of our results for each class and each measure taken.

5.6 Answering RQ2

Generally, the results of our second experiment are structurally very similar across the different classes, as well as metrics logged. Raw data from both the *baseline* and the $RESTARTS^{8}_{5\%}$ for all classes (that generated at least one test case) and for all metrics are made available in the accompanying online resources.

There we provide the *.zip of all raw data gathered, as well as a *.txt file with aggregated raw data, which was used in the evaluation below in this section.

We note that in the baseline experimental runs, 14 of the original 107 selected classes are rendered unusable for statistical evaluation, due to an excessively high number of internal EvoSuite errors. Of the original set of classes in the $RESTARTS^{8}_{5\%}$ experiments, 15 are failures due to internal errors of the tool. Of all failures 12 classes fail too often for both approaches, 3 only when using DynaMOSA, and 4 only using *Bet and Run*. Of those 12 classes failing for both approaches 1 class completely fails to generate *any* test cases for either approach (class *Evaluation* of project *weka*). In total, this problem leads to 17 of the 107 classes not being eligible for statistical evaluation, apart from looking at which of the approaches was able to execute EvoSuite more stably.

5.7 Statistical Analysis and Discussion

In this section, we will present the results of the statistical analysis for the complete experiments of DynaMOSA and $RESTARTS^{8}_{5\%}$. We obtained the p-values using the non-parametric *Wilcoxon Rank Sum* hypothesis test, as provided by *SciPy*. Note that of the 107 initially selected classes only 90 were eligible for statistical evaluation due to internal EvoSuite errors. Table 1 shows per metrics collected the number of classes for which *Bet and Run* (BAR) is (i) identical to, (ii) worse than, and (iii) better than the baseline (BL). In each column, within the brackets we also show the number of statistically significant results of the total number of classes within that field. For instance, the first line is to be read as: For the metric *Fitness Score* there were a total of 28 classes that led to identical results in both approaches. Of those 28, 28 are statistically significant. Further there were 19 classes where the baseline outperformed the *Bet and Run* approach, of which 2 were statistically significant. Finally, there were 43 instances, where *Bet and Run* outperformed the baseline approach, of which only 4 classes were statistically significant.

Table 1 shows that neither approach was significantly better than the other overall. In fact, in most cases the p-value either does not reach statistical significance or even indicates that both approaches reach statistically the exact same result. Over all 9 metrics and all 90 evaluated classes there were only 25 results that show any statistically significant *difference* between the two approaches. Of those, *Bet and Run* was 18 times significantly better than the baseline, and 7 times significantly worse (cf. Table 2). In contrast, over all 9 metrics and all 90 classes, we see 460 instances in which both approaches yield the exact same result. This means, when looking at *all* 810 results (i.e., 9 metrics $\times$ 90 classes) 785 instances are either only marginally different, or yield the exact same result.

We argue that this is indicative of a limitation of the approach: In a real-world setting in the TCG context, runs typically are strictly limited in their computation times. This strict and more importantly *short* time budget for the genetic algorithm appears to mitigate the plateauing problem quite effectively. Furthermore, the use of the state-of-the-art DynaMOSA algorithm with its optimization mechanisms that work on minimizing the initial number of coverage

Table 1. Shown here are the aggregated results of the statistical analysis over all metrics when comparing the baseline DynaMOSA (BL) and the *Bet and Run* $RESTARTS^{8}_{5\%}$ strategy (BAR), where BAR = BL means both approaches show statistically the same result (p-value = 1.0), BAR < BL means the *Bet and Run* approach was worse than the baseline, and BAR > BL means *Bet and Run* was better than the Baseline. Any value inside of brackets indicate the number of statistically significant results in that field. Note that of the 107 initially selected classes only 90 were eligible for statistical evaluation due to internal EvoSuite errors.

Metric	BAR = BL	BAR < BL	BAR > BL
Fitness score	28 (28)	19 (2)	43 (4)
Line coverage	39 (39)	13 (1)	38 (3)
Branch coverage	35 (35)	16 (2)	39 (4)
CBranch coverage	36 (36)	16 (2)	38 (4)
Method coverage	78 (78)	6 (0)	6 (0)
Method no exception coverage	69 (69)	11 (0)	10 (1)
Exception coverage	80 (80)	5 (0)	5 (0)
Output coverage	57 (57)	12 (0)	21 (1)
Weak mutation coverage	38 (38)	19 (0)	33 (1)

targets (i.e., only search solutions for branches that can actually be covered at this time) help to further decrease the potential of getting stuck in a local optimum.

Panichella et al. state in their 2018 paper [18] that DynaMOSA quickly increases test suite quality, once it starts to generate tests. We can confirm this behavior: In our early experiments for RQ1, we saw that after a short initialization phase the algorithm managed to quickly improve fitness scores across all metrics for the population. We believe this quick improvement phase at the beginning is due to the fact that DynaMOSA only works on targets that can possibly be covered currently, instead of working immediately on every single target (i.e., the mechanism that the algorithm does not start working on targets that are still restricted by higher level, yet uncovered targets). This behavior is also a possible explanation for what we stated above: Neither $RESTARTS^{40}_{1\%}$, nor $RESTARTS^{20}_{2\%}$ managed to generate any tests. When we are using these settings, no candidate instance can be chosen to survive past the decision time t_k, as t_k was too short to get past the above-mentioned initialization phase. We note that while $RESTARTS^{40}_{1\%}$ was shown in Friedrich et al. [10] to be the best strategy, we cannot confirm this finding here. Moreover, we also cannot safely confirm quality increases in the test suite generated by EvoSuite. In fact, we would argue that the additional computational overhead is indeed an argument against usage of that particular restart strategy in the field of *Test Case Generation*, when also looking at the achieved results.

We have, however, seen a significant increase (p = 0.002930) in tool stability over all selected classes, when employing the *Bet and Run* strategy

Table 2. Shown here are the classes that reached a significant difference when comparing the baseline DynaMOSA and the *Bet and Run* $RESTARTS^{8}_{5\%}$ strategy.

Metric	Stat. sig. worse classes	Stat. sig. better classes
Fitness score	99_newzgrabber: Downloader 12_dsachat: Handler	19_jmca: JMCAAnalyzer 61_noen: ProbeInformation 82_ipcalculator: BinaryCalculate 86_at-robots2-j: RobotRenderer
Line coverage	12_dsachat: Handler	19_jmca: JMCAAnalyzer 82_ipcalculator: BinaryCalculate 86_at-robots2-j: RobotRenderer
Branch coverage	39_diffi: IndexedString 12_dsachat: Handler	19_jmca: JMCAAnalyzer 61_noen: ProbeInformation 82_ipcalculator: BinaryCalculate 86_at-robots2-j: RobotRenderer
CBranch coverage	39_diffi: IndexedString 12_dsachat: Handler	19_jmca: JMCAAnalyzer 61_noen: ProbeInformation 82_ipcalculator: BinaryCalculate 86_at-robots2-j: RobotRenderer
Method coverage	–	–
Met. no exc. cov.	–	86_at-robots2-j: RobotRenderer
Exception cov.	–	–
Output coverage	–	50_biff: Scanner
Weak mut. cov.	–	61_noen: ProbeInformation

$RESTARTS^{8}_{5\%}$. While the *baseline* showed a total of 578 internal errors of the tool, the *Bet and Run* approach showed 500 errors across all classes. This difference of exactly 78 errors occurred over all 3210 experimental runs per approach (30 repetitions × 107 classes). That means in 15.6% of all experimental runs, the EvoSuite tool encountered errors during the $RESTARTS^{8}_{5\%}$ experiments, in comparison with 18.0% in the normal behavior EvoSuite experiments. This difference can be explained by the transience of internal EvoSuite errors, where a simple restart may fix the problem thanks to the selection of a new random seed value. As these transient internal tool errors usually happen very early during computation and since the *Bet and Run* restart strategy never selects any seed value that showed an error, this is expected behavior of the employed strategy. We assume the rather high number of internal errors by EvoSuite is due to the fact that we employed a self-compiled (and thus potentially unstable) tool snapshot, instead of an actual release of the software. We therefore argue, that unless tool stability is of particular importance to the use case, the statistically significant gains in this metric *do not* outweigh computational overhead needed for those increases.

Answer to RQ2: Apart from statistically significant increases in tool stability, we cannot confirm any gains in the remaining nine metrics when using *Bet and Run* with the best parameters found in literature. In fact, we question the effectiveness of *Bet and Run* in the field of TCG, as time constraints and highly specialized algorithms (e.g., DynaMOSA) in this field effectively mitigate the typical plateauing problem of metaheuristics.

5.8 Threats to Validity and Future Work

In this paper, we rely on a search-based approach with a restart strategy to solve the Test Case Generation Problem. As such we see the following potential threats to the validity of our work.

Internal: In all genetic algorithms, a fair amount of (pseudo-)randomness is involved in the generation of their respective results. In DynaMOSA, the randomness problem is somewhat alleviated by the use of the archiving function – a test case covering some new target is never lost due to future mutations again. Additionally, we try to mitigate this threat by (i) sampling 30 runs per selected class and evaluating all results only in the average case; as well as (ii) providing online the seed values for all runs, both in the normal EvoSuite and the *Bet and Run* experiments (in the latter we provide the seeds that showed the most promise at time t_k). The seed values are made available in the aggregated result data files of the online resources. Furthermore, there exists the possibility of faults within our strategy implementation. We tried to minimize this threat by use of standard components (such as the EvoSuite implementation of DynaMOSA or the *SciPy* libraries for the statistical testing) wherever possible. All implementations, codes, and scripts are made available for inspection, review, and validation with the online resources accompanying this study.

External: Our evaluation is based upon the benchmark data set "*SF110 Corpus of Classes*" as provided by Fraser et al. [7]. While Fraser et al. argue that they selected 100 statistically representative *SourceForge* projects and even further enhanced this set with another 10 "popular" projects, the possibility that the heterogeneous nature of software is indeed not fully accounted for still exists: The restart strategy might behave differently for a different set of *real* classes. We leave for future work to revisit the use of the restart strategy with a different data set.

Construct: The first threat in this category that we can see is the possibility of the experiments not yet having encountered any plateaus. This threat could only be mitigated by redoing the entire set of experiments with a much higher timeout. However, this would go against real world usage of the EvoSuite tool. In practical settings, the tool is used to *quickly* generate a test suite for project classes – extremely high timeouts are thus the exact opposite of the intended use.

We will, however, leave for future work a confirmation study with those increased timeouts: It is a matter of running the provided implementation with a new set of parameters. A second threat of this family is that while we selected restart strategy parameters that were shown to be "good" in the study by Friedrich et al. [10], "good" in the context of TCG might be something else entirely. In our case, it may be the case that most of t_k is in fact used for tool initialization, and no significant test generation is performed so as to correctly estimate the "best" performing one. Similarly to the first threat to construct validity, a parameter tuning study with a new set of parameters is left for future work, as even more parameters for the restart strategy would go beyond the scope of this initial study. And thirdly, by forcing us to restart the most promising instance completely, the implementation limited us to choose between either giving the remaining instance in the run phase a maximum run-time of (i) $t_f + t_k$, or (ii) t_f. We chose the latter, so that both *Bet and Run* and the baseline do not use more than t_{total}. This leads to the evaluated instance having $p\%$ less run-time, while the total run-time of the *Bet and Run* approach is 100% of t_{total}. Had we used approach (i), however, *Bet and Run* would have had a total run-time of $t_{total} + t_k$.

6 Conclusion

In this study, we provided a tool to run and evaluate the generic restart strategy *Bet and Run* in the context of Test Case Generation using EvoSuite and the state-of-the-art DynaMOSA algorithm. To the best of our knowledge this was the first study that applies the generic *Bet and Run* approach to this field. Our work indicates that use of a restart strategy instantiated with the best parameters found in the literature does *not* generally lead to gains in the quality metrics: Not a single metric was improved by the *Bet and Run* restart strategy when compared with the EvoSuite/DynaMOSA as the baseline. In fact, for most selected classes, both approaches showed statistically the exact same results.

Only the stability of runs showed statistically significant improvements when *Bet and Run* was employed, as the number of internal EvoSuite errors decreased in the final generation process. However, from what we saw in our experiments, most of the internal EvoSuite errors are transient: In a real-world setting where the tool did not generate a test suite immediately, it is more than likely that a second (manually started) run with the same parameters would fix the problem.

Our results indicate, that (contrary to the promising results in the 2017 study by Friedrich et al. [10]) the restart strategy *Bet and Run* is *not* suited for improving the quality of automatically generated test suites using EvoSuite and its state-of-the-art DynaMOSA implementation considering the best *Bet and Run* parameters found in literature.

References

1. Baars, A., et al.: Symbolic search-based testing. In: Proceedings of the 2011 26th International Conference on Automated Software Engineering, ASE 2011, pp. 53–62. IEEE (2011)
2. Bianchi, L., Dorigo, M., Gambardella, L.M., Gutjahr, W.J.: A survey on metaheuristics for stochastic combinatorial optimization. Nat. Comput. Int. J. **8**(2), 239–287 (2009)
3. Chan, K.P., Chen, T.Y., Towey, D.: Forgetting test cases. In: Proceedings of the 30th International Computer Software and Applications Conference, COMPSAC 2006, vol. 01, pp. 485–494. IEEE (2006)
4. Fischetti, M., Monaci, M.: Exploiting erraticism in search. Oper. Res. **62**(1), 114–122 (2014)
5. Fraser, G., Arcuri, A.: EvoSuite: automatic test suite generation for object-oriented software. In: 19th Symposium on the Foundations of Software Engineering (FSE) and 13th European Software Engineering Conference (ESEC), pp. 416–419. ACM (2011)
6. Fraser, G., Arcuri, A.: Whole test suite generation. IEEE Trans. Software Eng. **39**(2), 276–291 (2013)
7. Fraser, G., Arcuri, A.: A large-scale evaluation of automated unit test generation using EvoSuite. ACM Trans. Softw. Eng. Methodol. **24**(2), 81–842 (2014)
8. Fraser, G., Arcuri, A.: 1600 faults in 100 projects: automatically finding faults while achieving high coverage with EvoSuite. Empirical Softw. Eng. **20**(3), 611–639 (2015)
9. Fraser, G., Arcuri, A., McMinn, P.: Test suite generation with memetic algorithms. In: Proceedings of the 15th Annual Conference on Genetic and Evolutionary Computation, GECCO 2013, pp. 1437–1444. ACM (2013)
10. Friedrich, T., Kötzing, T., Wagner, M.: A generic bet-and-run strategy for speeding up stochastic local search. In: Proceedings of the Thirty-First AAAI Conference on Artificial Intelligence, pp. 801–807. AAAI Press (2017)
11. Kempka, J., McMinn, P., Sudholt, D.: Design and analysis of different alternating variable searches for search-based software testing. Theoret. Comput. Sci. **605**, 1–20 (2015)
12. Lourenço, H.R., Martin, O.C., Stützle, T.: Iterated local search: framework and applications. In: Gendreau, M., Potvin, J.Y. (eds.) Handbook of Metaheuristics, pp. 363–397. Springer, Boston (2010). https://doi.org/10.1007/978-1-4419-1665-5_12
13. Martí, R.: Multi-start methods. In: Glover, F., Kochenberger, G.A. (eds.) Handbook of Metaheuristics, pp. 355–368. Springer, Boston (2003). https://doi.org/10.1007/0-306-48056-5_12
14. Mathesen, L., Yaghoubi, S., Pedrielli, G., Fainekos, G.: Falsification of cyber-physical systems with robustness uncertainty quantification through stochastic optimization with adaptive restart. In: 15th International Conference on Automation Science and Engineering, CASE, pp. 991–997. IEEE (2019)
15. McCabe, T.J.: A complexity measure. IEEE Trans. Software Eng. **2**(4), 308–320 (1976)
16. McMinn, P.: Search-based software test data generation: a survey. Softw. Testing Verif. Reliab. **14**(2), 105–156 (2004)
17. Orso, A., Rothermel, G.: Software testing: a research travelogue (2000–2014). In: Future of Software Engineering Proceedings, FOSE 2014, pp. 117–132. ACM (2014)

18. Panichella, A., Kifetew, F.M., Tonella, P.: Automated test case generation as a many-objective optimisation problem with dynamic selection of the targets. IEEE Trans. Software Eng. **44**(2), 122–158 (2018)
19. Rojas, J.M., Vivanti, M., Arcuri, A., Fraser, G.: A detailed investigation of the effectiveness of whole test suite generation. Empirical Softw. Eng. **22**(2), 852–893 (2017)
20. Shamshiri, S., Rojas, J.M., Fraser, G., McMinn, P.: Random or genetic algorithm search for object-oriented test suite generation? In: Proceedings of the Genetic and Evolutionary Computation Conference, GECCO, pp. 1367–1374. ACM (2015)
21. Tonella, P.: Evolutionary testing of classes. In: Avrunin, G.S., Rothermel, G. (eds.) Proceedings of the International Symposium on Software Testing and Analysis, ISSTA, pp. 119–128. ACM (2004)
22. Vogel, T., Tran, C., Grunske, L.: Does diversity improve the test suite generation for mobile applications? In: Nejati, S., Gay, G. (eds.) SSBSE 2019. LNCS, vol. 11664, pp. 58–74. Springer, Cham (2019). https://doi.org/10.1007/978-3-030-27455-9_5

Bytecode-Based Multiple Condition Coverage: An Initial Investigation

Srujana Bollina[1] and Gregory Gay[2](✉)

[1] University of South Carolina, Columbia, SC, USA
sbollina@email.sc.edu
[2] Chalmers and the University of Gothenburg, Gothenburg, Sweden
greg@greggay.com

Abstract. Masking occurs when one condition prevents another from influencing the output of a Boolean expression. Adequacy criteria such as Multiple Condition Coverage (MCC) overcome masking within one expression, but offer no guarantees about subsequent expressions. As a result, a Boolean expression written as a single complex statement will yield more effective test cases than when written as a series of simple expressions. Many approaches to automated test case generation for Java operate not on the source code, but on bytecode. The transformation to bytecode simplifies complex expressions into multiple expressions, introducing masking. We propose Bytecode-MCC, a new adequacy criterion designed to group bytecode expressions and reformulate them into complex expressions. Bytecode-MCC should produce test obligations that are more likely to reveal faults in program logic than tests covering the simplified bytecode.

A preliminary study shows potential improvements from attaining Bytecode-MCC coverage. However, Bytecode-MCC is difficult to optimize, and means of increasing coverage are needed before the technique can make a difference in practice. We propose potential methods to improve coverage.

Keywords: Search-based test generation · Adequacy criteria · Coverage criteria

1 Introduction

For any reasonably complex software project, testing alone cannot prove the absence of faults. As we cannot know what faults exist a priori, dozens of *adequacy criteria*—ranging from the measurement of structural coverage to the detection of synthetic faults [8]—have been proposed to judge testing efforts. In theory, if the goals set forth by such criteria are fulfilled, tests should be *adequate* at detecting faults related to the focus of that criterion. Adequacy criteria such as Statement or Branch Coverage have proven popular in both research and practice, as they are easy to measure, offer clear guidance to developers,

A. Aleti and A. Panichella (Eds.): SSBSE 2020, LNCS 12420, pp. 220–236, 2020.
https://doi.org/10.1007/978-3-030-59762-7_16

and present an indicator of progress [9]. Adequacy criteria also play an important role in search-based test generation, as they offer optimization targets that shape the resulting test suite [16].

Masking occurs when one condition—an atomic Boolean variable or subexpression—prevents another condition from influencing the output of the expression. Even if a fault in a Boolean expression is triggered, other parts of that expression—or future expressions encountered along the path of execution—can prevent that fault from triggering an observable failure during test execution.

Sophisticated logic-based adequacy criteria such as Multiple Condition Coverage (MCC) or Multiple Condition/Decision Coverage (MC/DC) are designed to overcome masking within a single expression. However, they can offer no guarantees about masking in subsequent expressions. As a result, such criteria are sensitive to how expressions are written [8]. A Boolean expression written as a single complex statement will more effective test cases than the same expression written as multiple simple expressions, as the adequacy criterion will not prevent masking between expressions.

Many approaches to automated analysis and test case generation for Java operate not on the source code, but on the resulting bytecode [4,19]. The transformation from source to bytecode translates complex expressions into multiple simple expressions, introducing the risk of masking between expressions. This could limit the fault-finding potential of bytecode-based adequacy criteria. To overcome this limitation, we propose a new variant of Multiple Condition Coverage.

Our approach, **Bytecode-MCC**, is a new test coverage criteria that prescribes a set of test obligations—goals that must be satisfied by test cases—for a class-under-test. Bytecode-MCC groups related Boolean expressions from the bytecode, reformulates the grouping into a single complex expression, and calculates all possible combinations of conditions within the constructed expression. Bytecode-MCC should produce test obligations that—when satisfied—are more likely to reveal faults in the program logic than tests providing simple coverage over the simplified bytecode.

Bytecode-MCC can be used to measure the power of existing test suites or as a target for automated test generation. To examine both scenarios, we have implemented an algorithm to generate test obligations and measure coverage in the EvoSuite search-based test generation framework [4]. We have also implemented a fitness function within EvoSuite intended to enable the automated generation of test suites.

We conducted a preliminary study examining the effectiveness of test generation targeting Bytecode-MCC on 109 faults from Defects4J—a database of real faults from Java projects [10]. Results attained for the "Time" system, where targeting the combination of Bytecode-MCC and Branch Coverage yields an average of 92% Bytecode-MCC coverage, yield an average 32.50%–35.00% likelihood of fault detection—well over the overall average. This suggests the potential of approaches that can attain high Bytecode-MCC coverage.

However, the results for other systems are more negative. Bytecode-MCC is difficult to optimize, and our fitness function does not offer sufficient feedback to guide test generation. Additional search budget does not guarantee higher levels of coverage. This suggests that Bytecode-MCC may be best used as a method of judging test suite quality, rather than as a direct generation target. Simultaneously targeting Bytecode-MCC and Branch Coverage improves coverage of Bytecode-MCC and improves the likelihood of fault detection, as the fitness function for Branch Coverage offers more feedback to the search process. Therefore, other fitness functions may offer the means to satisfy Bytecode-MCC. Bytecode-MCC has potential to yield effective test suites if the identified limitations can be overcome. We propose suggestions on how to proceed in future work and make our implementation available.

2 Background

Adequacy Criteria: Adequacy criteria are important in providing developers with the guidance they need to test efficiently, as they identify inadequacies in the test suite. For example, if a given test does not reach and execute a statement, it is inadequate for finding faults in that statement.

Each adequacy criterion prescribes a series of *test obligations*—goals that must be met for testing to be considered "adequate" with respect to that criterion. Often, such criteria are structured around particular program elements and faults associated with those elements, such as statements, branches of control flow, or Boolean conditions [8]. When a criterion has been satisfied, the system is considered to be adequately tested with respect to that element. Adequacy criteria have seen widespread use, as they offer objective, measurable checklists [9].

In this study, we are concerned with adequacy criteria defined over Boolean *decisions*, complete Boolean expressions within a program. Decisions can be broken into simple *conditions*—atomic Boolean variables or subexpressions—connected with operators such as `and`, `or`, `xor`, and `not`.

- **Decision Coverage:** This simple criterion requires that all decision statements evaluate to both possible outcomes—`true` and `false`. Given the expression (A *or* B), the test suite `(TT), (FF)` attains decision coverage over that expression.
- **Branch Coverage:** The source code of a program can be broken into basic blocks—sets of statements executed sequentially. Branches are decision statements that can decide which basic blocks are executed, such as `if`, `loop`, and `switch` statements. Branch Coverage requires that the test suite cover each outcome of all branches. Improving branch coverage is a common objective in test generation [13].
- **Multiple Condition Coverage (MCC):** MCC requires test cases that guarantee all possible combinations of condition outcomes within the decision to be executed at least once. Given expression (A *or* B), MCC coverage requires the value combinations `(TF), (TT), (FF), (FT)`. MCC is more

expensive to attain than Decision Coverage, but offers greater potential fault-detection capability. Note that, in the presence of short-circuit evaluation, infeasible outcomes are not required. In the previous example, short-circuit evaluation would reduce the required test suite to (FF), (FT), (T-).

Search-Based Software Test Generation: Selection of test inputs is generally a costly manual task. However, given a measurable testing goal, input selection can be framed as a *search* for the input that achieves that goal. Automation of input selection can potentially reduce human effort and the time required for testing [13].

Meta-heuristic search provides a possible solution for test input generation. Given scoring functions denoting *closeness to the attainment of those goals*—called *fitness functions*—optimization algorithms can sample from a large and complex set of options as guided by a chosen strategy (the *metaheuristic*). Metaheuristics are often inspired by natural phenomena. For example, genetic algorithms evolve a group of candidate solutions by filtering out bad "genes" and promoting fit solutions [4]. Due to the non-linear nature of software, resulting from branching control structures, the search space of a real-world program is large and complex. Metaheuristic search—by strategically sampling from that space—can scale effectively to large problems. Such approaches have been applied to a wide variety of testing scenarios [2]. Adequacy criteria are ideal as test generation targets, as such criteria can be straightforwardly translated into the fitness functions used to guide the search [16].

3 Bytecode-Based Multiple Condition Coverage

Masking occurs when a condition, within a decision statement, has no effect on the value of the decision as a whole. As an example, consider the trivial program fragments to the right. The program fragments have different structures, but are functionally equivalent.

Version 1: Complex Implementation

```
out_1 = (in_1 or in_2) and in_3;
```

Version 2: Simple Implementation

```
expr_1 = in_1 or in_2;
out_1 = expr_1 and in_3;
```

Version 1 presents the full, complex expression. Version 2 is defined using intermediate variable `expr_1`. Given a decision of the form `in_1 or in_2`, the truth value of `in_1` is irrelevant if `in_2` is true, so we state that `in_1` is *masked out*. Masking can have negative consequences on the testing process by preventing the effect of a fault from propagating to a visible failure.

MCC is able to overcome masking within a single expression by requiring that all possible combinations of condition values be attempted, meaning that non-masking test cases must exist. However, MCC is sensitive to how expressions are written. Variable `in_3` can have a masking effect—when it is `false`, it determines the value of the decision it is in. In the complex implementation, MCC would require test cases that overcome this masking effect, showing the effect that `in_1` and `in_2` have on the overall decision. In the simple, multi-line case, we only require that `in_3` be evaluated with the overall expression `expr_1`.

Suppose this code fragment is faulty and the correct expression should have been `in_1 and in_2`. Tests over the simplified implementation may miss this fault, while any test set providing coverage of the complex implementation would reveal this fault. This can have significant ramifications with respect to fault finding of test suites [7,8,18]. The simplified version can be more trivially satisfied, with fewer test cases, than cases where the code is structured into fewer, more complex expressions. The complex version will have more complex test obligations and will generally require more test cases, but those test cases will generally have more fault revealing power [8].

Many approaches to automated analysis and test case generation for Java operate not on the source code, but on the bytecode [4,19]. Bytecode is often easier to instrument than the source code—for instance, it can be obtained without the source code being present—and bytecode-based techniques are often more efficient and scalable than source code-based techniques [19]. Many state-of-the-art techniques compute coverage and generate test cases by monitoring the instrumented bytecode [4].

```
package tutorial;

class Sample
{
    public void myMethod(int a, int b, int c, boolean status) {

        if (status && (a != 0 || b > c || a != b ) ) {
                // true Case
            if (c <= a) {
                // continue here...
                } else {
                    //return "
                }
        } else {
            // false case
        }
    }

}
```

```
1 Method Name: <init>()V
2 I0 (0) LABEL L554976255
3 I1 (0) LINE  4
4 I2 (0) ALOAD 0
5 I3 (1) INVOKESPECIAL java/lang/Object.<init>()V
6 I4 (4) ALOAD 0
7 I5 (5) INVOKESTATIC org/evosuite/runtime/System.registerOb
8 I6 (8) RETURN
9 I7 (9) LABEL L411754329
10 Method Name: myMethod(IIIZ)V
11 I0 (0) LABEL L2061041761
12 I1 (0) LINE  8
13 I2 (0) ILOAD 4
14 I3 (2) UNKNOWN Branch I3 IFEQ, jump to L1862090560
15 I4 (5) ILOAD 1
16 I5 (6) UNKNOWN Branch I5 IFNE, jump to L550284715
17 I6 (9) ILOAD 2
18 I7 (10) ILOAD 3
19 I8 (11) UNKNOWN Branch I8 IF_ICMPGT, jump to L550284715
20 I9 (14) ILOAD 1
21 I10 (15) ILOAD 2
22 I11 (16) UNKNOWN Branch I11 IF_ICMPEQ, jump to L1862090560
23 I12 (19) LABEL L550284715
24 I13 (19) LINE  10
25 I14 (19) ILOAD 3
26 I15 (20) ILOAD 1
27 I16 (21) UNKNOWN Branch I16 IF_ICMPGT, jump to L1862090560
28 I17 (24) LABEL L1862090560
29 I18 (24) LINE  18
30 I19 (24) RETURN
31 I20 (25) LABEL L160656677
```

Fig. 1. A complex Boolean expression. Source code is shown on the left, and its equivalent bytecode is shown on the right.

The transformation from source to bytecode requires a similar simplification. Consider the example depicted in Fig. 1, where the source code is shown on the left and the bytecode is shown on the right. The complex statement on the left is translated into a series of simple expressions. As a result of this transformation, the risk of masking is introduced between expressions. As all expressions are maximally simplified, a straight-forward implementation of MCC would be equivalent to Branch Coverage over each individual statement.

Given concerns over the fault-revealing power of tests generated over simplified representations of Boolean expressions [7,8]—as well as concerns over whether code coverage attained over bytecode accurately predicts coverage over the source code [12]—test generation approaches able to account for this simplification may yield more effective and representative testing results.

To overcome the limitations imposed by the translation to a simplified program structure, we propose a new variant of MCC for bytecode, **Bytecode-MCC**. Bytecode-MCC groups related Boolean expressions, reformulates the grouping into a single complex expression, and calculates all possible combinations of conditions within the constructed expression. Bytecode-MCC should produce test obligations that—when satisfied—are likely to reveal faults in program logic.

Bytecode-MCC can be used to assess existing test suites as well as as a target for automated test generation. To examine both scenarios, we have implemented an algorithm to generate test obligations and measure coverage in the EvoSuite test generation framework [4]. We have also implemented a fitness function intended to enable the automated creation of Bytecode-MCC-satisfying test suites.

Our implementation of Bytecode-MCC as a fitness function and coverage measurement mechanism in EvoSuite is available from https://github.com/Srujanab09/evosuite.

3.1 Test Obligation Generation

To formulate the test obligations for Bytecode-MCC, we perform the following process:

1. Search the bytecode for Boolean expressions.
2. When an expression is detected, begin building a group of related expressions.
3. Add any subsequent Boolean expressions in the same bytecode label—a basic block of sequentially executed expressions—to the grouping.
4. When a new label is reached, add any new Boolean expressions to that grouping.
5. Stop when a label is reached with no Boolean expressions.
6. Formulate a truth table containing all evaluations of the gathered expressions.
7. Translate each row of the truth table into a test obligation.

For a given class and method, we inspect the bytecode to gather related Boolean expressions. In the bytecode, expressions are grouped into labels. A label indicates the start of a series of sequentially-executed expressions, and is a point that another control-altering expression can jump to. While monitoring

```
1  package tutorial;
2
3  class Sample
4  {
5
6      public void myMethod(int a, int b, int c, int d) {
7          if (a > 555) {
8              if (b < 9) {
9                  if ( c >= 788) {
10                     if (d != 909) {
11                 // continue here...
12                     } else {
13                         //return "D failed"
14                     }
15                 } else {
16                     //return "C failed";
17                 }
18             } else {
19                 //return "B failed";
20             }
21         } else {
22             //return "A failed";
23         }
24
25     }
26
27
28 }
```

```
1  Method Name: <init>()V
2  I0 (0) LABEL L554976255
3  I1 (0) LINE  4
4  I2 (0) ALOAD 0
5  I3 (1) INVOKESPECIAL java/lang/Object.<init>()V
6  I4 (4) ALOAD 0
7  I5 (5) INVOKESTATIC org/evosuite/runtime/System.registerOb:
8  I6 (8) RETURN
9  I7 (9) LABEL L411754329
10 Method Name: myMethod(IIII)V
11 I0 (0) LABEL L1899798671
12 I1 (0) LINE  8
13 I2 (0) ILOAD 1
14 I3 (1) INT 555 Type=null, Opcode=SIPUSH
15 I4 (4) UNKNOWN Branch I4 IF_ICMPLE, jump to L1871531303
16 I5 (7) LABEL L717117575
17 I6 (7) LINE  9
18 I7 (7) ILOAD 2
19 I8 (8) INT 9 Type=null, Opcode=BIPUSH
20 I9 (10) UNKNOWN Branch I9 IF_ICMPGE, jump to L1871531303
21 I10 (13) LABEL L1866234461
22 I11 (13) LINE  10
23 I12 (13) ILOAD 3
24 I13 (14) INT 788 Type=null, Opcode=SIPUSH
25 I14 (17) UNKNOWN Branch I14 IF_ICMPLT, jump to L1871531303
26 I15 (20) LABEL L2064685037
27 I16 (20) LINE  11
28 I17 (20) ILOAD 4
29 I18 (22) INT 909 Type=null, Opcode=SIPUSH
30 I19 (25) UNKNOWN Branch I19 IF_ICMPEQ, jump to L1871531303
31 I20 (28) LABEL L1871531303
32 I21 (28) LINE  26
33 I22 (28) RETURN
34 I23 (29) LABEL L2121337309
```

Fig. 2. A simple Java class. Source code is on the left and bytecode is on the right.

the bytecode, we start a grouping when we detect a Boolean expression. Each Boolean expression in bytecode is represented using a form of `if-statement` where a `true` outcome causes a jump to another label. We add this *if-statement* to our grouping, noting the label that is jumped to if the statement evaluates to `true` and where we resume execution if the statement evaluates to `false`. We then continue to iterate over the code in the current label, if any, adding additional `if-statements` to the table. We continue parsing any labels jumped to by recorded statements for additional `if-statements`, and subsequent labels. Once we reach a label without additional *if-statements*, we stop collecting. For the sample code in Fig. 2, we extract the following grouping:

```
I0 (0) LABEL L1899798671
I4 (4) UNKNOWN Branch I4 IF_ICMPLE, jump to L1871531303
I5 (7) LABEL L717117575
I9 (10) UNKNOWN Branch I9 IF_ICMPGE, jump to L1871531303
I10 (13) LABEL L1866234461
I14 (17) UNKNOWN Branch I14 IF_ICMPLT, jump to L1871531303
I15 (20) LABEL L2064685037
I19 (25) UNKNOWN Branch I19 IF_ICMPEQ, jump to L1871531303
I20 (28) LABEL L1871531303
I23 (29) LABEL L2121337309
```

Next, we can connect the grouped statements through the order they are executed based on their evaluation: (1) We record the current label, where the expression resides. (2) We record the label that is jumped to if the expression evaluates to `true`. (3) We record where execution resumes if the expression evaluates to `false`. This is either a continuation of the current label, or a new label that is reached immediately after the current expression. For the grouping above, we extract the following:

Expression	Location	True jump location	False jump location
I4	L1899798671	L1871531303	L717117575
I9	L717117575	L1871531303	L1866234461
I14	L1866234461	L1871531303	L2064685037
I19	L2064685037	L1871531303	L1871531303

I4	I9	I14	I19	Outcome jump location
True	–	–	–	L2089187484
False	True	–	–	L2089187484
False	False	True	–	L2089187484
False	False	False	True	L2089187484
False	False	False	False	L2089187484

This information indicates the order in which expressions are evaluated, and the outcome once they are evaluated. Using this information, we can form a truth table containing all possible paths through the gathered expressions. Each row of this truth table corresponds to a concrete test obligation that we impose for the Bytecode-MCC criterion. In order to achieve Bytecode-MCC, we need to cover all of the rows of the table. The truth table for the gathered expressions is shown to the right. From this table, the test obligations for the simple class in Fig. 2 are: $(I4 = True)$, $(I4 = False \wedge I9 = True)$, $((I4 = False \wedge I9 = False) \wedge I14 = True)$, $(((I4 = False \wedge I9 = False) \wedge I14 = False) \wedge I19 = True)$, and $(((I4 = False \wedge I9 = False) \wedge I14 = False) \wedge I19 = False)$

3.2 Automated Test Generation to Satisfy Bytecode-MCC

Effective approaches to search-based generation require a fitness function that reports not just the percentage of goals covered, but *how close* the suite is to covering the remaining goals [15]. This feedback allows the search to efficiently maximize coverage of the chosen criterion.

In the case of Branch Coverage, the fitness function calculates the *branch distance* from the point where the execution path diverged from a targeted expression outcome. If an undesired outcome is reached, the function describes how "close" the targeted predicate was to the desired outcome. The fitness value of a test suite is measured by executing all of its tests while tracking the distances $d(b, Suite)$ for each branch.

$$F_{BC}(Suite) = \sum_{b \in B} v(d(b, Suite)) \tag{1}$$

Note that $v(...)$ is a normalization of the distance $d(b, Suite)$ between 0–1. The value of $d(b, Suite)$, then, is calculated as follows:

$$d(b, Suite) = \begin{cases} 0 & \text{if the branch is covered,} \\ v(d_{min}(b, Suite)) & \text{if the predicate has been executed at least twice,} \\ 1 & \text{otherwise.} \end{cases} \tag{2}$$

The cost function used to attain the distance value follows a standard formulation based on the branch predicate [13]. Note that an expression must be executed at least twice, because we must cover the `true` and `false` outcomes of each expression.

In order to measure coverage of Bytecode-MCC and generate test cases intended to satisfy the produced obligations, we can make use of the same branch distance calculation. To obtain the fitness of a test suite, we calculate the branch distances for each expression (and desired outcome) involved in each obligation. Then, the fitness for an individual obligation is the sum of fitness values of all expressions (and desired outcomes) present in the obligation.

For each Boolean expression, we can calculate the minimal branch distance achieved by that suite. For each obligation, we calculate the branch distance for each targeted expression and outcome, then score that obligation as the sum of the branch distances for its targeted expression and outcome combinations. As execution comes closer to satisfying the obligation, the fitness should converge to zero. This fitness formulation can be used as a test generation target, or to measure coverage of existing test suites.

4 Study

We hypothesize that the simplified nature of bytecode instructions limits the effectiveness of tests by introducing the potential for masking, and that Bytecode-MCC-satisfying tests will be effective at overcoming this masking effect. Specifically, we wish to address the following research questions:

1. Does the Bytecode-MCC fitness function attain high coverage of the Bytecode-MCC test obligations?
2. Are test suites generated targeting Bytecode-MCC more effective at detecting faults than suites targeting Branch Coverage?
3. Does targeting the Bytecode-MCC and Branch Coverage fitness functions simultaneously yield higher levels of Bytecode-MCC coverage?
4. Does targeting the Bytecode-MCC and Branch Coverage fitness functions simultaneously yield higher levels of fault detection?

To address these questions, we have performed the following experiment:

1. **Collected Case Examples:** We have used 109 real faults, from five Java projects, as test generation targets.
2. **Generated Test Cases:** For each fault, we generated 10 suites targeting Bytecode-MCC, Branch Coverage, and a combination of both Bytecode-MCC and Branch Coverage for each class-under-test (CUT) using EvoSuite. We perform this process with both a two-minute and a ten-minute search budget per CUT.
3. **Removed Non-Compiling and Flaky Tests:** Any tests that do not compile, or that return inconsistent results, are automatically removed.

4. **Assessed Effectiveness:** For each fault and fitness target, we measure likelihood of fault detection (proportion of suites that detect the fault to the number generated).
5. **Measured Bytecode-MCC Coverage:** For each generated suite, we measure the attained Bytecode-MCC Coverage over the CUT.

Case Examples: Defects4J is a database of real faults extracted from Java projects [10]. The version used in this research, 1.20, consists of 395 faults from six projects: Chart (26 faults), Closure (133 faults), Lang (65 faults), Math (106 faults), Time (27 faults), and Mockito (38 faults). As our focus is on complex Boolean expressions, we selected examples where the source code contains either a large number of Boolean expressions (at least 30), complex Boolean expressions (at least three conditions), or both. Following this filtering, we selected a subset of 109 faults: Chart (1), Closure (66), Lang (28), Math (11), and Time (4). For each fault, Defects4J provides access to the faulty and fixed versions of the code, developer-written test cases that expose the faults, and a list of classes and lines of code modified by the patch that fixes the fault.

Each fault is required to meet three properties. First, a pair of code versions must exist that differ only by the minimum changes required to address the fault. The "fixed" version must be explicitly labeled as a fix to an issue, and changes imposed by the fix must be to source code, not to other project artifacts. Second, the fault must be reproducible—at least one test must pass on the fixed version and fail on the faulty version. Third, the fix must be isolated from unrelated code changes such as refactoring.

Test Suite Generation: We generate tests using EvoSuite targeting both Bytecode-MCC and Branch Coverage[1]. EvoSuite can also simultaneously target multiple criteria, with fitness evaluated as a single combined score. Therefore, we have also targeted a combination of Bytecode-MCC and Branch Coverage to evaluate whether the combination can achieve higher Bytecode-MCC coverage or detect more faults.

Test suites are generated that target the classes reported as relevant to the fault by Defects4J. Tests are generated using the fixed version of the CUT and applied to the faulty version because EvoSuite generates its own assertions for use as oracles. In practice, this translates to a regression testing scenario, where tests are generated using a version of the system understood to be "correct" in order to guard against future issues [17]. Tests that fail on the faulty version, then, detect behavioral differences between the two versions[2].

Two search budgets were used—two minutes and ten minutes per class. This allows us to examine whether an increased search budget benefits coverage or fault detection efficacy. These values are typical of other testing experiments [16]. To control experiment cost, we deactivated assertion filtering—all

[1] Specifically, the `onlybranch` fitness function, which omits branchless methods. This was chosen as our implementation of Bytecode-MCC also omits branchless methods.
[2] Note that this is identical practice to other studies using EvoSuite with Defects4J, i.e. [16,17].

possible regression assertions are included. All other settings were kept at their default values. As results may vary, we performed 10 trials for each fault, criterion, and search budget.

Generation tools may generate flaky (unstable) tests [17]. We automatically removed non-compiling test suites and tests that return inconsistent results over five trials. On average, less than one percent of the tests are removed from each suite.

5 Results and Discussion

The goal of our preliminary study is to determine whether search-based test generation is able to satisfy the test obligations of Bytecode-MCC within a typical search budget. We also wish to evaluate the fault-detection performance of the suites generated under that budget, regardless of the attained level of coverage.

5.1 Attained Bytecode-MCC Coverage

Table 1 lists the average Bytecode-MCC coverage attained given two-minute and ten-minute search budgets when targeting Bytecode-MCC alone and when targeting both Branch and Bytecode-MCC. From Table 1, we can see that the attained coverage is generally quite low. Overall, only 23.31% of obligations are covered on average under a two-minute budget, and only 25.53% under the ten-minute budget. On a per-system basis, the average ranges from 8.70% (Chart) - 69.58% (Time) under the two-minute budget and 15.40% (Chart) - 70.68% (Time) under the ten-minute budget.

Table 1. Average Bytecode-MCC coverage (%) attained by test suites.

System	Two-minute budget		Ten-minute budget	
	MCC	MCC/BC	MCC	MCC/BC
Overall	23.31	40.47	25.53	43.36
Chart	8.70	35.20	15.40	39.70
Closure	13.52	20.81	16.33	23.97
Lang	31.97	66.07	32.55	68.89
Math	41.00	67.04	44.22	69.22
Time	69.58	92.88	70.68	93.33

We can compare this to the attained Branch Coverage when targeting Branch Coverage as the optimization target, as detailed in Table 2. While these suites fail to attain 100% coverage of their targeted goal, these figures are much higher. Given two minutes for generation, these suites attain 71.39% more coverage of their stated goal (Branch Coverage) on average than suites targeting Bytecode-MCC (Table 1). Under a ten-minute budget, this increases to 86.72%.

Table 2. Average Branch Coverage (%) attained by test suites targeting Branch Coverage.

System	Two-minute budget	Ten-minute budget
Overall	39.95	47.67
Chart	33.10	54.41
Closure	13.30	21.36
Lang	81.59	87.99
Math	73.00	77.36
Time	68.50	86.27

What this shows is that Bytecode-MCC is a more difficult criterion to satisfy than Branch Coverage. Given the same period of time, we can naturally expect higher attainment of Branch Coverage than Bytecode-MCC coverage. Therefore, "typical" generation time frames like two minutes may not be enough to attain reasonable levels of Bytecode-MCC coverage. However, moving from two minutes to ten minutes offers only a 9.52% average improvement in attained Bytecode-MCC coverage, compared to an average improvement of 19.32% in Branch Coverage. The limited improvement suggests that an increased budget alone may not be enough to overcome the difficulty of satisfying Bytecode-MCC obligations.

This idea is further reinforced by examining the Bytecode-MCC coverage results when Branch Coverage and Bytecode-MCC are targeted simultaneously, as listed in Table 1 for each system and budget. Overall, targeting Branch and Bytecode-MCC coverage simultaneously yields a 73.62% increase in attained Bytecode-MCC coverage under a two-minute budget over targeting Bytecode-MCC on its own, and a 69.84% improvement under a ten-minute budget. Targeting Branch Coverage in addition to Bytecode-MCC offers easier-to-cover intermediate goals that, ultimately, result in improved Bytecode-MCC coverage. Coverage is still lower than desired, but the situation is improved over single-target optimization of Bytecode-MCC by introducing feedback (using Branch Coverage) that the test generator can work with.

5.2 Fault Detection

Table 3 lists the average likelihood of fault detection for suite generated to target Bytecode-MCC, Branch Coverage, and a combination of Bytecode-MCC and Branch Coverage, divided by system and overall, for each search budget. Overall, Branch-targeting suites have a 21.20–22.13% likelihood of detection. This is consistent with previous experiments using this set of faults, and reflects the complex nature of the studied faults [16]. Overall, Bytecode-MCC-targeting tests only have a 4.27% average likelihood of detection under a two-minute budget, and a 3.47% average likelihood of detection under a ten-minute budget—far lower than when Branch Coverage is targeted.

This drop is likely due to the low coverage of Bytecode-MCC when it is the sole optimization target. Results improve when Bytecode-MCC and Branch Coverage are targeted simultaneously. Targeting both yields an overall average likelihood of detection of 16.67% (two-minute budget) and 19.33% (ten-minute budget). Still, this is lower than when Branch Coverage is targeted alone. Previous research indicates that

Table 3. Average likelihood of fault detection (%) for each generation target and budget, broken down by system and overall.

System	Two-minute budget			Ten-minute budget		
	MCC	Branch	MCC/BC	MCC	Branch	MCC/BC
Overall	4.27	21.20	16.67	3.47	22.13	19.33
Chart	20.00	100.00	90.00	1.00	100.00	90.00
Closure	0.00	1.33	1.00	0.00	3.67	2.33
Lang	8.28	41.72	28.28	7.24	43.79	32.41
Math	5.46	20.91	16.36	3.64	16.36	19.09
Time	0.00	2.50	32.50	0.00	0.00	35.00

multi-objective optimization can be more difficult than single-objective optimization [16], and it is likely that the additional burden of satisfying Bytecode-MCC results in lower Branch Coverage as well when both are targeted.

However, if we examine results on a per-system basis, we can see that Bytecode-MCC satisfaction may have some promise for improvement in fault-detection. For the Time examples, targeting the combination of Branch Coverage and Bytecode-MCC yields over 92% Bytecode-MCC coverage on average. The combination also has an average likelihood of detection of 32.50–35.00%—well over the overall average. In this case, targeting the combination makes it possible to detect faults completely missed when targeting Branch Coverage alone.

On average, the Time examples contain more complex Boolean expressions than the other systems, with an average of 3.25 conditions per decision (compared to an overall average of 2.29). These are not trivial examples, and the performance when targeting the combination of Branch Coverage and Bytecode-MCC is promising. If systems contain complex Boolean expressions and high Bytecode-MCC coverage can be achieved, then we may also see improvements in fault detection. However, it is also clear that we must first find the means to improve attained Bytecode-MCC coverage.

5.3 Discussion

Even if Bytecode-MCC attainment is theoretically able to overcome issues with masking, we cannot test its abilities without first finding ways to improve coverage. The Time examples were the only ones where Bytecode-MCC coverage was reasonably high—particularly with the boost offered by simultaneously targeting Branch Coverage. While those showed promising improvements in fault detection as well, such improvements require increased ability to attain coverage.

Some criteria are inherently more difficult to satisfy than others [18]. It will be more difficult—and will require more test cases—to satisfy MCC over Branch Coverage. It may not be reasonable to expect equal coverage of Branch Coverage and Bytecode-MCC given the same time budget. Still, there may be means of improving coverage.

Reformulating the Fitness Function: A complicating factor in search-based test generation comes from the fitness function and its ability to offer feedback. When attempting to achieve Branch Coverage, the branch distance is used instead because it offers clear feedback, suggesting whether one solution is closer to covering the remaining obligations than another. It is possible that a fitness formulation other than the one employed in this work would yield better results. In the proposed function, each Bytecode-MCC obligation is a combination of smaller Boolean conditions. Fitness is measured by scoring each condition independently and linearly combining the resulting scores. Progress towards covering any of the individual conditions will yield a better fitness score. This, in theory, should offer reasonable feedback. However, there may exist cases where the independent subgoals conflict, and the choice of input may improve the coverage of one condition while increasing the distance for another condition.

A linear combination of condition distances may not be an ideal mechanism for judging fitness for Bytecode-MCC obligations, and other fitness formulations may yield better results. For example, it may be better to weight conditions based on the order they must be satisfied in. Alternatively, rather than combining the distances into a single score, each obligation could be treated as a set of distance scores to be optimized independently. This would be a more complex approach, but could potentially yield better results in cases where goals conflict.

Use Bytecode-MCC to Measure Adequacy Instead of a Direct Generation Target: Some criteria could yield powerful test cases, but lack sufficient feedback mechanisms to drive a search towards high levels of coverage. For example, Exception Coverage rewards test suites that throw many exceptions. However, there is no feedback mechanism that suggests "closeness" to throwing more exceptions [1]. This criterion yields poor suites when targeted as the sole fitness function, but offers great utility as a means of judging adequacy, and as a stopping criterion to determine when to finish testing. Likewise, we could target other fitness functions, but use Bytecode-MCC to assess the final test suites as a means to determine when to stop test generation.

Research has suggested that targeting unrelated fitness functions like Branch Coverage, or combining additional fitness functions with uninformative ones—i.e., Branch and Exception Coverage—results in higher Exception Coverage of the final suite [16]. Likewise, we could choose alternative optimization targets, then measure the attained Bytecode-MCC of the resulting tests. If we can find fitness functions that yield higher levels of Bytecode-MCC, we will be better able to evaluate the potential of the criterion for overcoming masking and improving the fault-detection potential of test suites.

Recent work explored the use of reinforcement learning to improve attainment of Exception Coverage [1]. The proposed approach was able to strategically adjust the targeted fitness functions over time in service of improving Exception Coverage. A similar adaptive fitness function selection approach could be used to discover combinations of fitness functions that attain high coverage of Bytecode-MCC.

We may wish to also consider other forms of test generation, beyond search-based generation. For example, (dynamic) symbolic execution techniques use sophisticated solvers to attain input designed to drive program execution towards particular paths [2]. Such approaches suffer from limitations in terms of the type of programs and language features they can handle, and in terms of scalability [5]. However, they can be very effective at producing the input needed to traverse specific paths—which is required for Bytecode-MCC satisfaction. The use of symbolic execution—or approaches that combine search and symbolic execution—may be required to achieve high levels of Bytecode-MCC coverage.

6 Threats to Validity

Internal Validity: Because EvoSuite's test generation process is non-deterministic, we have generated ten test suites for each combination of fault, budget, and fitness function. It is possible that larger sample sizes may yield different results. However, we believe that this is a sufficient number to draw stable conclusions.

External Validity: Our study has focused on only five systems. We believe that such systems are representative of, at minimum, other small to medium-sized open-source Java systems. We believe that we have chosen enough examples to gain a basic understanding of Bytecode-MCC, and that our results are generalizable to sufficiently similar projects. In this study, we have implemented Bytecode-MCC within EvoSuite. Results may differ using a different test generation algorithm. However, we believe that EvoSuite is sufficiently powerful to explore our proposed ideas.

7 Related Work

Hayhurst et al. observed sensitivity to statement structure, stating that "if a complex decision statement is decomposed into a set of less complex (but logically equivalent) decision statements, providing MC/DC for the parts is not always equivalent to providing MC/DC for the whole" [11]. Gargantini et al. have also observed the sensitivity of structural coverage metrics to modification of the code structure and proposed a method of automatically measuring the resilience of a piece of code to modification [6]. Chilenski further made the observation that "If the number of tests M is fixed at N + 1 (N being the number of conditions), the probability of distinguishing between incorrect functions grows exponentially with N, N > 3" [3]. This observation is based on the number of tests, but notes that testing power grows with statement complexity.

In past work, we empirically demonstrated the effects of expression structure on coverage and suite effectiveness, clearly illustrating the negative impact of statement simplification on a series of industrial case examples [8]. Our results supported the previous observations. We also proposed a set of "observability" extensions for source-based coverage criteria to overcome masking between-expressions [14,18]. Our proposed method, Bytecode-MCC acts in a similar—but more limited—manner to the notion of "observability", requiring that masking be overcome in closely-connected statements. To date, ours is the first approach to address masking in search-based test generation or when considering bytecode representations of programs.

8 Conclusions

Masking occurs when one condition prevents another from influencing the output of a Boolean expression. Adequacy criteria such as Multiple Condition Coverage (MCC) overcome masking within one expression, but offers no guarantees about subsequent expressions. As a result, a Boolean expression written as a single complex statement will yield more effective test cases than when written as a series of simple expressions. Many approaches to automated test case generation for Java operate not on the source code, but on bytecode. The transformation to bytecode simplifies complex expressions into multiple expressions, introducing masking. We propose Bytecode-MCC, a new adequacy criterion designed to group bytecode expressions and reformulate them into complex expressions. Bytecode-MCC should produce test obligations that are more likely to reveal faults in program logic than tests covering the simplified bytecode.

A preliminary study conducted over 109 faults from Defects4J indicate the potential of the technique. Results attained for the "Time" system, where targeting the combination of Bytecode-MCC and Branch Coverage yields high Bytecode-MCC coverage, show fault detection well above the overall average. However, there are multiple research challenges to be overcome. Bytecode-MCC is more difficult to achieve than Branch Coverage, and its fitness function does not offer sufficient feedback to guide test generation. This suggests that Bytecode-MCC may be best used as a method of judging test suite quality, rather than as a direct generation target. Simultaneously targeting Bytecode-MCC and Branch Coverage improves coverage of Bytecode-MCC and the likelihood of fault detection. It may be possible to identify other fitness functions that are effective at attaining Bytecode-MCC.

In future work, we will explore methods of improving Bytecode-MCC coverage. In particular, we plan to: (1) Explore alternative formulations of the fitness function for Bytecode-MCC, such as applying weights based on the order that sub-obligations must be solved. (2) Examine the use of Bytecode-MCC as a way to judge test suites generated targeting other criteria, as well as its use as a stopping condition for test generation. (3) Investigate the use of reinforcement learning to automatically identify alternative generation targets that will yield higher attainment of Bytecode-MCC than direct targeting of Bytecode-MCC during test generation. (4) Vary the algorithms used to generate Bytecode-MCC-covering test suites.

References

1. Almulla, H., Gay, G.: Learning how to search: generating exception-triggering tests through adaptive fitness function selection. In: 13th IEEE International Conference on Software Testing, Validation and Verification (2020)
2. Anand, S., et al.: An orchestrated survey on automated software test case generation. J. Syst. Softw. **86**(8), 1978–2001 (2013)
3. Chilenski, J.: An investigation of three forms of the modified condition decision coverage (MCDC) criterion. Technical report DOT/FAA/AR-01/18, Office of Aviation Research, Washington, D.C., April 2001
4. Fraser, G., Staats, M., McMinn, P., Arcuri, A., Padberg, F.: Does automated white-box test generation really help software testers? In: Proceedings of the 2013 International Symposium on Software Testing and Analysis, ISSTA, pp. 291–301. ACM, New York (2013). https://doi.org/10.1145/2483760.2483774
5. Galeotti, J.P., Fraser, G., Arcuri, A.: Improving search-based test suite generation with dynamic symbolic execution. In: 2013 IEEE 24th International Symposium on Software Reliability Engineering (ISSRE), pp. 360–369, November 2013

6. Gargantini, A., Guarnieri, M., Magri, E.: AURORA: automatic robustness coverage analysis tool. In: 2013 IEEE Sixth International Conference on Software Testing, Verification and Validation (ICST), pp. 463–470. IEEE (2013)
7. Gay, G., Staats, M., Whalen, M., Heimdahl, M.: The risks of coverage-directed test case generation. IEEE Trans. Softw. Eng. PP(99) (2015)
8. Gay, G., Rajan, A., Staats, M., Whalen, M., Heimdahl, M.P.E.: The effect of program and model structure on the effectiveness of MC/DC test adequacy coverage. ACM Trans. Softw. Eng. Methodol. **25**(3), 25:1–25:34 (2016). https://doi.org/10.1145/2934672
9. Groce, A., Alipour, M.A., Gopinath, R.: Coverage and its discontents. In: Proceedings of the 2014 ACM International Symposium on New Ideas, New Paradigms, and Reflections on Programming & Software, Onward! 2014, pp. 255–268. ACM, New York (2014). https://doi.org/10.1145/2661136.2661157
10. Just, R., Jalali, D., Ernst, M.D.: Defects4J: a database of existing faults to enable controlled testing studies for Java programs. In: Proceedings of the 2014 International Symposium on Software Testing and Analysis, ISSTA 2014, pp. 437–440. ACM, New York (2014). https://doi.org/10.1145/2610384.2628055
11. Kelly J., H., Dan S., V., John J., C., Leanna K., R.: A practical tutorial on modified condition/decision coverage. Technical report (2001)
12. Li, N., Meng, X., Offutt, J., Deng, L.: Is bytecode instrumentation as good as source code instrumentation: an empirical study with industrial tools (experience report). In: 2013 IEEE 24th International Symposium on Software Reliability Engineering (ISSRE), pp. 380–389, November 2013
13. McMinn, P.: Search-based software test data generation: a survey. Softw. Testing Verif. Reliab. **14**, 105–156 (2004)
14. Meng, Y., Gay, G., Whalen, M.: Ensuring the observability of structural test obligations. IEEE Trans. Softw. Eng. 1–1 (2018). http://greggay.com/pdf/18omcdc.pdf
15. Rojas, J.M., Campos, J., Vivanti, M., Fraser, G., Arcuri, A.: Combining multiple coverage criteria in search-based unit test generation. In: Barros, M., Labiche, Y. (eds.) SSBSE 2015. LNCS, vol. 9275, pp. 93–108. Springer, Cham (2015). https://doi.org/10.1007/978-3-319-22183-0_7
16. Salahirad, A., Almulla, H., Gay, G.: Choosing the fitness function for the job: automated generation of test suites that detect real faults. Softw. Testing Verifi. Reliab. 29(4–5), e1701 (2019). https://doi.org/10.1002/stvr.1701. e1701 stvr.1701
17. Shamshiri, S., Just, R., Rojas, J.M., Fraser, G., McMinn, P., Arcuri, A.: Do automatically generated unit tests find real faults? An empirical study of effectiveness and challenges. In: Proceedings of the 30th IEEE/ACM International Conference on Automated Software Engineering (ASE), ASE 2015. ACM, New York (2015)
18. Whalen, M., Gay, G., You, D., Heimdahl, M., Staats, M.: Observable modified condition/decision coverage. In: Proceedings of the 2013 International Conference on Software Engineering. ACM, May 2013
19. Whalen, M.W., Person, S., Rungta, N., Staats, M., Grijincu, D.: A flexible and non-intrusive approach for computing complex structural coverage metrics. In: Proceedings of the 37th International Conference on Software Engineering, vol. 1, pp. 506–516. IEEE Press (2015)

Challenge Solutions

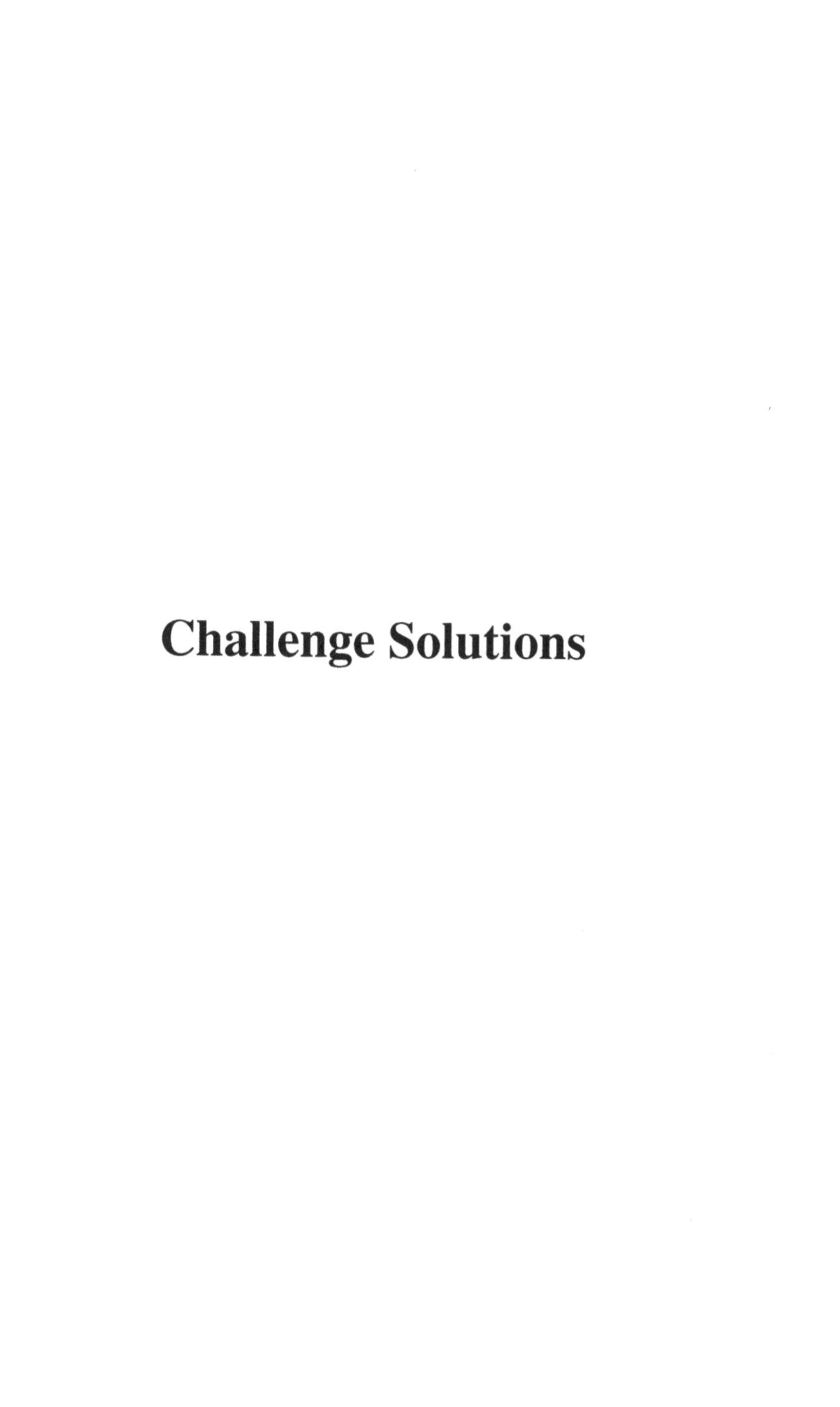

An Application of Model Seeding to Search-Based Unit Test Generation for Gson

Mitchell Olsthoorn(✉), Pouria Derakhshanfar, and Xavier Devroey

Delft University of Technology, Delft, The Netherlands
{m.j.g.olsthoorn,p.derakhshanfar,x.d.m.devroey}@tudelft.nl

Abstract. Model seeding is a strategy for injecting additional information in a search-based test generation process in the form of models, representing usages of the classes of the software under test. These models are used during the search-process to generate logical sequences of calls whenever an instance of a specific class is required. Model seeding was originally proposed for search-based crash reproduction. We adapted it to unit test generation using EvoSuite and applied it to Gson, a Java library to convert Java objects from and to JSON. Although our study shows mixed results, it identifies potential future research directions.

Keywords: Model seeding · Search-based software testing · Case study

1 Introduction

Over the years, several techniques have been developed to generate unit tests by applying search-based algorithms to source code. Among the existing tools, EvoSuite is one of the references in the state-of-the-art for Java unit test generation [5]. It has been developed and maintained over the years and received several contributions to improve code coverage and mutation score of the generated tests or to generate tests for specific purposes. Despite the numerous improvements, one of the challenges still faced by EvoSuite is the generation of complex objects with logical (*i.e.,* not random) sequences of method calls. There exist several strategies to address this challenge partially. Among those, *seeding* [7] consists of the injection of additional information that will be used during the search. For instance, constant values collected from the source code, and the usage (as-is) of objects collected from an existing test suite.

In their recent study, Derakhshanfar *et al.* [3] propose to abstract the behavior of the different classes of the system under test (SUT) using a *transition system model.* For each class, one transition system describes the sequences of method calls previously observed on the instances of that class. They seeded those models to a search-based crash reproduction algorithm to generate complex objects and found that it improved the overall crash reproduction rate. Crash reproduction

A. Aleti and A. Panichella (Eds.): SSBSE 2020, LNCS 12420, pp. 239–245, 2020.
https://doi.org/10.1007/978-3-030-59762-7_17

does not seek to cover all the elements of a class under test (CUT) but rather to generate a test exercising a specific behavior causing the software to crash.

For unit test generation, the coverage of different elements of a CUT also requires specific objects that might be difficult to generate randomly. In this paper, we applied model seeding for unit test generation using EvoSuite on a set of eight classes from the Gson library. We compare *model seeding* to the *default* configuration of EvoSuite *w.r.t.* the branch coverage and mutation scores achieved by the generated tests.

2 Evaluation Setup

Classes Under Test. Gson is a *Java serialization* and *deserialization library to convert Java Objects into JSON and back.*[1] It is used as a dependency by more than 222 000 projects on GitHub. We used Gson v.2.8.7 (5924 LOC) and selected 8 classes with at least one method (with the exception of the `toString` and `equals` method) with a cyclomatic complexity above 3: `Gson`, `JsonReader`, `JsonTreeReader`, `JsonTreeWriter`, `JsonWriter`, `LinkedHashTreeMap`, `LinkedTreeMap`, and `TypeAdapters`. The overall branch coverage of the existing manually written tests is 79%, and the overall mutation score is 75%.

Learning the Models. We followed Derakhshanfar *et al.*'s [3] approach and generated our models using the existing source code and tests of the Gson library. For each class used in the project, each time an object is created, we collected the sequence of methods called on this object. For that, we statically analyzed the source code of Gson and dynamically executed (a heavily instrumented version of the) existing test cases. The models are then *learned* from the collected call sequences using a 2-gram inference. Learning the models is a one-time operation. Models are then seeded to the different executions of EvoSuite. In total, we collected 328 models for 328 different classes. The average number of states is 7 and the average number of transitions is 15. We rely on the implementation of Derakhshanfar *et al.* [3] to collect call sequences and learn the different models, and on EvoSuite-RAMP,[2] a customized version of EvoSuite [5] for unit test generation.

Configurations. Model seeding works either *online* or *offline*. In the offline mode, during the initialization of the search, for each model, EvoSuite-RAMP creates a fixed number of objects by selecting *abstract behaviors* from the model. For each selected abstract behavior, it instantiates the object, calls the corresponding methods, and adds the result to an *object pool*. Whenever an object is required, the search process *copies* (with a defined probability `p_object_pool`) one object and its method calls from this object pool. Additionally, during the initialization of the population, EvoSuite-RAMP can also copy (with a defined probability `seed_clone`) an instance of the CUT (as-is) from the object pool

[1] https://github.com/google/gson.
[2] https://github.com/STAMP-project/evosuite-ramp.

and use it as a plain (initial) test. In the online mode, objects are created during the search process using the same procedure. The main difference with the offline mode is that the objects are created *on demand*, slightly overloading the search process. For our evaluation, we used the *online* mode as it does not overload the initialization and generates only objects required by the search (and therefore leaves more budget for the search itself). We used probabilities `p_object_pool` = 0.3, and `seed_clone` = 0.3, following Derakhshanfar *et al.* [3].

To select abstract behaviors, EvoSuite-RAMP supports *random* selection, corresponding to random walks in the models, and *dissimilarity* selection, trying to increase diversity in the selected behaviors. For our evaluation, we used *random selection* as it gave slightly better results in our initial trial.

In our evaluation, we compare unit test generation with model seeding activated (*model s.*) to the default EvoSuite configuration (*default*). For both configurations, we used *DynaMOSA* [6] with the default set of objectives (*i.e.*, branch, line, weak mutation, input, output, method, and exception coverage) and a search budget of 180 s. Additionally, we deactivated model seeding after 90 s to increase exploration. We ran our evaluation (1600 runs) on a server with 12 CPU cores @ 3.50 GHz. The total execution time for unit test generation took around 40 min. Our replication package is available on Zenodo [4].

Data Analysis. For each class under test, we compare the generated test suites *w.r.t.* their branch coverage (reported by EvoSuite-RAMP) and their mutation score, computed using Pit v1.4.3 [1] with `ALL` mutation operators activated. The total execution time for the mutation analysis of the 1600 generated test suites took around 2 days. We used the the non-parametric Wilcoxon Rank Sum test ($\alpha = 0.05$) for Type I error, and the Vargha-Delaney statistic $\hat{A}_{12}$ to evaluate the effect size between *model s.* and *default*.

3 Results

Figure 1 presents the branch coverage and mutation score of the test suites generated using the *default* and *model s.* configurations. On average, the highest branch coverage is achieved by the *default* configuration for the class `JsonTreeReader` with 92.04%. The lowest branch coverage is, on average, also achieved by the *default* configuration for the class `TypeAdapters` with 50.04%. For the mutation score, the highest average mutation score is achieved by the *model s.* configuration for the `TypeAdapters` class with a score of 93.75%, and the lowest average mutation score is also achieved by *model s.* configuration on the `LinkedHashTreeMap` class with an average of 35.91%.

For each class, we compared the coverage and mutation score of the generated test suites. The configuration *model s.* achieved significant better results ($\alpha \leq 0.05$) for three classes (with two small and one large $\hat{A}_{12}$ magnitudes). It performs worse compared to the *default* configuration for two classes (with one small and one medium $\hat{A}_{12}$ magnitudes). For the mutation score, the *model s.* configuration performed significantly better ($\alpha \leq 0.05$) on three classes (with two small and

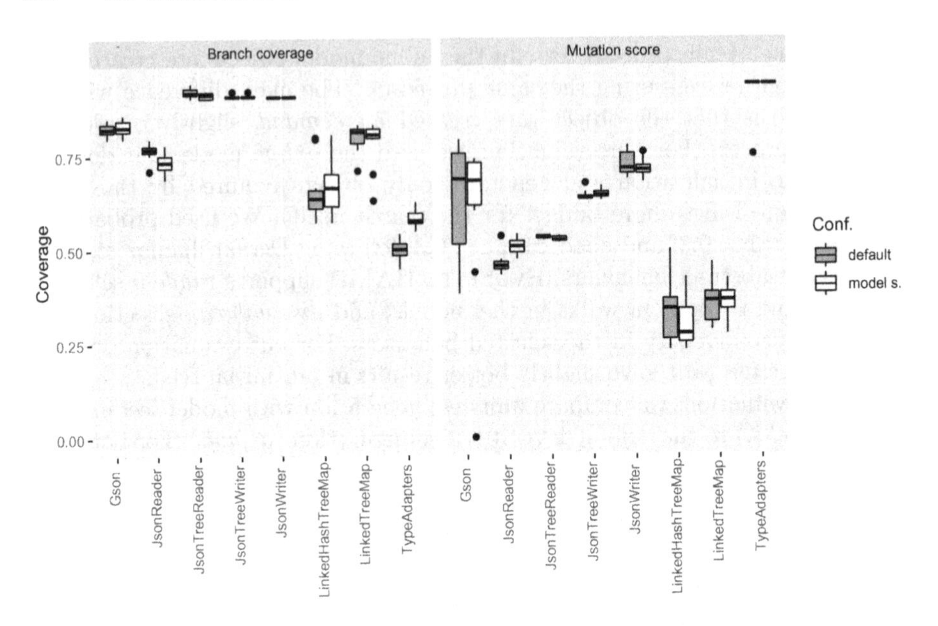

Fig. 1. Coverage of the tests generated using the *default* and *model s.* configurations.

one large $\hat{A}_{12}$ magnitudes). The *default* configuration performed better for two classes (with one small and one medium $\hat{A}_{12}$ magnitudes).

In general, our results are mixed. The *model s.* configuration can lead to an improvement of the mutation score with, in general, a lower variability than the *default* configuration. The most interesting class of our evaluation is the `JsonReader` class for which *model s.* achieves a significantly worse branch coverage (p-value $= 4.98 \times 10^{-15}$) with a large magnitude ($\hat{A}_{12} = 0.180$) than *default*, but, in the same time, also achieves a significantly better (p-value $= 9.26 \times 10^{-25}$) mutation score, also with a large magnitude ($\hat{A}_{12} = 0.92$). We focus our discussion on the `JsonReader` class.

4 Discussion and Future Work

Code Complexity and Model Generation. From analyzing the project using CodeMR,[3] we see that the `JsonReader` class is the most complex class of the project with a *very-high* complexity rate and a Weighted Method Count (WMC) of 359 for 891 lines of code (LOC). Complex code is a well-known challenge for search-based testing algorithms.

At the same time, the model generated from the collected call sequences for `JsonReader` is highly connected with an average degree (*i.e.*, the average number of incoming and outgoing transitions per state) of 9.0, 252 transitions for only 28 states, and a BFS height of 6 (*i.e.*, number of levels when navigating the

[3] https://www.codemr.co.uk.

model using a breadth-first search algorithm). This permissiveness of the model tends to indicate that the usages are not well captured and that the model can provide only limited guidance. Future research will investigate the usages of other learning approaches (including higher-values of n for the n-gram inference) to better reflect the usages of the classes. Additionally, the models are created from the source code and the existing tests. We followed the procedure defined by Derakhshanfar *et al.* [3] with the same assumption that the existing tests are representative of valid usages of the classes (for crash reproduction). However, this assumption might not be right for unit testing. Therefore, future research will investigate other sources of call sequences, like projects using GSON, as well as including information about object and parameter values for those calls.

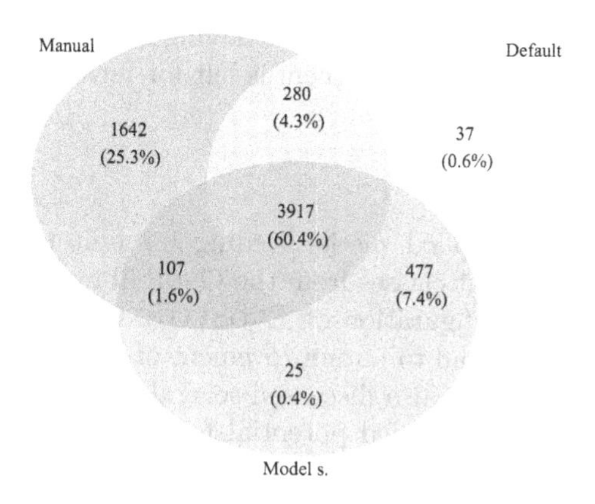

Fig. 2. Combined mutation analysis of the `JsonReader` class with the number and percentage of mutants killed by the *manually* written, *model s.*, and *default* test suites.

Mutation Analysis. To further investigate the mutants killed by the generated tests, we performed a combined mutation analysis on the `JsonReader` class. For that, we used all the tests generated out of the 10 rounds of execution and the manually written tests from the `JsonReaderTest` and `JsonReaderPathTest` test suites. Figure 2 presents the number of mutants killed, grouped by the source of the test suite. Mutants killed by more than one test suite coming from different sources are placed at intersections in the diagram.

We see that between 7.8% and 8% of the mutants are killed only by a *generated test suite*. Additionally, we see that between 62% and 64.7% of the mutants are killed by both manually written and automatically generated tests, which tends to confirm that effort on testing can be reduced using automated approaches. Figure 2 also shows that the largest amount of mutants only killed by EvoSuite (7.4%) are killed both by the *default* and the *model s.* configurations. This tends to indicate that the randomness of the evolution process helps to explore new areas of the search space, compared to manually written tests. We also see that 25 mutants are killed only by the *model s.* configuration, and

37 mutants are killed only by the *default* configuration. Finally, from Figs. 1 and 2, we see that the *default* configuration achieves a significantly lower mutation score, compared to *model s.*, but kills a larger diversity of mutants (at least once) when the 10 test suites are merged together.

Test Case Understandability. In this case study, we only consider the functional properties (*i.e.,* branch coverage and mutation score) of the generated tests. Recent studies have investigated other aspects of generated tests, like the readability and understandability by a developer [2]. We believe that, by generating objects with common usages, model seeding can contribute to improving test case readability and understandability. From the manual analysis of the test cases killing 25 mutants only killed by the *model s.* configuration, we could retrace the usages of the `JsonReader` class observed in the test case in the usage model of the class. The confirmation that having such usages in the test cases helps in reading and understanding them is left for future work.

5 Conclusion

In this case study, we applied model seeding for unit test generation using EvoSuite-RAMP on eight classes from the Gson library. We compared *model seeding* to the *default* configuration of EvoSuite. Overall, results are mixed. Using model seeding can lead to an improvement of branch coverage and mutation score in some cases. We also discussed several aspects of model seeding for unit test generation and identified potential future research directions regarding the collection of call sequences and generation of the models, and the usage of model seeding to improve the understandability of automatically generated tests.

Acknowledgement. This research was partially funded by the EU Horizon 2020 ICT-10-2016-RIA "STAMP" project (No. 731529).

References

1. Coles, H., Laurent, T., Henard, C., Papadakis, M., Ventresque, A.: PIT: a practical mutation testing tool for Java. In: ISSTA 2016, pp. 449–452. ACM (2016). https://doi.org/10.1145/2931037.2948707
2. Daka, E., Campos, J., Fraser, G., Dorn, J., Weimer, W.: Modeling readability to improve unit tests. pp. 107–118 (2015). https://doi.org/10.1145/2786805.2786838
3. Derakhshanfar, P., Devroey, X., Perrouin, G., Zaidman, A., Deursen, A.: Search-based crash reproduction using behavioural model seeding. Softw. Test. Verif. Reliab. **30**(3), e1733 (2020). https://doi.org/10.1002/stvr.1733
4. Derakhshanfar, P., Olsthoorn, M., Devroey, X.: Replication package of An Application of Model Seeding to Search-based Unit Test Generation for Gson (2020). https://doi.org/10.5281/zenodo.3963956
5. Fraser, G., Arcuri, A.: EvoSuite: automatic test suite generation for object-oriented software. In: ESEC/FSE 2011, p. 416. ACM (2011). https://doi.org/10.1145/2025113.2025179

6. Panichella, A., Kifetew, F.M., Tonella, P.: Automated test case generation as a many-objective optimisation problem with dynamic selection of the targets. IEEE Trans. Softw. Eng. **44**(2), 122–158 (2018). https://doi.org/10.1109/TSE.2017.2663435
7. Rojas, J.M., Fraser, G., Arcuri, A.: Seeding strategies in search-based unit test generation. Softw. Test. Verif. Reliab. **26**(5), 366–401 (2016). https://doi.org/10.1002/stvr.1601d

Generating Diverse Test Suites for Gson Through Adaptive Fitness Function Selection

Hussein Almulla[1] and Gregory Gay[2](✉)

[1] University of South Carolina, Columbia, SC, USA
halmulla@email.sc.edu
[2] Chalmers and the University of Gothenburg, Gothenburg, Sweden
greg@greggay.com

Abstract. Many fitness functions—such as those targeting test suite diversity—do not yield sufficient feedback to drive test generation. We propose that diversity can instead be improved through *adaptive fitness function selection* (AFFS), an approach that varies the fitness functions used throughout the generation process in order to strategically increase diversity. We have evaluated our AFFS framework, EvoSuiteFIT, on a set of 18 real faults from Gson, a JSON (de)serialization library. Ultimately, we find that AFFS creates test suites that are more diverse than those created using static fitness functions. We also observe that increased diversity may lead to small improvements in the likelihood of fault detection.

Keywords: Search-based test generation · Fitness function · Reinforcement learning

1 Introduction

In search-based test generation, testers seek input that attains their testing *goal*. An optimization algorithm systematically samples the space of possible test input in search of a solution to that goal, guided by feedback from one or more **fitness functions**—numeric scoring functions that judge the optimality of the chosen input [8].

During this search, the fitness functions embody the high-level goals of the tester [1]. Feedback from the fitness function will shape the resulting test suite, and we choose fitness functions based on their suitability for those goals. For example, a common goal is to attain Branch Coverage—to ensure that all outcomes of control-diverging statements have been executed. Search-based test generation generally attains this goal using a fitness function based on the branch distance—a fine-grained measurement of how close we came to covering each branch outcome. If our goal is Branch Coverage, we know how to attain it through search-based test generation. For other goals, we may not be so fortunate. Many goals *do not* have an effective fitness function formulation.

A. Aleti and A. Panichella (Eds.): SSBSE 2020, LNCS 12420, pp. 246–252, 2020.
https://doi.org/10.1007/978-3-030-59762-7_18

One such goal is *test suite diversity*. When testing, it is generally impossible to try every input. It follows, then, that different test cases are more effective than similar ones [2,9]. This intuition has led to effective automated test generation, prioritization, and reduction [2]. While numerous diversity metrics exist—for example, the *Levenshtein distance* [9]—these metrics serve as poor fitness functions, as they may no present sufficient actionable feedback to optimize.

This does *not* mean that test suite diversity cannot be attained. Rather, we do not *yet* know what fitness functions will be effective. There are many fitness functions used in search-based test generation for attainment of other goals. Careful selection of one or more of *those* functions—paired with, or even excluding, a diversity metric—could provide that missing feedback. In fact, we may even attain higher diversity by reevaluating our choice of fitness functions over time based on the test suite population.

We previously introduced *adaptive fitness function selection* (AFFS)—a hyperheuristic that adapts the test generating process, using reinforcement learning [7], to adjust the fitness functions in service of optimizing attainment of a higher-level goal [1]. We used AFFS to increase the number of exceptions thrown by test suites [1]. In this study, we extend our AFFS framework, EvoSuiteFIT, to a new goal—increasing test suite diversity. We have implemented the Levenshtein distance as both a fitness function and as a target for reinforcement learning in EvoSuiteFIT, and made improvements to the two implemented RL approaches, UCB and DSG-Sarsa [7].

As a case study to evaluate the ability of AFFS to increase test suite diversity, we perform a case study on the Gson library[1]. Gson is an open-source library for serializing and deserializing JSON input, and is an essential tool of Java and Android development [5]. A set of 18 real-world faults from Gson are available in the Defects4J fault database [3]. Previous work has found that this framework is a challenging target for test generation [3]. We evaluate EvoSuiteFIT on these examples in terms of attained test suite diversity and fault detection, comparing against two static baselines.

Ultimately, we find that EvoSuiteFIT creates test suites that are more diverse than those created using static fitness functions. While results on fault detection are inconclusive, we observe that diversity may lead to small improvements in the likelihood of fault detection. We make EvoSuiteFIT available for use in research and practice.

2 Adaptive Fitness Function Selection

Careful selection and reevaluation of fitness functions could result in test suites that are more diverse than those generated targeting a single diversity metric alone or a naively chosen set of fitness functions. Identifying this set of fitness functions is a secondary search problem—one that can be tackled as an additional step within the normal test generation process using reinforcement learning (RL). This process is known as *adaptive fitness function selection* (AFFS) [1].

[1] https://github.com/google/gson.

Due to space constraints, we present only a brief overview of AFFS[2]. Adjusting the set of fitness functions can be considered as an instance of the n-armed bandit problem [7]. Given a measurement of diversity, each action—a choice of fitness functions—has an expected reward—increase in diversity—when it is selected. The modified test generation process is illustrated in Fig. 1. At a defined interval, the RL agent will reevaluate the set of fitness functions and refine its estimation of their ability to increase diversity. Because the population of test suites at round N depends on the population from round $N-1$, RL can not only choose effective fitness functions, but can strategically adjust that choice based on the test suite state.

AFFS has been implemented within the standard Genetic Algorithm in the EvoSuite test generation framework [8]. Two RL algorithms—Upper Confidence Bound (UCB) and Differential Semi-Gradient Sarsa (DSG-Sarsa) [7]—have been implemented. UCB is a classic reinforcement learning approach that works well for many problems, while DSG-Sarsa is an approximate approach adapted for problems with large state spaces [7]—i.e., test generation. We refer to the overall framework as **EvoSuiteFIT**[3].

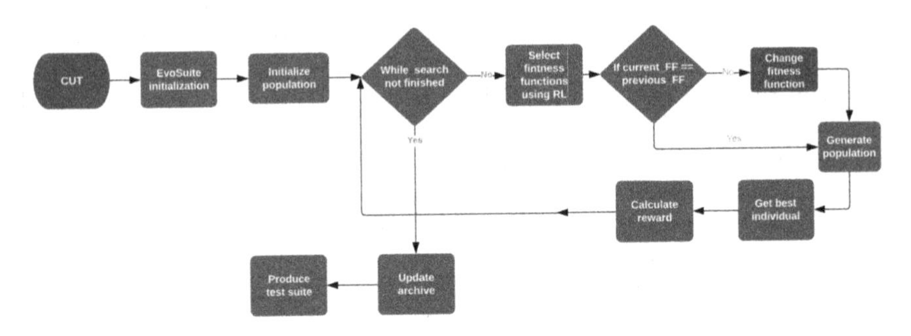

Fig. 1. Overview of the test generation process with AFFS.

Diversity-Based Fitness Function: We have implemented a fitness function to measure test suite diversity based on the Levenshtein distance [9]. The Levenshtein distance is the minimal cost of the sum of individual operations—insertions, deletions, and substitutions—needed to convert one string to another (i.e., one test to another). The distance between two tests (ta and tb) can be calculated as follows [9]:

$$lev_{ta,tb}(i,j) = \begin{cases} max(i,j) & \text{if } min(i,j) == 0 \\ min \begin{cases} lev_{ta,tb}(i-1,j)+1 \\ lev_{ta,tb}(i,j-1)+1 \\ lev_{ta,tb}(i-1,j-1)+1_{(ta_i \neq tb_j)} \end{cases} & otherwise \end{cases} \quad (1)$$

where i and j are the letters of the strings representing t_a and t_b. To calculate the diversity of a test suite (TS), we calculate the sum of the Levenshtein

[2] For full details, refer to our prior work [1].

[3] EvoSuiteFIT is available from https://github.com/HusseinAlmulla/evosuite/tree/evosuitefit.

distance between each pair of test cases: $div(TS) = \sum_{ta,tb}^{TS} lev_{ta,tb}$. To attain a normalized value between 0–1 for use in a multi-fitness function environment, we then calculate and attempt to minimize the final fitness as $\frac{1}{1+div(TS)}$.

RL Implementation in EvoSuiteFIT: Every five generations, EvoSuiteFIT will select one to four fitness functions from the following: Diversity, Exception Count, Branch Coverage, Direct Branch Coverage, Method Coverage, Method Coverage (Top-Level, No Exception), Output Coverage, and Weak Mutation Coverage. Rojas et al. provide more details on each [8]. To constrain the number of combinations, we (1) use only the combinations that include the diversity score, and (2), remove a small number of semi-overlapping combinations (i.e., Branch and Direct Branch). Ultimately, the RL agent can choose from 44 combinations of fitness functions. To seed reward estimates, EvoSuiteFIT will make sure that all the actions have been tried (in a random order) before it starts using the UCB or DSG-Sarsa selection mechanisms. After selecting fitness functions, EvoSuiteFIT will proceed through the normal population evolution mechanisms. After five generations, the reformulated population is used to calculate the reward and update the expected reward. This allows sufficient population evolution to judge the effect of changing fitness functions. Over time, the combination that gains the highest reward will be more likely to be selected again until reaching convergence.

Related Work: Hyperheuristic search—often based on reinforcement learning—has been employed to improve search-based test generation. For instance, it has been used to tune the metaheuristic during Combinatorial Interaction Testing [6] or for addressing the test ordering problem [4]. In all of these cases, the hyperheuristic is used to tune the algorithm, and not the fitness functions. Although AFFS has been performed in other domains, such as production scheduling, we were the first to apply this concept in test generation [1]. We extend our earlier approach to test suite diversity.

3 Case Study

We have assessed EvoSuiteFIT using 18 real faults from the Gson project in order to address the following research questions: **(1)** *Is EvoSuiteFIT able to yield more diverse test suites than static fitness function choices?* **(2)** *Does increased test suite diversity lead to greater likelihood of fault detection?*

In order to investigate these questions, we have performed the following experiment:

(1) Collected Case Examples: We use 18 real faults from the Gson project from the Defects4J fault database. We target the affected classes for each fault for test generation.

(2) Generated Test Suites: For each class, we generate 10 suites per approach (the two reinforcement learning algorithms—UCB and DSG-Sarsa—and two baselines—generation guided by diversity score alone and a combination of all eight fitness functions). A search budget of 10 min is used per suite.

(3) Removed Non-Compiling and Flaky Tests: Any tests that do not compile, or that return inconsistent results, are removed.
(4) Assessed Results: We measure the diversity of each test suite and the likelihood of fault detection (proportion of failing suites to the number generated) for each fault.

Case Examples: Defects4J is an extensible database of real faults extracted from Java projects[4]. For each fault, Defects4J provides access to the faulty and fixed versions of the code, developer-written test cases that expose each fault, and a list of classes and lines of code modified to fix the fault. The current version contains 18 faults mined from the Gson framework [3]. We target the classes affected by faults for test generation.

Baselines: We have generated test suites using both reinforcement learning approaches, UCB and DSG-Sarsa. In addition, we generate tests for two baseline approaches representing current practice. The first is the diversity score alone. This would be the likely starting point for a tester interested in improving suite diversity. The second is the combination of all eight functions that are used in this study. This configuration represents a "best guess" at what would produce effective test suites, and would be considered a reasonable approach in the absence of a known, informative fitness function.

Test Generation: Tests are generated using the fixed version of the class and applied to the faulty version because EvoSuite generates assertions for use as oracles. Tests that fail on the faulty version display behavioral differences between the two versions[5]. To perform a fair comparison between approaches, each is allocated a ten minute search budget. To control experiment cost, we deactivated assertion filtering—all possible regression assertions are included. All other settings were kept at their default values. As results may vary, we performed 10 trials for each approach and fault. We automatically remove tests that return inconsistent results over five executions and non-compiling test suites. On average, less than one percent of tests are removed from each suite.

4 Results and Discussion

Below, we illustrate the final diversity fitness score attained by each test suite, the median fitness for each technique, and the median likelihood of fault detection for the four detected faults (faults 4, 9, 12, and 17). Because we minimize fitness, *lower* scores indicate *higher* diversity.

[4] Available from http://defects4j.org.
[5] This is the same practice followed by other studies using EvoSuite, i.e. [1,3].

	DSG-Sarsa	UCB	Combination	Diversity-only
Diversity	1.47E−06	9.99E−07	1.93E−06	3.91E−06
Fault detection likelihood	0.60	0.75	0.60	0.05

We can clearly see that the diversity score is a difficult fitness function to optimize on its own. When the sole target of test generation, the diversity score gives the worst final results. There is also a wide variance (distance between first and third quartile) in its results. Diversity alone not offer actionable feedback to the test generation algorithm. The combination of all eight fitness functions attains more stable results, with a better median. By adding additional fitness functions, we offer the test generation framework detailed feedback, leading to improvements in the final diversity.

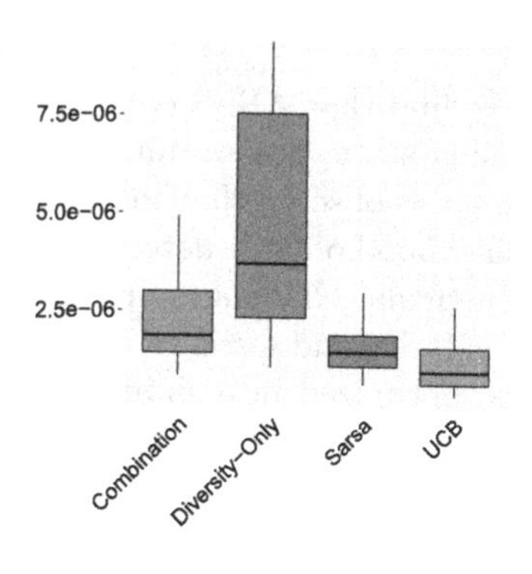

Both RL approaches outperform the static approaches. DSG-Sarsa attains a 62% median improvement over diversity-only and 24% improvement over the combination. UCB is even better, attaining a 75% median improvement over diversity-only and 49% over the combination. UCB also attains a 33% median improvement over DSG-Sarsa. Both RL approaches also demonstrate lower variance in the results they attain.

A one-sided (strictly lesser) Mann-Whitney-Wilcoxon rank-sum ($\alpha = 0.05$) confirms (p-values < 0.01) that UCB outperforms all other approaches and that DSG-Sarsa outperforms the static approaches. The Vargha-Delaney A measure indicates that UCB outperforms DSG-Sarsa with small, the combination with medium, and diversity-only with large effect size. DSG-Sarsa outperforms the combination with small and diversity-only with large effect size. ***Both EvoSuiteFIT techniques—particularly UCB—increase test suite diversity over static fitness function choices with significance.***

There is limited evidence that we can use to assess the impact of diversity on the likelihood of fault detection. Only four faults are detected by any of the approaches—faults 4, 9, 12, and 17. We identified several factors impacting fault detection in Gson, including faults that are easier to expose through system-level testing, complex datatypes, and faults exposed through stronger test requirements [3]. Increased diversity may not overcome these broader issues. However, our observations suggest a positive impact from increased diversity.

Examining fault detection across the four faults, we see that (1) all approaches outperform diversity-alone, and (2) the approach with the greatest diversity—UCB—also attains the highest likelihood of fault detection. In addition, Fault 17 is uniquely detected by DSG-Sarsa. This fault is based around serialization of date values[6]. Increased diversity could lead to a wider range of

[6] See https://github.com/google/gson/issues/1096.

attempted input and method calls. ***AFFS may improve likelihood of fault detection. However, more examples are needed to draw clear conclusions on the impact of diversity.***

5 Conclusions

We find that AFFS creates test suites that are more diverse than those created using static fitness functions. While results on fault detection are inconclusive, we also observe that increased diversity may lead to small improvements in the likelihood of fault detection. However, additional research is needed to tackle the particular challenges presented by Gson. Future work on AFFS will extend this study beyond Gson to a greater pool of faults and examine other goals that can be optimized in a similar manner. We make the EvoSuiteFIT available for use in test generation research or practice.

References

1. Almulla, H., Gay, G.: Learning how to search: generating exception-triggering tests through adaptive fitness function selection. In: 13th IEEE International Conference on Software Testing, Validation and Verification (2020)
2. De Oliveira Neto, F.G., Feldt, R., Erlenhov, L., Nunes, J.B.D.S.: Visualizing test diversity to support test optimisation. In: 2018 25th Asia-Pacific Software Engineering Conference (APSEC), pp. 149–158 (2018)
3. Colanzi, T.E., McMinn, P. (eds.): SSBSE 2018. LNCS, vol. 11036. Springer, Cham (2018). https://doi.org/10.1007/978-3-319-99241-9
4. Guizzo, G., Fritsche, G.M., Vergilio, S.R., Pozo, A.T.R.: A hyper-heuristic for the multi-objective integration and test order problem. In: Proceedings of the 2015 Annual Conference on Genetic and Evolutionary Computation, pp. 1343–1350. GECCO 2015, ACM, New York (2015). https://doi.org/10.1145/2739480.2754725
5. Idan, H.: The top 100 java libraries in 2017 - based on 259,885 source files (2017). https://blog.takipi.com/the-top-100-java-libraries-in-2017-based-on-259885-source-files/
6. Jia, Y., Cohen, M.B., Harman, M., Petke, J.: Learning combinatorial interaction test generation strategies using hyperheuristic search. In: Proceedings of the 37th International Conference on Software Engineering, ICSE 2015, pp. 540–550. IEEE Press, Piscataway (2015). http://dl.acm.org/citation.cfm?id=2818754.2818821
7. Sutton, R.S., Barto, A.G.: Reinforcement Learning: An Introduction, 2nd edn. MIT Press, Cambridge (2018)
8. Rojas, J.M., Campos, J., Vivanti, M., Fraser, G., Arcuri, A.: Combining multiple coverage criteria in search-based unit test generation. In: Barros, M., Labiche, Y. (eds.) SSBSE 2015. LNCS, vol. 9275, pp. 93–108. Springer, Cham (2015). https://doi.org/10.1007/978-3-319-22183-0_7
9. Shahbazi, A.: Diversity-based automated test case generation. Ph.D. thesis, University of Alberta (2015)

Challenge Cases

Defects4J as a Challenge Case for the Search-Based Software Engineering Community

Gregory Gay[1(✉)] and René Just[2]

[1] Chalmers and the University of Gothenburg, Gothenburg, Sweden
greg@greggay.com
[2] University of Washington, Seattle, WA, USA
rjust@cs.washington.edu

Abstract. Defects4J is a collection of reproducible bugs, extracted from real-world Java software systems, together with a supporting infrastructure for using these bugs. Defects4J has been widely used to evaluate software engineering research, including research on automated test generation, program repair, and fault localization. Defects4J has recently grown substantially, both in number of software systems and number of bugs. This report proposes that Defects4J can serve as a benchmark for Search-Based Software Engineering (SBSE) research as well as a catalyst for new innovations. Specifically, it outlines the current Defects4J dataset and infrastructure, and details how it can serve as a challenge case to support SBSE research and to expand Defects4J itself.

Keywords: Software faults · Research infrastructure · Research benchmarks

1 Introduction

Each year, the Symposium on Search-Based Software Engineering (SSBSE) hosts a Challenge Track. This track presents a series of challenge cases, often centered around particular software systems or research domains, and tasks researchers with applying their tools, techniques, and algorithms to those challenge cases. The Challenge Track has attracted great attention and competition, and has been a powerful mechanism for highlighting the applicability of state-of-the-art SBSE research to complex, real-world problems [5–7,13,17,23].

This report proposes that Defects4J can serve as a compelling challenge case for future editions of SSBSE. Defects4J is a collection of reproducible bugs, extracted from real-world Java software systems, together with a supporting infrastructure for using these bugs [9]. The current version (v2.0.0) contains 835 bugs from 17 Java software systems as well as supporting infrastructure for conducting experiments in software testing and debugging research. For example, Defects4J has been used to evaluate automated test generation [18,20], automated program

A. Aleti and A. Panichella (Eds.): SSBSE 2020, LNCS 12420, pp. 255–261, 2020.
https://doi.org/10.1007/978-3-030-59762-7_19

repair [11,12], and fault localization [16] research. Furthermore, past contributions to the SSBSE Challenge Track have expanded the Defects4J dataset [2,5,6].

The inclusion of Defects4J in the SSBSE Challenge Track can serve as a benchmark for SBSE research as well as a catalyst for new innovations. This report outlines the current version of Defects4J (Sect. 2), and details how it can support SBSE research and inspire extensions to Defects4J itself (Sect. 3).

2 Defects4J

Defects4J is an extensible collection of reproducible bugs from Java software systems, together with a supporting infrastructure, and aims at advancing software engineering research [9]. Defects4J is available at: https://defects4j.org.

The Bugs: The current version of Defects4J (v2.0.0) targets Java 8 and consists of 835 reproducible bugs from 17 projects: Chart (26 bugs), Cli (39), Closure (174), Codec (18), Collections (4), Compress (47), Csv (16), Gson (18), JacksonCore (26), JacksonDatabind (112), JacksonXml (6), Jsoup (93), JxPath (22), Lang (64), Math (106), Mockito (38), and Time (26). The 835 bugs span more than a decade of development history, and the 17 projects span a wide range of domains, including compilers, parsers, testing infrastructure, and a variety of libraries.

Each bug in Defects4J has the following three properties:

1. Each bug consists of a buggy and a fixed source code version. The fixed version is explicitly labeled as a fix to an issue reported in the project's issue tracker, and the changes imposed by the fix must be to source code, not to other project artifacts such as configuration or build files.
2. Each bug is reproducible: all tests pass on the fixed version and at least one of those tests fails on the buggy version, thereby exposing the bug.
3. Each bug is isolated: the buggy and the fixed version differ only by a minimal set of changes, all of which are related to the bug. That is, the difference is free of unrelated code changes, such as refactoring or feature additions.

For each bug, Defects4J provides the following artifacts and metadata:

- A pair of source code versions—the buggy and the fixed version.
- A set of classes and source-code lines modified by the patch that fixes the bug.
- A set of developer-written tests that expose the bug—called "trigger tests".
- A stacktrace for each trigger test, when executed on the buggy version.
- A set of classes loaded by the classloader during execution of the trigger tests.
- A set of tests that are relevant to the bug—tests that load at least one class modified by the patch that fixes the bug.

Supporting Infrastructure: Defect4J offers a command-line utility to execute a set of common tasks for each bug, including the following: print information about a bug, checkout a buggy or fixed source code version, compile a source code version, execute tests and (optionally) monitor the classloader, perform coverage or mutation analysis, export metadata such as classpaths, directories, or sets of tests, and a utility that queries the metadata to support automated analyses.

By default, Defects4J commands use the developer-written tests that come with each project. However, each command can be executed for an arbitrary JUnit test suite, including those created by automated test generation tools. Defects4J offers a uniform interface for automated test generation. Concrete instantiations of that interface are provided for EvoSuite [4], a search-based test generator, and Randoop [14], a feedback-directed random test generator. Defects4J provides a template to ease incorporation of additional test generators into the infrastructure. Coverage and mutation analyses are provided through Cobertura [1] and the Major mutation framework [8], respectively.

Expanding Defects4J: Defects4J can be expanded along different dimensions. First, testing and debugging tools can be integrated into the supporting infrastructure, through well-defined interfaces.

Second, a semi-automated process[1] supports mining candidate bugs from an existing project's version control system and issue tracker. This process requires the creation of a meta build file that can compile any source code version of the project (generally by calling that version's existing build script). An automated step mines candidate bugs by cross-referencing the project's version control history with its issue tracker, identifying commits that fix a reported and closed issue. This step also compares each fix commit with its predecessor commit to determine whether at least one trigger test exists that reliably passes on the fixed version and fails on the buggy version. Each reproducible bug is then subject to a manual minimization process that eliminates irrelevant code changes (i.e., refactoring or feature additions). Finally, an automated step adds all reproducible, minimized bugs to the dataset and computes their metadata.

3 Research Challenges

Defects4J can serve as a challenging and diverse benchmark for SBSE research as well as a catalyst for new innovations. Past research has successfully used Defects4J to evaluate software testing and debugging approaches. For example, past research used Defects4J to assess the effectiveness of automated test generation and corresponding fitness functions [20], automated program repair [11], and fault localization [16].

This section outlines three concrete areas in which Defects4J can serve as a challenge case. These areas and their corresponding challenges do not form an exhaustive set, but are intended to provide inspiration. Defects4J can be used

---

[1] The entire process is documented at https://github.com/rjust/defects4j/tree/master/framework/bug-mining/README.md.

to validate and extend work presented in prior editions of SSBSE or to explore new SBSE-based approaches.

3.1 Empirical Validation

Genetic Improvement: Search-based approaches have been used to improve system performance [17]. These performance improvements should be semantics preserving—that is, not alter the functional behavior of a system. Defects4J can support the empirical validation of performance improvement research by providing a supporting infrastructure for integrating tools and a thorough set of tests for assessing generated patches.

Hyper-Parameter Tuning: Search-based approaches often have many parameters that can, and need to, be tuned to increase effectiveness on particular problem instances. Examples include crossover and mutation rates of a genetic algorithm [22]. A fair and comprehensive assessment of automated tuning processes requires a well-defined dataset. Defects4J provides such a dataset, fosters replicability of experiments, and supports validation of in-project and cross-project generalization of hyper-parameter settings.

Longitudinal Studies and Software Evolution: Software evolves over time, and Defects4J's artifacts capture this evolution. With a diverse set of projects and multiple artifacts per project, spanning multiple years of development history, Defects4J supports longitudinal studies that, e.g., investigate the effectiveness and generalizability of SBSE approaches over time.

Test Suite Diversity: Prior work hypothesized that diverse test suites are more effective than those that contain similar tests [21]. Defects4J supports controlled experiments that can assess the impact of increased or decreased diversity on fault detection. The use of a common dataset also offers common grounds for researchers to compare different diversification techniques.

3.2 Novel SBSE Techniques

Predictive Modeling: Search-based approaches can be used to tackle prediction problems, such as defect prediction [19]. The 835 bugs in Defects4J correspond to over 1000 buggy Java classes. Features of those classes and the isolated bugs themselves can be used to develop or train predictive models.

State Space Exploration in Program Repair: Search-based approaches for automated program repair often have difficulty traversing the search space due to a costly fitness evaluation. Recent work addressed this challenge by capturing dependencies between source code statements and high-level differences between patch variants [3]. Defects4J is a natural benchmark for novel program repair research, as the dataset includes both buggy and fixed versions of a variety of complex bugs.

Topic Modeling: Recent work examined the extraction of information from textual artifacts, such as bug reports [10,15]. Defects4J provides many examples for topic modeling and can serve as a basis for linking code and textual artifacts, since each bug in Defects4J is linked to a bug report and the projects and classes themselves often include detailed documentation.

The topics above capture only a portion of the research presented at SSBSE 2019. Defects4J can also support other research areas, such as **crash reproduction** (Defects4J provides trigger tests, detailed stacktraces, and isolated bugs, all of which can be used to assess crash-reproduction approaches) or **mutation testing** (Defects4J's real bugs can serve as templates for evolving new mutation operators and generation techniques).

3.3 Extending Defects4J

In addition to providing a challenging benchmark for assessing SBSE research, extending Defects4J also poses interesting challenges that may inspire new approaches. In particular, researchers may wish to consider the following:

- **Automated build script creation and repair:** Adding bugs to Defects4J currently requires the manual creation of a meta build file. Genetic programming could be used to automate this step. Some of the associated challenges include gathering dependencies and inferring necessary properties to build each source code version.
- **New projects and bugs:** The set of software systems in Defects4J is diverse, but many domains are not accounted for. New projects and additional bugs would increase the range of research that could be supported. Of particular interest are AI-based systems, concurrent systems, and systems in difficult-to-test domains.
- **New test generation approaches:** Defects4J includes a standardized interface for automated test generation. This interface allows researchers to quickly integrate and evaluate new approaches.
- **Interface for automated patch generation:** Similar to the standardized interface for test generation, Defects4J would benefit from a well-defined interface for tools that generate patches, including tools for automated program repair and genetic improvement. For example, tools for automated program repair are routinely benchmarked on Defects4J [11]. A standardized interface for executing such tools would facilitate reproducability and comparability.
- **Support for newer Java versions:** The current version of Defects4J supports Java 8. Newer version of Java may result in unexpected compilation errors, test failures, or metadata inconsistencies. Novel approaches for automatically migrating source code and tests to newer Java versions would be a welcome contribution.

In addition to the examples above, researchers may identify other needs and challenges that require attention to support different research areas or experimental protocols.

4 Conclusions

This report proposes that Defects4J can serve as a challenge case for SBSE research as well as a catalyst for new innovations. Among the topics explored in work published at the 2019 Symposium on Search-Based Software Engineering, Defects4J could benefit research in genetic improvement, hyper-parameter tuning, test suite diversity and generation, predictive modeling, state space exploration, and topic modeling. Furthermore, extending Defects4J poses additional challenges that can be tackled through SBSE.

Acknowledgements. This material is based upon work supported by the National Science Foundation under grants CNS-1823172 and CCF-1942055.

References

1. Cobertura. https://cobertura.github.io/cobertura/
2. Almulla, H., Salahirad, A., Gay, G.: Using search-based test generation to discover real faults in Guava. In: Menzies, T., Petke, J. (eds.) SSBSE 2017. LNCS, vol. 10452, pp. 153–160. Springer, Cham (2017). https://doi.org/10.1007/978-3-319-66299-2_13
3. Dantas, A., de Souza, E.F., Souza, J., Camilo-Junior, C.G.: Code naturalness to assist search space exploration in search-based program repair methods. In: Nejati, S., Gay, G. (eds.) SSBSE 2019. LNCS, vol. 11664, pp. 164–170. Springer, Cham (2019). https://doi.org/10.1007/978-3-030-27455-9_12
4. Fraser, G., Staats, M., McMinn, P., Arcuri, A., Padberg, F.: Does automated unit test generation really help software testers? A controlled empirical study. ACM Trans. Softw. Eng. Methodol. **24**(4), 23:1–23:49 (2015). https://doi.org/10.1145/2699688
5. Gay, G.: Challenges in using search-based test generation to identify real faults in mockito. In: Sarro, F., Deb, K. (eds.) SSBSE 2016. LNCS, vol. 9962, pp. 231–237. Springer, Cham (2016). https://doi.org/10.1007/978-3-319-47106-8_17
6. Colanzi, T.E., McMinn, P. (eds.): SSBSE 2018. LNCS, vol. 11036. Springer, Cham (2018). https://doi.org/10.1007/978-3-319-99241-9
7. Harman, M., Jia, Y., Langdon, W.B.: Babel pidgin: SBSE can grow and graft entirely new functionality into a real world system. In: Le Goues, C., Yoo, S. (eds.) SSBSE 2014. LNCS, vol. 8636, pp. 247–252. Springer, Cham (2014). https://doi.org/10.1007/978-3-319-09940-8_20
8. Just, R.: The major mutation framework: efficient and scalable mutation analysis for java. In: Proceedings of the 2014 International Symposium on Software Testing and Analysis, ISSTA 2014, pp. 433–436. ACM, New York (2014). https://doi.org/10.1145/2610384.2628053
9. Just, R., Jalali, D., Ernst, M.D.: Defects4J: a database of existing faults to enable controlled testing studies for Java programs. In: Proceedings of the 2014 International Symposium on Software Testing and Analysis, ISSTA 2014, pp. 437–440. ACM, New York (2014). https://doi.org/10.1145/2610384.2628055
10. Just, R., Parnin, C., Drosos, I., Ernst, M.D.: Comparing developer-provided to user-provided tests for fault localization and automated program repair. In: Proceedings of the International Symposium on Software Testing and Analysis (ISSTA), pp. 287–297 (2018)

11. Martinez, M., Durieux, T., Sommerard, R., Xuan, J., Monperrus, M.: Automatic repair of real bugs in java: a large-scale experiment on the defects4j dataset. Empirical Softw. Eng. **22**(4), 1936–1964 (2016). https://doi.org/10.1007/s10664-016-9470-4
12. Motwani, M., Soto, M., Brun, Y., Just, R., Le Goues, C.: Quality of automated program repair on real-world defects. IEEE Trans. Softw. Eng. **PP**, 1 (2020)
13. de Oliveira Barros, M., de Almeida Farzat, F.: What can a big program teach us about optimization? In: Ruhe, G., Zhang, Y. (eds.) SSBSE 2013. LNCS, vol. 8084, pp. 275–281. Springer, Heidelberg (2013). https://doi.org/10.1007/978-3-642-39742-4_24
14. Pacheco, C., Ernst, M.D.: Randoop: feedback-directed random testing for java. In: Companion to the 22nd ACM SIGPLAN Conference on Object-Oriented Programming Systems and Applications Companion, OOPSLA 2007, pp. 815–816. ACM, New York (2007). https://doi.org/10.1145/1297846.1297902
15. Panichella, A.: A systematic comparison of search algorithms for topic modelling—a study on duplicate bug report identification. In: Nejati, S., Gay, G. (eds.) SSBSE 2019. LNCS, vol. 11664, pp. 11–26. Springer, Cham (2019). https://doi.org/10.1007/978-3-030-27455-9_2
16. Pearson, S., et al.: Evaluating and improving fault localization. In: 2017 IEEE/ACM 39th International Conference on Software Engineering (ICSE), pp. 609–620 (2017)
17. Petke, J., Brownlee, A.E.I.: Software improvement with Gin: a case study. In: Nejati, S., Gay, G. (eds.) SSBSE 2019. LNCS, vol. 11664, pp. 183–189. Springer, Cham (2019). https://doi.org/10.1007/978-3-030-27455-9_14
18. Rueda, U., Just, R., Galeotti, J.P., Vos, T.E.: Unit testing tool competition: round four. In: Proceedings of the International Workshop on Search-Based Software Testing (SBST), pp. 19–28, May 2016
19. Sarro, F.: Search-based predictive modelling for software engineering: how far have we gone? In: Nejati, S., Gay, G. (eds.) SSBSE 2019. LNCS, vol. 11664, pp. 3–7. Springer, Cham (2019). https://doi.org/10.1007/978-3-030-27455-9_1
20. Shamshiri, S., Just, R., Rojas, J.M., Fraser, G., McMinn, P., Arcuri, A.: Do automatically generated unit tests find real faults? An empirical study of effectiveness and challenges. In: Proceedings of the 30th IEEE/ACM International Conference on Automated Software Engineering (ASE), ASE 2015. ACM, New York (2015)
21. Vogel, T., Tran, C., Grunske, L.: Does diversity improve the test suite generation for mobile applications? In: Nejati, S., Gay, G. (eds.) SSBSE 2019. LNCS, vol. 11664, pp. 58–74. Springer, Cham (2019). https://doi.org/10.1007/978-3-030-27455-9_5
22. Zamani, S., Hemmati, H.: Revisiting hyper-parameter tuning for search-based test data generation. In: Nejati, S., Gay, G. (eds.) SSBSE 2019. LNCS, vol. 11664, pp. 137–152. Springer, Cham (2019). https://doi.org/10.1007/978-3-030-27455-9_10
23. Zhang, Y., Harman, M., Jia, Y., Sarro, F.: Inferring test models from Kate's bug reports using multi-objective search. In: Barros, M., Labiche, Y. (eds.) SSBSE 2015. LNCS, vol. 9275, pp. 301–307. Springer, Cham (2015). https://doi.org/10.1007/978-3-319-22183-0_27

Author Index

Printed in the United States
By Bookmasters